Introduction to Metabolism

8

Metabolism comprises all of the chemical reactions that take place in a living system, be it a cell, a tissue, an organ, or an organism. Metabolic reactions are almost all enzyme-catalyzed and include transformations of energy and nutrients, syntheses and degradations, and excretions of waste products. Chemical changes concerned with the production, storage, and utilization of metabolic energy for biosynthesis are known as **intermediary metabolism.** Typically, intermediary metabolism includes all aspects of metabolism (including digestion) except those involved in the transfer of genetic information—replication, transcription, and translation.

At the level of an individual organism, metabolic reactions form the basis of all the functions—growth, movement, reproduction, and the like—characteristic of living systems. On a larger scale, the combined metabolism of microorganisms, plants, and animals leads to the great natural cycles of the biosphere (Figure 8.1).

The Sun is the ultimate source of energy for metabolism in all organisms; it constitutes the source of life on Earth. Plants trap solar energy during photosynthesis, the single most important chemical reaction in the biosphere. Photosynthesis results in conversion of carbon dioxide to carbohydrates, which then provide carbon skeletons for the synthesis of proteins, lipids, and nucleic acids. Nonphotosynthetic organisms feed on plants and use plant products for their growth and reproduction.

8.1. CATABOLISM AND ANABOLISM

We divide metabolism into two parts, **catabolism,** or degradative reactions, and **anabolism,** or synthetic reactions. Catabolism consists of three stages (Figure 8.2). In the first stage, polymeric nutrients break down to small monomeric building blocks. In animals, this stage comprises digestion, as a result of which carbohydrates, proteins, lipids, and nucleic acids are degraded to, respectively, monosaccharides (principally glucose), amino acids, fatty acids and glycerol, and nucleotides.

The second stage produces only a few different types of molecules. These products have even simpler structures than the building blocks from which they derive and include the two key compounds *pyruvate* and *acetate*. Acetate enters metabolism in the form of *acetyl coenzyme A*. Protein catabolism also yields ammonia, formed by deamination of amino acids.

In the third stage, acetyl coenzyme A enters the *cit-*

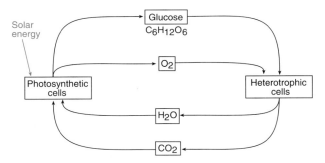

Figure 8.1. The carbon and oxygen cycles of the biosphere.

ric acid cycle, where its acetyl group is oxidized to carbon dioxide and water. All told, catabolism yields only three major end products—CO_2, H_2O, and NH_3—in addition to the capture of chemical energy in the form of ATP. The number of end products is small because of the *convergent* nature of catabolic reactions; multiple pathways of nutrient degradation come together and ultimately enter one major metabolic system.

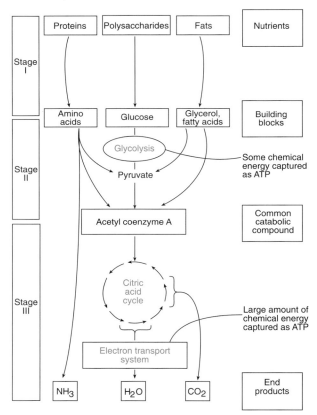

Figure 8.2. The three stages of catabolism for the major energy-yielding nutrients. Nucleic acids are also degraded in three stages, but they have been omitted from the figure because their breakdown does not contribute significantly to the cell's energy supply.

By contrast, anabolic reactions have *divergent* characteristics; a simple precursor gives rise to numerous larger molecules by reactions that fan out from the initial step (Figure 8.3).

Both catabolism and anabolism involve hundreds of different reactions, all interconnected in a giant network (Figure 8.4). As you can see, most metabolic intermediates can be converted to several compounds by reactions that proceed along different paths. Not surprisingly, therefore, interconversions between major nutrients occur readily. Carbohydrates can yield proteins, proteins can form lipids, and so on. We will discuss specific examples of such interrelationships later on.

Studying this multitude of reactions may appear to be an insurmountable task, but that is not so. Without having to discuss each and every reaction, you will find that it is possible to understand the operation of the major metabolic systems by focusing on a few groups of closely linked reactions. In the following chapters, we deal with these unifying concepts of metabolism.

8.1.1. Fermentation

Sets of some catabolic reactions constitute fermentations. We define **fermentation** as an energy-yielding catabolic pathway that proceeds without net oxidation; oxidation of one intermediate is balanced by reduction of another. In fermentation, organic compounds act as both donors and acceptors of electrons. Fermentation reactions do not require molecular oxygen and yield energy in the form of *ATP,* an *energy-rich compound,* discussed in Section 9.2. Two examples will illustrate the process.

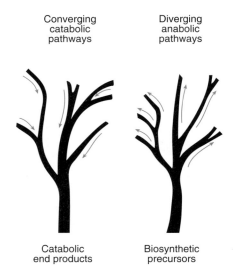

Figure 8.3. Converging and diverging metabolic pathways.

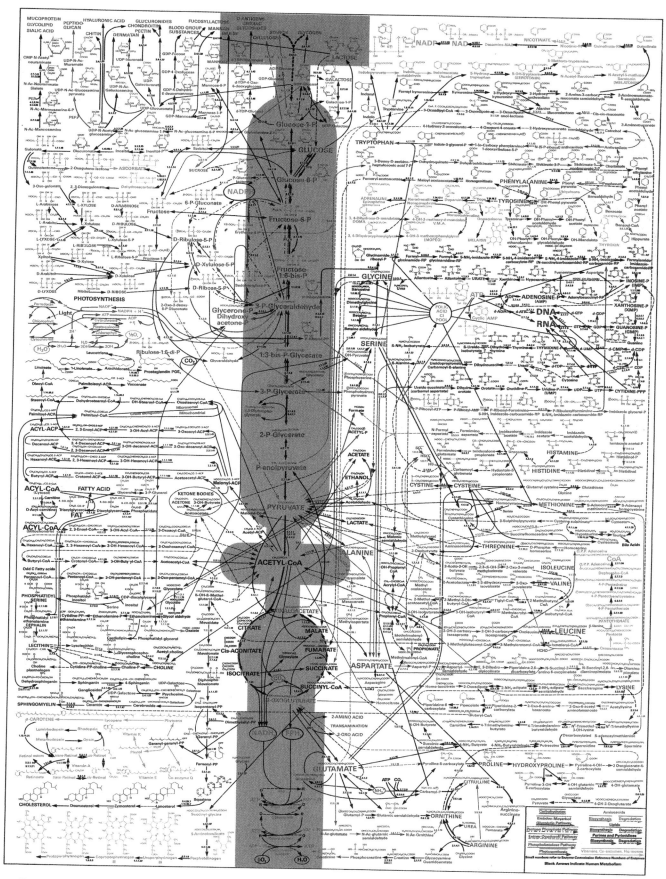

Figure 8.4.   Schematic map of the major metabolic pathways in a typical cell. The shaded region encompasses the main reactions of glucose metabolism. (Designed by D. E. Nicholson. Published by BDH Ltd., Poole, Dorset, England.)

8.1.1A. Alcoholic Fermentation. In **alcoholic fermentation,** yeast converts glucose to ethanol. *Alcohol dehydrogenase* catalyzes the last reaction in this sequence:

$$\underset{\text{Acetaldehyde}}{CH_3-\overset{\overset{\displaystyle O}{\|}}{CH}} + NADH + H^+ \rightarrow \underset{\text{Ethanol}}{CH_3CH_2OH} + NAD^+$$

Acetaldehyde is *reduced* to ethanol while NADH is *oxidized* to NAD$^+$. The overall reaction of alcoholic fermentation includes the synthesis of ATP from ADP:

$$\underset{\text{Glucose}}{C_6H_{12}O_6} + 2P_i^{2-} + 2ADP^{3-} + 2H^+ \rightarrow 2\underset{\text{Ethanol}}{CH_3CH_2OH} +$$

$$2CO_2 + 2ATP^{4-} + 2H_2O$$

8.1.1B. Lactate Fermentation. **Lactate fermentation** involves the catabolism of glucose to lactate (the anionic form of lactic acid). In microorganisms, lactate is formed by lactic acid bacteria. In humans, lactate forms under relatively anaerobic conditions, such as those that occur during strenuous exercise. *Lactate dehydrogenase* catalyzes the last reaction in this process:

$$\underset{\text{Pyruvate}}{CH_3-\overset{\overset{\displaystyle O}{\|}}{C}-COO^-} + NADH + H^+ \rightarrow$$

$$\underset{\text{Lactate}}{CH_3-CHOH-COO^-} + NAD^+$$

Pyruvate is *reduced* to lactate while NADH is *oxidized* to NAD$^+$. The overall reaction of lactate fermentation also includes the synthesis of ATP from ADP:

$$\text{Glucose} + 2P_i^{2-} + 2ADP^{3-} + 2H^+ \rightarrow 2 \text{ lactate} +$$

$$2ATP^{4-} + 2H_2O$$

8.2. REGULATION OF METABOLISM

Because almost all metabolic reactions are enzyme-catalyzed, the primary regulation of metabolism occurs by controlling enzymatic activity. This control, as you saw earlier, involves many different factors and mechanisms (Figure 4.6). Recall that a typical cell contains thousands of different kinds of substrates and enzymes (Table 1.1). Regulating the multitude of intracellular reactions that use

these substrates and enzymes amounts to an astounding feat. When you consider that in large multicellular organisms there exists the added element of coordinating the reactions of billions of cells, the regulation of metabolism emerges as a staggering evolutionary achievement.

In addition to regulating individual enzymes, metabolic control involves three major elements—the existence and regulation of specific *metabolic pathways,* the occurrence of *compartmentation,* and the action of *hormones.*

8.2.1. Metabolic Pathways

A **metabolic pathway** consists of a sequence of steps by which a metabolite is synthesized, degraded, or transformed. The sequence proceeds from some key intermediate to a specific terminal product. Pathways may be *linear, branched,* or *cyclic* (Figure 8.5).

Metabolic pathways typically involve a large number of steps, best described by steady-state rather than by equilibrium conditions (Section 4.5). Multiple steps mean that alterations of intermediates generally occur in small increments. In this way, catabolism releases energy in amounts that can be used for efficient ATP synthesis. Metabolic pathways frequently commence by effectively *trapping* the first key intermediate inside the cell or by *activating* it to a chemically more reactive form. All metabolic pathways have three important characteristics in common.

First, *metabolic pathways are irreversible.* Catabolic pathways constitute "downhill" sequences in which one or more reactions are strongly *exergonic,* resulting in the release of free energy ($\Delta G'$ is negative). Hence, the reverse pathway, the anabolic sequence, constitutes an "uphill" process in which these same reactions are strongly *endergonic* ($\Delta G'$ is positive). However, endergonic reactions are thermodynamically not feasible and cannot pro-

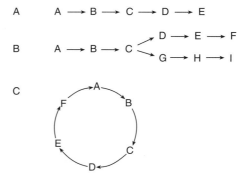

Figure 8.5.   Types of metabolic pathways: (A) linear; (B) branched; (C) cyclic.

ceed as written. Accordingly, these reactions serve as energy barriers for the anabolic pathway. Because of them, the pathway cannot occur even though some of its steps may be readily reversible.

In order for the anabolic sequence to become feasible, the pathway must be altered (Figure 8.6). Each highly endergonic step must be changed by replacing it with one or several different reactions. Anabolic and catabolic sequences must differ in the reactants and products of these altered steps since only then will the steps have different free energy changes associated with them. Recall that a reaction's free energy change is a function of the reactants and the products but does not depend on the path whereby reactants are converted to products.

Second, *metabolic pathways have committed steps.* A **committed step** is a unique step of a pathway. It may be a step producing a metabolite that has no other role than to serve as an intermediate in this pathway. In most instances, the committed step constitutes a highly exergonic and essentially irreversible reaction. The committed step ensures that, once it has taken place, all subsequent reactions in the sequence will also take place; it "commits" the intermediate it produces to continue down the pathway. Committed steps occur early in metabolic pathways and frequently represent rate-determining steps.

Lastly, *metabolic pathways are regulated.* Regulation of pathways may be accomplished by multienzyme systems, allosteric enzymes, activators and inhibitors, and the like. More than one factor may be operative at a given step, and frequently more than one step of the pathway is subject to control. A committed step usually represents the most important control element of the pathway. Because catabolic and anabolic sequences differ, they can be regulated independently of each other. Separate regulation

of degradation and synthesis of key intermediates allows for fine-tuned control of metabolism.

8.2.2. Compartmentation

Compartmentation represents a second important element in the regulation of overall metabolism. The term denotes an unequal distribution of a metabolite, an enzyme, a pathway, or some other biomolecule or system within a cell or organism. Compartmentation affects metabolic reactions in several ways. In eukaryotes, it frequently accomplishes complete segregation of entire metabolic pathways in specific subcellular locations. For example, fatty acid catabolism occurs in the mitochondria, but fatty acid biosynthesis takes place in the cytoplasm. Likewise, production of ATP occurs in the mitochondria, but utilization of ATP takes place largely in the cytoplasm.

Separation of degradative and synthetic processes is advantageous because otherwise the two opposing activities would cancel each other out, in part or in entirety. We term such a set of opposing reactions a *futile cycle* (see Figure 10.35); it achieves nothing except the dissipation of free energy and, possibly, the generation of some heat.

Compartmentation also regulates enzymatic activity via the permeability properties of the compartment. By being selectively permeable, a compartment membrane controls the entry of substrates into and the exit of products from the compartment. Relative concentrations of substrates and products in turn influence enzymatic activities.

Lastly, compartmentation is linked to the action of several hormones that affect the transfer of metabolites across cell and subcellular membranes. In animals, all of these different aspects of compartmentation are related in a complex fashion. Compartmentation involves not only specific organs such as kidney, liver, stomach, and intestine, each of which is responsible for particular aspects of metabolism, but also specific subcellular locations of biochemical functions.

Mitochondria exhibit a high degree of compartmentation. Mitochondria possess two membranes, an inner and an outer one, separated by an intermembrane space (Figure 8.7). The inner membrane has many infoldings called *cristae.* Filling the spaces between cristae and surrounded by the inner membrane is the *matrix,* a gel-like substance. Thus, there exist four compartments in a mitochondrion: *outer membrane, intermembrane space, inner membrane,* and *matrix.* Each compartment contains specific metabolic systems and one or more enzymes that are located only in that compartment. We call such enzymes,

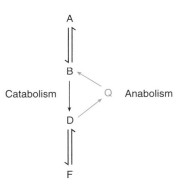

Figure 8.6. Irreversibility of catabolic pathways. The catabolic sequence A → B → D → F has two readily reversible steps but one exergonic and irreversible step (B → D). To overcome the energy barrier presented by this step, the anabolic sequence must proceed via a different path, at least at this step. Introduction of the Q bypass allows anabolism to proceed from F to A.

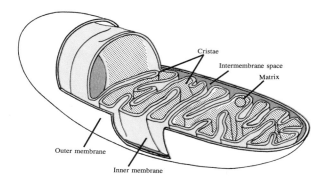

Figure 8.7.   Schematic structure of a mitochondrion. (Adapted, with permission, from S. L. Wolfe, *Biology of the Cell,* © 1972 by Wadsworth Publishing Co., Belmont, California.)

located *exclusively* in certain parts of a cell, **marker enzymes** (Table 8.1). Because of its exclusive location, a marker enzyme can serve to identify an experimentally obtained cellular fraction. Thus, if a fraction possesses cytochrome oxidase activity, you could assume that it contains inner mitochondrial membrane fragments.

8.2.3. Hormones

Hormone action constitutes a third key factor in the regulation of overall metabolism. According to their classic definition, hormones comprise regulatory substances in plants and animals that are synthesized by specialized cells, are active at low concentrations, and affect either all cells of an organism or only certain target cells in specific organs. Biochemists have expanded this definition to include any substance that carries a signal to produce some change at the cellular level. We now classify hormones based on the distance over which they act: an *au-*

Table 8.1. Location of Some Metabolic Activities and Marker Enzymes in Mitochondria

Location	Metabolic activities or marker enzymes
Outer membrane	Monoamine oxidase[a]; phospholipid synthesis; fatty acid elongation and desaturation
Intermembrane space	Adenylate kinase[a]; nucleoside diphosphate kinase
Inner membrane	Electron transport system; oxidative phosphorylation; succinate dehydrogenase[a]; cytochrome oxidase[a]; ATP synthase
Matrix	Citric acid cycle; β-oxidation of fatty acids; citrate synthase[a]; fumarase[a]; pyruvate dehydrogenase complex; urea cycle

[a]Marker enzyme.

tocrine hormone acts on the same cell that produced it (e.g., *interleukin*); a *paracrine hormone* acts on cells in close proximity to the one that produced it (e.g., *prostaglandins*); and an *endocrine hormone* acts on cells remote from the one that produced it (e.g., *insulin*).

Animal hormones are secreted principally by ductless glands and transported via the circulation to target tissues or organs. There they exert their effects either directly or indirectly and help regulate overall physiological processes like metabolism, growth, and reproduction. Plant hormones are organic compounds that control growth or some other function at a site removed from their place of production in the plant.

Chemically speaking, there exist four main classes of hormones: polypeptides, steroids, amino acid derivatives (amines), and eicosanoids. Table 8.2 lists major human hormones.

Many hormonal systems in animals originate in the brain, where an environmental or internal stimulus triggers the first signal (Figure 8.8). From the central nervous system the signal may be transmitted as an electrical pulse or as a chemical signal—or both—and then lead to release of a hormone. Frequently, a number of hormones act on each other in a *cascade mechanism* (Figure 4.24), thereby *amplifying* the initial signal. Reaction of the *ultimate hormone* with target cells produces particular metabolic changes. Hormones act on these target cells by binding to specific sites, called **receptors,** located on the cell membrane or in the interior of responsive cells.

Biochemists recognize four classes of hormone receptors that differ in their location and in the mechanism of action of the corresponding hormone. Some receptors are located on the cell periphery and bind hormones there. Binding leads to synthesis and/or release of a compound inside the cell that evokes a metabolic change by acting on a specific enzyme. We refer to the compound released inside the cell as a **second messenger** because we consider the hormone binding to the cell membrane as the **first messenger.** We will discuss specific hormones and second messengers as we encounter them in the text.

The second class of receptors comprises receptors that are located on cytoplasmic or nuclear proteins and bind hormones capable of traversing the cell or nuclear membrane. The remaining two classes of receptors constitute transmembrane proteins. One type of transmembrane protein has a hormone binding site on its extracellular side and a catalytic site on its intracellular side; the cytoplasmic site becomes activated when the hormone binds to the external site. The second type of transmembrane protein functions as an ion channel. Upon binding of the hormone to the cell periphery, the channel opens and specific ions flow across the membrane.

Table 8.2. Selected Human Hormones

Hormone	Source	Major effects
Polypeptides		
Corticotropin-releasing factor (CRF)	Hypothalamus	Stimulates ACTH release
Gonadotropin-releasing factor (GnRF)	Hypothalamus	Stimulates LH and FSH release
Growth hormone-releasing factor (GRF)	Hypothalamus	Stimulates GH release
Thyrotropin-releasing factor (TRF)	Hypothalamus	Stimulates TSH release
Adrenocorticotropic hormone (ACTH)	Anterior pituitary	Stimulates release of adrenocorticosteroids (glucocorticoids and mineralocorticoids)
β-Endorphin	Anterior pituitary	Exerts opioid effects on central nervous system
Follicle-stimulating hormone (FSH)	Anterior pituitary	Stimulates ovulation and estrogen synthesis in ovaries; stimulates spermatogenesis in testes
Growth hormone (somatotropin, GH)	Anterior pituitary	Stimulates bone growth and release of insulin and glucagon
Leu-enkephalin	Anterior pituitary	Exerts opioid effects on central nervous system
Luteinizing hormone (LH)	Anterior pituitary	Stimulates estrogen and progesterone synthesis in ovaries; stimulates androgen synthesis in testes
Met-enkephalin	Anterior pituitary	Exerts opioid effects on central nervous system
Prolactin (PRL)	Anterior pituitary	Stimulates lactation of mammary gland
Thyrotropin (TSH)	Anterior pituitary	Stimulates T_3 and T_4 release
Oxytocin	Posterior pituitary	Stimulates uterine contractions
Vasopressin	Posterior pituitary	Stimulates water resorption by the kidney and increases blood pressure
Glucagon	Pancreas (α cells)	Stimulates glucose release by glycogen breakdown
Insulin	Pancreas (β cells)	Stimulates glucose uptake by cells from the blood
Steroids		
Glucocorticoids (e.g., cortisol)	Adrenal cortex	Decrease inflammation, increase resistance to stress
Mineralocorticoids (e.g., aldosterone)	Adrenal cortex	Maintain salt and water balance
Androgens (e.g., testosterone)	Gonads and adrenal cortex	Promote development of secondary sex characteristics, particularly in males
Estrogens (e.g., estrone)	Gonads and adrenal cortex	Promote development of secondary sex characteristics, particularly in females
Amino acid derivatives		
Epinephrine (adrenaline)	Adrenal medulla	Increases heart rate and blood pressure
Norepinephrine	Adrenal medulla	Decreases peripheral circulation
Thyroxine (T_4)	Thyroid	Stimulates metabolism
Triiodothyronine (T_3)	Thyroid	Stimulates metabolism
Eicosanoids		
Prostaglandins	Cell membranes	Stimulate smooth muscle contraction; regulate inflammatory reactions

8.2.4. Genetic Diseases

Although metabolism is carefully regulated, abnormalities do occur. Some abnormalities result from major dietary changes or specific pathological states. Other abnormalities result from *mutations* in the genetic makeup of an organism and constitute **genetic** or **hereditary diseases.** Each genetic disease reflects a *deficiency* of a specific enzyme or other protein.

The deficiency may result from one of two causes. First, the specific enzyme or other protein *may not be synthesized at all.* In this case, the gene coding for the enzyme or protein is either missing or not expressed. Second, the

enzyme or other protein may be synthesized, but *in an altered and ineffective form.* For example, an enzyme may be formed with a nonfunctional active site.

In some instances, we know the nature of the defect resulting in the genetic disease. We might be able to distinguish between the two causes of deficiency by determining whether or not antibodies prepared against the protein produce an antigen–antibody reaction with cell-free extracts expected to contain the protein. A positive antigen–antibody reaction indicates that the enzyme or other protein (the antigen) is probably present but does not react properly with its substrate.

Whatever the underlying reason, an enzyme or pro-

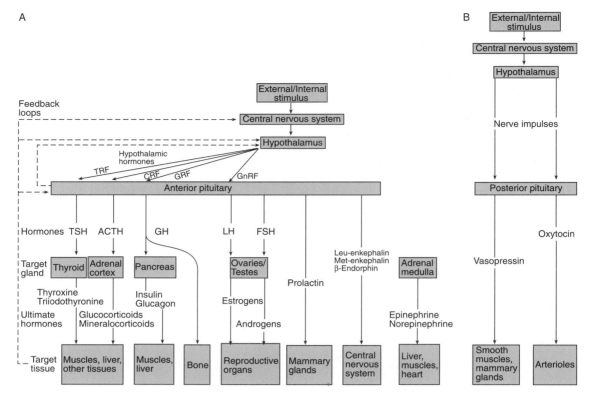

Figure 8.8.   The hormonal control network in animals involves both the anterior (A) and the posterior (B) pituitary gland.

Table 8.3. Some Human Genetic Diseases

Disease	Symptoms	Defective enzyme or protein
Acanthocytosis	Abnormal erythrocytes with projecting spines	Low-density lipoprotein
Albinism	Lack of skin pigmentation	Tyrosinase
Analbuminemia	Impaired synthesis of serum albumin	Serum albumin
Argininemia	Elevated levels of arginine in blood and urine	Arginase
Diabetes	Deranged carbohydrate metabolism	Insulin
Dwarfism	Premature arrest of skeletal growth	Growth hormone
Galactosemia	Inability to convert galactose to glucose metabolites	Galactokinase
Gaucher's disease	Accumulation of cerebrosides in tissues	Glucocerebrosidase
Glycogen storage disease II	Accumulation of normal glycogen in all organs	α-1,4-Glucosidase
Glycogen storage disease IV	Glycogen with very long unbranched chains in liver and spleen	Branching enzyme
Hemolytic anemia	Excessive destruction of red blood cells	Glucose 6-phosphate dehydrogenase
Lactose intolerance	Inability to metabolize lactose	Lactase
Methylmalonic acidemia	Massive ketosis	Methylmalonyl CoA mutase
Niemann–Pick disease	Sphingomyelin accumulation and mental retardation	Sphingomyelinase
Phenylketonuria	Inability to hydroxylate phenylalanine, associated with mental retardation	Phenylalanine hydroxylase
Sickle-cell anemia	Hemolysis of sickle-shaped red blood cells	Hemoglobin
Tangier disease	Almost complete absence of high-density lipoprotein (HDL) and accumulation of cholesterol	High-density lipoprotein
Tay–Sachs disease	Accumulation of gangliosides, blindness, and brain deterioration	N-Acetylhexosaminidase

tein deficiency results in a **metabolic block** so that a specific reaction cannot proceed. Because of the block, products of the reaction do not form, reactants accumulate, and a set of disease symptoms may develop. Attempts are currently under way to "cure" various genetic diseases by replacing the defective or missing gene with a normal gene from some other source. Such attempts, called *gene therapy,* require *genetic engineering* or *recombinant DNA technology* (see Appendix C).

Scientists have identified a large number of genetic diseases and characterized their specific enzymatic defects. Table 8.3 lists a number of human genetic diseases and their known defects.

8.3. EXPERIMENTAL APPROACHES TO METABOLISM

8.3.1. Levels of Cellular Organization

We can investigate metabolism at different levels of organization, from the intact organism (*in vivo,* i.e., within a living organism) to the isolated molecule (*in vitro,* i.e., outside a living organism). Biochemists face an unavoidable dilemma: the more closely a system reflects its natural state, the more complex its level of organization and the less the amount of specific information we can derive from it. Thus, an entire animal or plant constitutes a fully organized biological system but, unfortunately, lends itself to only a few, relatively simple studies. Conversely, a purified enzyme or other protein can be studied in great detail. However, in the process of isolating and purifying the enzyme or protein, organization at all levels had to be destroyed.

In one type of whole-organism study, we determine the effects on metabolism produced by surgically removing an endocrine gland. Another experiment with whole organisms consists of performing a **balance study.** Here we measure the difference between intake of a substance and output of the same or a different substance. *Nitrogen balance,* for example, represents the difference between the amount of nitrogen ingested and the amount of nitrogen excreted. *Respiratory quotient* represents the ratio of moles of carbon dioxide produced to moles of oxygen consumed.

Physicians also employ measurements performed on intact organisms when they diagnose and treat medical problems. Some typical examples are construction of a *glucose tolerance curve* (see Section 10.1); determinations of cholesterol, triglycerides, and urea in blood (BUN, blood urea nitrogen); and determinations of urea, uric acid, and phenylpyruvate in urine.

Bacteria are particularly useful for studies requiring an intact organism. We can easily grow bacteria in large numbers because they are small, require relatively simple growth media, and multiply rapidly. Thus, *Escherichia coli (E. coli),* commonly found in mammalian intestines, has a generation time of about 30 minutes in the laboratory so that the number of cells doubles every 30 minutes. We can grow *E. coli* cells on a defined medium containing only glucose, citrate, K_2HPO_4, $MgSO_4$, and $NaHNH_4PO_4$. Using a growth medium composed of only a few components is advantageous for metabolic studies because it simplifies the task of isolating and determining specific metabolites. Because of its short generation time, *E. coli* lends itself well to genetic studies; researchers can investigate a large number of generations in a limited time.

An organizational level less complex than an intact organism consists of an isolated and perfused organ. When we *perfuse* an organ, we pass blood, plasma, or other fluids through the organ's blood vessels. With this technique, we can measure the metabolism of a particular substance by the isolated organ. For example, passing a known amount of glucose through a perfused liver and analyzing the effluent allows us to determine the rate of glucose oxidation by the liver. Surgeons use perfusion to keep organs "alive" during organ transplantation.

A level of organization below that of a perfused organ comprises several types of tissue preparations. These include thin sections of tissue (*tissue slices*), specific types of cells maintained in an artificial medium (*tissue culture*), and tissue disrupted in a blender (*tissue homogenate*). Researchers have used tissue slices for many years to study aerobic metabolism by determining oxygen uptake and carbon dioxide evolution from manometric measurements.

Studies at the lowest level of cellular organization involve those of a purified enzyme or other protein. Such studies can provide a wealth of information about the isolated molecule. Keep in mind, though, that the properties determined may differ from those the molecule has inside the cell, where it may be located in a specific compartment, be adsorbed to a particular subcellular structure, or be subject to unique control mechanisms.

8.3.2. Radioactive Labels

Studies at all levels of cellular organization can be enhanced by using radioactive isotopes (see Appendix C). Unlabeled and radioactively labeled compounds have identical chemical properties and undergo the same reactions in metabolism. Using compounds that carry radioactive labels provides ways of tracing their metabolic fates in a cell or in an organism. We can illustrate this by asking the following question: Can the carbons of glucose, administered to a rat, be converted to carbons of exhaled CO_2?

$$\text{Glucose} \xrightarrow[\text{(Rat)}]{??} CO_2$$

Without using isotopes, you would find it very difficult to investigate this problem. However, by means of isotopes you can readily arrive at an answer to the question. To do so, you would administer some ^{14}C-labeled glucose to a rat, collect the exhaled CO_2 as a function of time, and analyze the gas for radioactivity. If the exhaled CO_2 becomes radioactive during the experiment, you can conclude that at least some of the carbons of glucose can be converted to those of CO_2. Glucose, in other words, can serve as a metabolic **precursor** of CO_2 (a precursor is a substance that precedes another in a metabolic pathway). Whether the conversion of glucose to CO_2 occurs by means of a single step or via a long series of reactions cannot be answered by your experiment. Note that even though glucose undergoes numerous reactions in metabolism, the use of radioisotopes allows you to study one particular reaction (or set of reactions) in a very complex system.

The use of isotopes has been invaluable in elucidating metabolic pathways. Continuing with the above example, suppose you have reason to suspect that *pyruvate* serves as an intermediate in the conversion of glucose to CO_2. To investigate your hypothesis, you repeat the experiment, administer radioactive glucose to a rat, isolate pyruvate, and determine whether or not it contains any ^{14}C. If the isolated pyruvate is radioactive, you conclude that it, too, must serve as a precursor of CO_2 so that you can expand the pathway to read:

$$\text{Glucose} \longrightarrow \text{pyruvate} \longrightarrow CO_2$$

To illustrate another use of isotopes, assume that you wanted to determine where proteins are being synthesized in the bacterial cell. To find the answer, you could add radioactively labeled amino acids to a growing culture of cells and allow the cells to incorporate the label for a very short time. You would then stop protein synthesis, harvest the cells, lyse them, collect various subcellular fractions (mitochondria, cell walls, etc.), and determine their radioactivity. You would find that the ribosome fraction has the greatest concentration of radioactive amino acids. You conclude that ribosomes very likely represent the sites of intracellular protein synthesis.

8.3.3. Mutants

Mutants provide another useful tool for studying metabolism. Bacterial mutants have similar genetic defects to the defects found in humans. However, whereas human genetic diseases are difficult to study because they occur at low frequencies, bacterial mutants can be produced in the laboratory at high frequencies and studied with ease. One way to produce mutants consists of exposing bacteria to mutagenic agents, such as X rays or specific chemicals. The resultant mutations in DNA lead to changes in bacterial metabolism.

Biochemists find *nutritional (auxotrophic) mutants* particularly useful for metabolic studies. An auxotroph is a microorganism that has a block in a metabolic pathway due to lack of a specific enzyme or presence of a defective enzyme. Such mutants require for their growth either the product of the enzymatic reaction or other metabolites not required by the normal, *wild-type* organism.

As an example, suppose you know that a given microorganism can synthesize the amino acid lysine when provided with compound A. There must, therefore, exist a pathway in this organism such that

$$\text{A} \dots\dots\dots\dots\dots\dots \longrightarrow \text{Lys}$$

Assume further that, when you expose the wild-type organism to mutagenic agents, several mutants (I–III) form. When you isolate the mutants and determine their growth requirements, you find that they will not grow on the medium of the wild type unless you supplement that medium with additional nutrients.

You discover that you can get mutant I to grow only by addition of lysine. When you add lysine, the organism grows and produces an excess of compound D. You may conclude that D serves as an intermediate in the pathway of lysine biosynthesis. Moreover, because it is likely that only one enzyme has been damaged in mutant I, you deduce that the conversion of D to lysine probably represents the last step in the pathway and that mutant I has a block (X) at this step.

$$\text{Mutant I} \quad \text{A} \dots\dots\dots\dots \longrightarrow \text{D} \overset{X}{\longrightarrow} \text{Lys}$$

You interpret the results obtained with the other mutants in a similar fashion. Mutant II grows when you add *either* lysine or D and produces an excess of compound C so that

$$\text{Mutant II} \quad \text{A} \dots\dots\dots\dots \text{C} \overset{X}{\longrightarrow} \text{D} \longrightarrow \text{Lys}$$

Mutant III grows when you add lysine, D, or C and produces an excess of compound B. It follows that

$$\text{Mutant III} \qquad \text{A.........B} \xrightarrow{\quad X \quad} \text{C} \longrightarrow \text{D} \longrightarrow \text{Lys}$$

You can see how you begin to elucidate the pathway from A to lysine. At this point, you must still determine whether the conversion of A to B proceeds in a single step or via a multiple-step sequence.

8.4. NUTRITIONAL ASPECTS

8.4.1. Digestion and Absorption

Digestion represents the first stage in metabolism in humans and other animals. It comprises the hydrolysis of food macromolecules to smaller molecules in the digestive tract, followed by their absorption across the intestinal membrane into the bloodstream or lymph and their transport to the tissues. We refer to the agents of digestion as *digestive fluids* and distinguish five such fluids (Table 8.4).

Hydrolysis during digestion is enzymatic in nature, except for the acid hydrolysis of proteins in the stomach. Strictly speaking, *bile* is not a *digestive* fluid since it does not contain enzymes. However, it does play an important role in lipid digestion. Bile empties into the intestine and acts as an *emulsifying agent,* hence its inclusion with the digestive fluids.

Saliva contains an *amylase* that catalyzes starch hydrolysis. Because of food's short exposure to the enzyme, only partial degradation of starch to oligosaccharides takes place in the mouth. *Gastric juice* has a pH of about 1–2 (it represents a 0.10–0.15M HCl solution), low enough to result in some hydrolysis of peptide bonds. Additionally, stomach secretions contain *pepsin,* an unusual enzyme that not only functions at the low pH of the stomach but actually has its optimum pH value in that range. Owing to the combined action of HCl and pepsin, a significant amount of protein digestion takes place in the stomach. Gastric juice also contains an acid-stable *lipase* that results in a small amount of lipid digestion.

Most of the digestion of the three major nutrients—carbohydrates, lipids, and proteins—occurs through the action of the *pancreatic* and *intestinal fluids.* Nucleic acids as well are digested primarily by action of these digestive fluids. Both fluids contain many hydrolytic enzymes that catalyze the degradation of oligo- and polysaccharides, oligo- and polypeptides, lipids, and oligo- and polynucleotides. Pancreatic fluid contains large amounts of protease zymogens that undergo conversion to active enzymes in the intestine.

The combined action of the digestive fluids and bile results in degradation of nutrient macromolecules to small building blocks—proteins to amino acids and di- and tripeptides; carbohydrates to mono- and oligosaccharides; and lipids to fatty acids, monoacylglycerols, cholesterol, and other lipid components.

Protein digestion also destroys the antigenic character of foreign proteins taken in through the diet. If foreign proteins were not degraded in this fashion and were absorbed intact into the bloodstream, humans would experience a powerful, and possibly fatal, immunological reaction called *anaphylactic shock.* When such a reaction occurs, the immune system overreacts, and antigen–antibody complexes trigger the release of harmful substances, notably histamine, by a class of body cells termed mast cells. The amino acids produced from dietary proteins are, of course, indistinguishable from those found in humans and do not trigger any immunological response.

Following the digestive process, nutrients are absorbed across the membranes of the epithelial cells of the intestine. Free amino acids are absorbed via a carrier-mediated transport. Small peptides are taken up by means of specific transport systems and are generally hydrolyzed by cytoplasmic enzymes of these cells to amino acids. The portal vein then transports all of the amino acids to the liver (Figure 8.9).

Oligosaccharides are hydrolyzed to monosaccharides by surface enzymes of the epithelial cells. All of the monosaccharides are absorbed by means of carrier-mediated transport and moved via the portal vein to the liver.

Bile salts solubilize lipid components by forming micelle-like structures that diffuse into the epithelial cells. This absorption is virtually complete for free fatty acids and monoacylglycerols but is less efficient for other lipids. Only 30–40% of dietary cholesterol, for example, is absorbed.

The fate of the absorbed fatty acids inside the cells depends on the chain lengths of the fatty acids. Fatty acids of medium chain length (C_6–C_{10}) pass through the cells into the portal blood and are carried directly to the liver.

Table 8.4. Digestion of Carbohydrates, Lipids, and Proteins

Digestive fluid	Compounds digested		
	Carbohydrates	Lipids	Proteins
Saliva	Some	None	None
Gastric juice	None	Some	Yes
Pancreatic fluid	Yes	Yes	Yes
Intestinal fluid	Yes	Yes	Yes
Bile	None	Yes	None

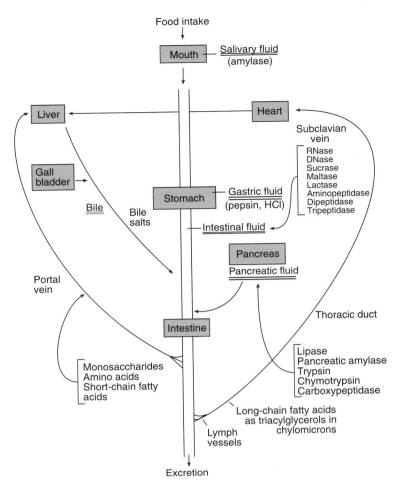

Figure 8.9.   Some components of the digestive system in animals.

Long-chain fatty acids ($>C_{12}$) are resynthesized into tri-acylglcyerols that form parts of *chylomicrons,* which leave the intestine via the lymphatic system and bypass the liver. We refer to the suspension of chylomicrons in the lymphatic system present after a meal as *chyle.* Intestinal lymph vessels drain into large body veins via the thoracic duct. Blood in large veins first reaches peripheral tissues, including adipose and muscle tissue, before coming in contact with the liver.

8.4.2. Dietary Nutrients

An organism's nutritional requirement usually reflects its biosynthetic capabilities. Many prokaryotes require only carbon, energy, and a few simple inorganic salts for growth. As you read earlier, *E. coli* can grow when provided with a few simple inorganic salts and glucose as a carbon and energy source. From these nutrients, the bacterium can synthesize all 20 amino acids and many other compounds. Animals, on the other hand, require a larger number of nutrients and have a more limited capacity for amino acid synthesis. Humans require some 50 different nutrients (Table 8.5) and can only synthesize 10 amino acids (see Table 2.4).

We classify nutrients required by humans as *macronutrients,* required in gram quantities per day, and *micronutrients,* needed only in milligram or microgram amounts. Carbohydrates, lipids, and proteins constitute macronutrients. Vitamins and minerals represent micronutrients. Minerals comprise *macro-* and *microminerals.* We require macrominerals (calcium, chlorine, magnesium, phosphorus, potassium, sodium, and sulfur) in the diet in relatively large amounts. Microminerals, or *trace elements,* are required only in small amounts.

Table 8.6 shows the *recommended dietary allowances* (*RDAs*) of some minerals and vitamins. These daily allowances are not minimal values needed to avoid clinical symptoms of deficiency but rather amounts that provide an ample safety margin. RDAs are established by the *Food and Nutrition Board* of the *National Research Council* (an agency of the *National Academy of Sciences*) for a normal individual engaged in average activity and living in a temperate climate. RDAs represent general guidelines that vary with age, sex, physiological condition, and activity.

For some vitamins and minerals, less information exists on which to base dietary allowances. Instead, recommended amounts are quoted in terms of ranges of *estimated safe and adequate daily dietary intakes (ESADDIs)*. Table 8.7 lists a number of examples.

Two other general nutritional recommendations have emerged in recent times: to decrease fat intake, especially intake of saturated fat, and to include *dietary fiber*—components of food that cannot be degraded by human digestive enzymes. Dietary fiber consists primarily of cellulose and other nondigestible cell-wall polymers of plants. Cellulose and hemicellulose enhance intestinal motility, increase stool bulk, and decrease the stool's transit time through the colon. Researchers believe that dietary fiber may reduce the risk of colon cancer because the fiber lowers the concentration of a potential carcinogen

Table 8.5. Nutrients Requied by Humans

Energy sources	Pyridoxine (vitamin B_6)
Carbohydrates[a]	Pantothenic acid
Fats	Folic acid
Proteins	Biotin
	Vitamin B_{12}
Essential amino acids	Ascorbic acid
Arginine (in young)	
Histidine	Water
Isoleucine	
Leucine	Minerals
Lysine	Arsenic[b]
Methionine	Calcium
Phenylalanine	Chlorine
Threonine	Chromium
Tryptophan	Copper
Valine	Fluorine
	Iodine
Essential fatty acids	Iron
Linoleic acid	Magnesium
Linolenic acid	Manganese
	Molybdenum
Vitamins (fat-soluble)	Nickel[b]
Vitamin A	Phosphorus
Vitamin D	Potassium
Vitamin E	Selenium
Vitamin K	Silicon
	Sodium
Vitamins (water-soluble)	Tin[b]
Thiamine (vitamin B_1)	Vanadium[b]
Riboflavin (vitamin B_2)	Zinc
Niacin	

[a]Recommended carbohydrate intake is 100 g/day.
[b]Believed to be required by humans.

Table 8.6. Recommended Dietary Allowances (RDAs) for the Average Human between Ages 19 and 24[a]

Nutrient	RDA	
	Male	Female
Protein	58 g	46 g
Fat-soluble vitamins		
Vitamin A	1000 μg RE[b]	800 μg RE[b]
Vitamin D	10 μg[c]	10 μg[c]
Vitamin E	10 mg α-TE[d]	8 mg α-TE[d]
Vitamin K	70 μg	60 μg
Water-soluble vitamins		
Vitamin C	60 mg	60 mg
Thiamine (vitamin B_1)	1.5 mg	1.1 mg
Riboflavin (vitamin B_2)	1.7 mg	1.3 mg
Vitamin B_6	2.0 mg	1.6 mg
Vitamin B_{12}	2.0 μg	2.0 μg
Niacin	19 mg NE[e]	15 mg NE[e]
Folate	200 μg	180 μg
Minerals		
Calcium	1200 mg	1200 mg
Phosphorus	1200 mg	1200 mg
Magnesium	350 mg	280 mg
Zinc	15 mg	12 mg
Iron	10 mg	15 mg
Iodine	150 μg	150 μg
Selenium	70 μg	55 μg
Energy[f]	12,134 kJ	9,205 kJ
	(2,900 kcal)	(2,200 kcal)

[a]Source: National Research Council (1989).
[b]RE, Retinol equivalent (1 RE = 1 μg of retinol or 6 μg of β-carotene).
[c]As cholecalciferol (10 μg of cholecalciferol = 400 IU of vitamin D).
[d]α-TE, α-Tocopherol equivalent (1 α-TE = 1 mg of d-α-tocopherol).
[e]NE, Niacin equivalent (1 NE = 1 mg of niacin or 60 mg of dietary tryptophan).
[f]These values may be compared with those required to maintain the *basal metabolic rate (BMR)*, the rate of energy utilization in the resting state. Energy requirements for BMR are about 7531 kJ (1800 kcal) for males and 5439 kJ (1300 kcal) for females.

Table 8.7. Estimated Safe and Adequate Daily Dietary Intakes (ESADDIs) of Selected Vitamins and Minerals[a]

Nutrient	ESADDI		
	Children (1–3 yr)	Adolescents (11+ yr)	Adults
Vitamins			
Biotin	20 μg	30–100 μg	30–100 μg
Pantothenic acid	3 mg	4–7 mg	4–7 mg
Minerals			
Copper	0.7–1.0 mg	1.5–2.5 mg	1.5–3.0 mg
Manganese	1.0–1.5 mg	2.0–5.0 mg	2.0–5.0 mg
Fluoride	0.5–1.5 μg	1.5–2.5 μg	1.5–4.0 μg
Chromium	20–80 μg	50–200 μg	50–200 μg
Molybdenum	25–50 mg	75–250 mg	75–250 mg

[a]Source: National Research Council (1989).

(by increasing stool bulk) and accelerates its passage through the colon. Some water-soluble types of fiber, such as pectin and gums, form viscous solutions in the stomach and the intestine. This decreases the rate of carbohydrate digestion and absorption. Consequently, the level of blood sugar rises more slowly after a meal. Soluble fibers also help to lower the level of serum cholesterol in most people. Researchers have proposed that this effect results from increased fecal excretion of bile acids that are major degradation products of cholesterol.

8.4.3. Vitamins

Vitamins are integral components of the diet of humans and other animals, and their importance for general health and growth is well known. We define a **vitamin** as an organic compound that (a) occurs in natural foods in extremely small concentrations and is distinct from carbo-hydrates, lipids, proteins, and nucleic acids; (b) is required by the organism (generally restricted to animals) in minute amounts for normal health and growth; (c) when absent from the diet, or improperly absorbed from the intestine, leads to development of a specific deficiency disease; and (d) cannot be synthesized by the organism or is not synthesized in sufficient amount and must, therefore, be obtained through the diet. The term *vitamin* was proposed by Casimir Funk in 1912 and was originally spelled "vita-mine" to designate a "vital amine" from rice husks that cured the disease beriberi. We now know that this "vital amine" is thiamine or vitamin B_1. Funk thought that all vitamins were amines, but other scientists showed later that that was incorrect, and the "e" was omitted from the name.

Because all vitamins occur naturally, there clearly exist organisms capable of synthesizing each vitamin. However, a compound that is a vitamin for one organism may not be a vitamin for another organism. We classify vi-

Table 8.8. Water-Soluble Vitamins and Their Coenzymes

Coenzyme	Reaction mediated	Vitamin precursor	Deficiency disease	Section
L-Ascorbic acid	Hydroxylation	L-Ascorbic acid (vitamin C)	Scurvy	5.1
Biotin attached to ε-NH_2 of Lys (biocytin)	Carboxylation	Biotin (vitamin B complex)	Dermatitis (humans)	13.5
Cobamide coenzymes	Alkylation	Cobalamin (vitamin B_{12})	Pernicious anemia	13.3
Coenzyme A	Acyl transfer	Pantothenic acid (vitamin B complex)	Dermatitis (chickens)	11.1
Flavin coenzymes	Oxidation–reduction	Riboflavin (vitamin B_2)	Growth retardation	11.1
Folate coenzymes	One-carbon group transfer	Folic acid (vitamin B complex)	Macrocytic anemia	14.4
Lipoic acid[a] attached to ε-NH_2 of Lys (lipoamide, lipoyllysine)	Acyl transfer	Lipoic acid[a] (vitamin B complex)	Growth deficiencies	11.1
Nicotinamide coenzymes	Oxidation–reduction	Niacin (nicotinic acid; vitamin B complex)	Pellagra	11.1
Pyridoxal phosphate	Amino group transfer (transamina-tion)	Pyridoxine, pyridoxal, and pyridoxamine (vitamin B_6)	Dermatitis (rats); neu-rological symptoms	14.2
Thiamine pyrophosphate	Aldehyde transfer and decarboxylation	Thiamine (vitamin B_1)	Beriberi	11.1

[a]Lipoic acid is a vitamin-like compound but not a vitamin.

tamins into two broad classes, *fat-soluble* and *water-soluble vitamins*

Fat-soluble vitamins are soluble in organic solvents and insoluble in water. They are structurally related but have diverse functions. The group includes vitamins A, D, E, and K and was discussed in Chapter 6.

Water-soluble vitamins include the B vitamins, or *vitamin B complex,* as well as some others, such as vitamin C. They have diverse structures but show functional similarity; all participate in transfer reactions involving protons, electrons, acyl groups, *one-carbon fragments* (HCHO, —CH$_2$OH, —CH$_3$, and the like), or other metabolic groups.

As you read in Section 4.4, water-soluble vitamins form parts of **coenzymes.** Some vitamins represent small parts of coenzymes; others constitute major portions. In one case, ascorbic acid, the vitamin and the coenzyme are one and the same. Sometimes the vitamin functions as the "business end" of the coenzyme, site of the chemical reaction. In other instances, the vitamin serves as a structural component but does not participate directly in the reaction. Whatever the particulars, each water-soluble vitamin is required for the biosynthesis of a coenzyme; each serves as a coenzyme precursor. Table 8.8 lists coenzymes and the types of reactions they mediate. We will examine the structures and functions of these coenzymes as we encounter them in later chapters.

Because of the high efficiency of enzymes, most intracellular enzyme and coenzyme concentrations are low. Accordingly, vitamins are required in only small amounts in the diet. For a healthy individual living under suitable conditions, a well-balanced diet probably provides all the necessary types and amounts of vitamins required. However, people suffering from specific ailments or living under special conditions may benefit from supplemental vitamin intake.

Usually, additional moderate intake of water-soluble vitamins poses no harm because they are readily excreted. Fat-soluble vitamins, on the other hand, tend to sequester in lipid-containing structures like membranes and fat cells and may accumulate to toxic levels. Therefore, present-day "megavitamin" regimens must be carried out with caution.

SUMMARY

Metabolism comprises a network of interconnected reactions that consists of two parts: catabolism, or degradative processes, and anabolism, or synthetic processes. Degradation and synthesis of intermediates proceeds via specific metabolic pathways. Because an exergonic reaction is endergonic in the opposite direction, a "downhill" catabolic pathway cannot be simply reversed. Instead, changes must occur at one or more steps of the sequence to circumvent these energy barriers. All metabolic pathways are regulated, and each has a committed step that ensures completion of the sequence of reactions.

Metabolism is regulated by control of individual enzymes; by control of metabolic pathways; by compartmentation of substrates, enzymes, and pathways; and by hormone action. Hormones bind to receptors on cell membranes or on cytoplasmic or nuclear proteins. Some hormones cause release of a second messenger inside the cell.

We can study metabolism at the level of whole organisms, perfused organs, tissue preparations, or purified proteins. The more closely a system reflects its natural state, the more complex its level of organization and the less the amount of specific information that we can derive from it. Biochemists use radioactively labeled compounds and bacterial mutants to elucidate metabolic pathways.

Five digestive fluids accomplish the digestion of macromolecules to smaller building blocks. Most digestion of the major fooodstuffs occurs via the pancreatic and intestinal fluids.

Humans require both macro- and micronutrients, including vitamins, in the diet. Water-soluble vitamins have functional similarity but diversity of structure. All form parts of coenzymes that function in group transfer reactions.

SELECTED READINGS

Beishir, L., *Microbiology in Practice,* 6th ed., HarperCollins, New York (1996).

Bender, D., *Nutritional Biochemistry of the Vitamins,* Cambridge University Press, Cambridge (1992).

Cowan, J. A., *Inorganic Biochemistry—An Introduction,* VCH Publishers, New York (1993).

Ford, T. C., and Graham, J. M., *An Introduction to Centrifugation,* Bios Scientific Publishers, Oxford (1991).

Freshney, R. I., *Culture of Animal Cells: A Manual of Basic Technique,* 3rd ed., Wiley, New York (1993).

Liscum, L., and Underwood, K. W., Intracellular cholesterol transport and compartmentation, *J. Biol. Chem.* 270:15443–15446 (1995).

Matthews, J. C., *Fundamentals of Receptor, Enzyme, and Transport Kinetics,* CRC Press, Boca Raton, Florida (1993).

Norman, A. W., and Litwack, G., *Hormones,* Academic Press, Orlando (1987).

Russell, J. B., and Cook, G. M., Energetics of bacterial growth: Balance of anabolic and catabolic reactions, *Microbiol. Rev.* 59:48–62 (1995).

Scriver, C. R. *The Metabolic and Molecular Bases of Inherited Disease,* 7th ed., McGraw-Hill, New York (1995).

Weindruch, R., Caloric restriction and aging, *Sci. Am.* 274:46–52 (1996).

REVIEW QUESTIONS

A. Define each of the following terms:

Marker enzyme	Digestion
Mitochondria	Balance study
Committed step	Precursor
Compartmentation	Fermentation
Receptor	Genetic (hereditary) disease

B. Differentiate between the two terms in each of the following pairs:

Catabolism/anabolism	Digestion/fermentation
First messenger/second messenger	Metabolism/intermediary metabolism
Alcoholic fermentation/ lactate fermentation	Metabolic block/metabolic pathway

C. (1) Describe the levels at which metabolism can be studied. How does the organization of the system being studied relate to the amount of information that you can derive from it? Why is this so?

(2) What are the three general characteristics of metabolic pathways?

(3) What are the three stages of catabolism?

(4) Outline the main features of the digestion of carbohydrates, lipids, and proteins, beginning with the intake of food through the mouth and ending with absorption of components across the membranes of intestinal cells.

(5) How is the hypothalamus–pituitary hormonal control network organized?

PROBLEMS

8.1. Why is it advantageous to control a metabolic pathway by means of its committed step?

8.2. The 12-hour human gastric secretion of HCl typically varies from about 150 to 1000 ml. Calculate the number of protons in these two volumes of output on the assumption that the secretion is $0.150M$ in HCl.

8.3. Assume that you have disrupted mitochondria and obtained three fractions designated A, B, and C. You then assay the three fractions for a number of different enzymes with the following results (for each enzyme the total activity in all three fractions is 100%):

	Percent of total activity		
	A	B	C
Monoamine oxidase	5	82	13
Cytochrome oxidase	10	11	79
Fumarase	85	7	8

On this basis, identify fractions A, B, and C as corresponding primarily to matrix, inner membrane, or outer membrane preparations.

8.4.* You can see that each mitochondrial preparation in the previous problem is contaminated by material from the other two fractions. Assuming that a unit of enzymatic activity reflects the same amount (in milligrams) of subcellular fraction for all three enzymes, estimate the approximate contamination of each fraction by the other two.

8.5.* A liver is perfused with a $1.50 \times 10^{-3}M$ solution of glucose. When the flow rate is adjusted to 0.500 ml/min, the exiting solution has a glucose concentration of $1.30 \times 10^{-3}M$. What is the rate of glucose metabolism by the liver in terms of milligrams of glucose per hour? The molecular weight of glucose is 180.

8.6. You incubate aliquots of a cell-free extract with a radioactively labeled compound (X) for various lengths of time. Following the incubation, you fractionate the aliquots and identify the labeled compounds. You obtain the following results:

Incubation time (min)	Labeled compounds found
5	K
10	K, G
25	K, G, T

Deduce the pathway whereby compound X is metabolized in this cell-free extract.

8.7. In the illustration on the use of radioactively labeled amino acids to identify the sites of protein synthesis (Section 8.3), the bacterial culture was exposed to the label for only a *short time*. Why does this need to be stressed? What would happen if the organisms are allowed to grow the entire time in the presence of labeled amino acids? (Hint: Once a polypeptide chain has been completed, it is released from the ribosome.)

8.8. A wild-type bacterium can synthesize compound X when compound Y is present in the growth medium. An auxotrophic mutant grows only if X is present in the medium, and during its growth compound Q accumulates. A second auxotrophic mutant can grow when *either* X or Q is present in the medium, and during its growth compound S accumulates. Sketch the metabolic pathway that these data indicate.

8.9.* A solution of a ^{14}C-labeled compound that is not metabolized contains 50.0 mmol/ml and has a specific activity of 2.40×10^3 cpm/μmol (cpm = counts per minute; see Appendix C). A patient receives an intravenous injection of 2.00 ml of this solution. After a short time, the radioactivity is evenly distributed throughout the blood. At that point, 5.00 ml of blood is collected from the patient. The protein in the blood is removed by precipitation (the precipitate is not radioactively labeled), and the supernatant contains a total of 4.00×10^5 cpm. What is the patient's total blood volume?

8.10. In a laboratory experiment, two rats are injected with labeled compounds that are not metabolized and that are excreted at identical rates. One rat receives 1.80 millicuries (see Appendix C) of a ^{14}C-labeled compound, and the other receives 20.0 millicuries of a ^{35}S-labeled compound. The half-life (see Appendix C) of ^{14}C is 5568 years, and that of ^{35}S is 87.1 days. Which animal will have the larger number of millicuries left in its system after 3, 6, and 9 months?

8.11.* What is the half-life of an isotope if 50 atoms out of 1000 disintegrate in one year? (See Appendix C.)

8.12. Damage to the pituitary gland results in a condition known as *diabetes insipidus* and is characterized by a massive flow of dilute urine and an unquenchable thirst. Why is this so?

8.13. Based on the data in Table 8.6, how many calories above those needed to maintain the basal metabolic rate do men and women aged 19–24 need each day?

Bioenergetics

9

Broadly speaking, metabolism deals with transformations of substances and transformations of energy. How usable energy is derived from nutrients and how it is used to drive metabolic processes are questions fundamental to understanding the workings of metabolism.

Historically, little progress was made in biochemistry until a key principle of biochemical energetics had been formulated. Antoine Lavoisier recognized in 1777 that cellular respiration was slower than combustion but not essentially different from it. Both processes accomplish the oxidation of foodstuffs to carbon dioxide and water. Lavoisier's observation led to the realization that biochemical reactions have the same characteristics as ordinary chemical reactions performed in the laboratory.

Specifically, investigators concluded that life plays by the rules of thermodynamics. A reaction in a living system has the same free energy change associated with it as the corresponding *in vitro* reaction. We now understand that this must be so because free energy, like other thermodynamic functions, *depends only on the final and initial states* of the system and not on the mechanism, or pathway, whereby the system proceeds from the initial to the final state. Hence, the conversion of glucose to CO_2 and water must have the same free energy change associated with it whether it is carried out by combustion in a calorimeter or by enzymatic reactions in a living cell.

9.1. FREE ENERGY

Free energy (**G** for **Gibbs free energy**) is the key thermodynamic function in biochemical systems. It allows us to deduce the direction that a spontaneous reaction will take and indicates the energetic yield or "cost" of a process. A reaction with a negative free energy change (an *exergonic* reaction) proceeds spontaneously (Figure 4.4); a reaction with a positive free energy change (an *endergonic* reaction) does not proceed spontaneously. An endergonic reaction is thermodynamically not feasible unless energy is applied to it.

We cannot determine the actual free energy of any process or substance; we can only determine the *change* in free energy (ΔG). The symbol delta (Δ) always represents the difference between *final* and *initial* states. Thus, ΔG of a chemical reaction represents the difference between the free energies of the *products* and the *reactants*:

$$\Delta G_{reaction} = G_{products} - G_{reactants}$$

The free energy change is related to two other thermodynamic functions—enthalpy and entropy—at constant temperature and pressure:

$$\Delta G = \Delta H - T\Delta S \qquad (9.1)$$

where ΔH is the change in **enthalpy** (or **heat content**) of the reaction, T is the absolute temperature in degrees Kelvin (K = 273.2 + °C), and ΔS is the change in **entropy,** a measure of randomness. You can infer from Eq. (9.1) that free energy constitutes the difference between the total heat content of a reaction and the energy lost by randomnization ($T\Delta S$); it represents the fraction of the total energy of a system that can be used to do work (hence, the term *free* energy).

The units of free energy are joules (J) per mole or kilojoules (kJ) per mole [calories (cal) per mole or kilocalories (kcal) per mole; 1 cal = 4.184 J]. As these units indicate, the free energy yield of a reaction depends on the *number of moles* of reactants that undergo reaction.

9.1.1. Standard and Actual Free Energy Changes

9.1.1A. Standard Free Energy Change.

We base all thermodynamic functions on arbitrarily selected reference states, called *standard states,* that represent particular sets of conditions. For solutions, we use as standard state a solute concentration of $1.0M$, a temperature of 25°C, and a pressure of 1.0 atm. We call the free energy change that corresponds to this state the **standard free energy change** and designate it $\Delta G°$.

You can calculate the standard free energy change of a reaction from its equilibrium constant (K'_{eq}) by means of the equation

$$\Delta G° = -RT \ln K'_{eq} \qquad (9.2)$$

where ln is the natural logarithm (ln = 2.303 $\log_{10}$), T is the absolute temperature, and R is the gas constant (8.314 J deg^{-1} mol^{-1}; 1.987 cal deg^{-1} mol^{-1}). Table 9.1 illustrates the quantitative relationship between the standard free energy change and the equilibrium constant.

You can see from Table 9.1 that the larger the equilibrium constant is (product concentrations greatly exceed reactant concentrations at equilibrium), the more negative the free energy change. Remember, however, that the *magnitude of the free energy* change tells you nothing about the *rate of the reaction*. A reaction with a highly negative free energy change does not necessarily proceed rapidly; the reaction rate depends on the magnitude of the *energy of activation* (Figure 4.4).

Table 9.1. Relationship between the Equilibrium Constant (K'_{eq}) and the Standard Free Energy Change ($\Delta G°$) at 25°C[a]

K'_{eq}	$\log K'_{eq}$	$\Delta G°$ (J mol^{-1})
0.001	−3	+17,117
0.01	−2	+11,410
0.1	−1	+5,707
1	0	0
10	1	−5,707
100	2	−11,410
1000	3	−17,117

[a]The same relationship holds between the biochemical equilibrium constant (K'_{bio}) and the biochemical standard free energy change ($\Delta G°'$), both of which are discussed in Section 9.1.2.

We can define $\Delta G°$ in two ways. One represents a *mathematical definition* and consists of Eq.(9.2). Given a reaction for which you know the equilibrium constant, you can compute the *value* of $\Delta G°$ from Eq.(9.2).

But what exactly does this calculated value of $\Delta G°$ mean in physical terms? To what specific reaction conditions does it refer? We might call the answer to these questions a *conceptual definition* of $\Delta G°$ and can phrase it as follows:

> The standard free energy change of a reaction is the free energy change associated with the reaction when all reactants and all products are at an initial concentration of $1.0M$ each, the temperature is 25°C, the pressure is 1.0 atm, and the reaction is allowed to proceed to equilibrium.

It is immaterial how, or whether, this set of conditions can be attained. A $1.0M$ concentration, for example, might exceed the solubility of a given compound. These specifications serve merely as a *reference for defining* the free energy change. *If* we mix reactants and products under these conditions, *then* the free energy change associated with the reaction has the value denoted as $\Delta G°$.

We must distinguish clearly between the two ways of defining $\Delta G°$. We base our mathematical evaluation of $\Delta G°$ on the equilibrium constant of the reaction. By contrast, the physical meaning of $\Delta G°$ does not pertain to a state of equilibrium but rather to the reaction proceeding under specified conditions. Let us illustrate these concepts by considering the ionization of acetic acid:

$$CH_3COOH \rightleftharpoons CH_3COO^- + H^+$$

This reaction has an equilibrium constant of 1.76×10^{-5} at 25°C. To calculate the *numerical value* of $\Delta G°$, we use Eq.(9.2):

$$\Delta G^\circ = -(8.31)\,(298.2)\,(2.303)\,\log(1.76 \times 10^{-5})$$

$$= 27{,}150 \text{ J mol}^{-1}$$

$$= 27.2 \text{ kJ mol}^{-1}$$

Now ask yourself what this calculated quantity *means* in physical terms. The answer is that if you were to mix CH_3COOH, CH_3COO^-, and H^+ in such a fashion that each species had an initial concentration of $1.0M$ in the reaction mixture, the temperature was kept at 25°C, and the pressure was kept at 1.0 atm, then the tendency of the reaction to proceed to equilibrium *under these conditions* is described by ΔG°. Because ΔG° is very positive, the reaction is strongly endergonic and will not proceed as written. (In fact, the reaction will proceed spontaneously in the opposite direction, from right to left.) At the low pH of zero, acetic acid will not ionize; the high proton concentration ($[H^+] = 1.0M$) depresses the ionization of the acid.

9.1.1B. Actual Free Energy Change.

Clearly, not all reactions commence with standard 1.0 molar concentrations of reactants and products. What free energy changes are associated with initial nonstandard reaction conditions? We describe such reactions by a free energy change that, to emphasize its distinction from ΔG°, we term an **actual free energy change (ΔG)**. It is defined by the equation

$$\Delta G = \Delta G^\circ + RT \ln \frac{[\text{products}]}{[\text{reactants}]} \tag{9.3}$$

where the ratio of products to reactants is that under *actual initial reaction conditions*. This ratio is *not an equilibrium constant,* so that the last term of Eq. (9.3) *cannot be* replaced by $RT \ln K'_{eq}$!

Only if the reaction is already at equilibrium are the initial concentrations of reactants and products identical to their equilibrium concentrations. *Then, and only then,* can the ratio in Eq. (9.3) be replaced by the equilibrium constant. At equilibrium (and since $\Delta G^\circ = -RT \ln K'_{eq}$), Eq. (9.3) becomes

$$\Delta G = \Delta G^\circ + RT \ln K'_{eq} = -RT \ln K'_{eq} + RT \ln K'_{eq} = 0$$

It follows that *at equilibrium, and only at equilibrium,* ΔG is equal to zero. We can define the actual free energy change in two ways, much as we did for the standard free energy change. Equation (9.3) represents the *mathematical definition*. The *conceptual definition* of ΔG states that ΔG constitutes the free energy change of a reaction in which reactants and products are mixed at the indicated

initial concentrations, at a temperature of 25°C, and at a pressure of 1.0 atm, and the reaction is allowed to proceed to equilibrium.

As you can see from Eq.(9.3), the actual free energy change depends on both the standard free energy change and the actual initial concentrations of reactants and products.

9.1.2. Biochemical Free Energy Changes

9.1.2A. Biochemical Standard Free Energy Change. While ΔG° adequately describes ordinary chemical reactions, it fails to do so for biochemical systems. The reason is that many biochemical reactions involve *protons* as either a reactant or a product. Because ΔG° refers to reaction conditions at which all reactants and products are at an initial $1.0M$ concentration, reactions involving protons would require a proton concentration of $[H^+] = 1.0M$, so that pH = 0.

A pH of zero constitutes a highly "unbiological" condition since most biochemical systems operate in the vicinity of neutrality (pH 7.0). Because ΔG° is not very meaningful under physiological conditions, we use a different standard that we term **biochemical standard free energy change** and designate $\Delta G^{\circ\prime}$ ("delta G zero prime"); it is based on a standard state that has a proton concentration of $10^{-7}M$ (pH 7.0).

We define the biochemical standard free energy change *mathematically* in analogous fashion to ΔG°. Specifically,

$$\Delta G^{\circ\prime} = -RT \ln K'_{bio} \tag{9.4}$$

where K'_{bio} represents the equilibrium constant calculated for the biochemical reaction conditions of pH 7.0. (A detailed treatment of K'_{bio} and its relationship to K'_{eq} is beyond the scope of this book.)

The *conceptual definition* of $\Delta G^{\circ\prime}$ is also analogous to that for ΔG°:

> The biochemical standard free energy change of a reaction is the free energy change associated with the reaction when all reactants and all products are at an initial concentration of $1.0M$ each, except for protons, the initial concentration of which, unless otherwise specified, is taken as $[H^+] = 10^{-7}M$ (pH 7.0); the temperature is 25°C; the pressure is 1.0 atm; and the reaction is allowed to proceed to equilibrium.

Table 9.2 lists biochemical standard free energy changes ($\Delta G^{\circ\prime}$) for a number of reactions.

Let us return to the ionization of acetic acid. $\Delta G^{\circ\prime}$ for this reaction has a value of -12.8 kJ mol^{-1}. This means that

**Table 9.2. Biochemical Standard Free Energy Changes ($\Delta G^{\circ\prime}$)
of Some Reactions of Biochemical Relevance (pH 7.0, 25°C)**

Reaction[a]	$\Delta G^{\circ\prime}$ (kJ mol^{-1})
Hydrolysis	
Fumarate^{2-} + H$_2$O $\rightleftharpoons$ malate^{2-}	−3.3
Glycylglycine + H$_2$O $\rightleftharpoons$ 2 glycine	−9.2
Glucose 6-P^{2-} + H$_2$O $\rightleftharpoons$ glucose + HPO$_4^{2-}$	−13.8
Fructose 1,6-bis-P^{4-} + H$_2$O $\rightleftharpoons$ fructose 6-P^{2-} + HPO$_4^{2-}$	−14.2
Glutamine + H$_2$O $\rightleftharpoons$ glutamate$^-$ + NH$_4^+$	−14.2
AMP^{2-} + H$_2$O $\rightleftharpoons$ adenosine + HPO$_4^{2-}$	−14.6
Maltose + H$_2$O $\rightleftharpoons$ 2 glucose	−15.5
Ethyl acetate + H$_2$O $\rightleftharpoons$ acetate$^-$ + ethanol + H$^+$	−20.1
Acetylcholine$^+$ + H$_2$O $\rightleftharpoons$ acetate$^-$ + choline$^+$ + H$^+$	−25.1
Sucrose + H$_2$O $\rightleftharpoons$ fructose + glucose	−27.6
ATP^{4-} + H$_2$O $\rightleftharpoons$ ADP^{3-} + HPO$_4^{2-}$ + H$^+$	−30.5
ADP^{3-} + H$_2$O $\rightleftharpoons$ AMP^{2-} + HPO$_4^{2-}$ + H$^+$	−30.5
ATP^{4-} + H$_2$O $\rightleftharpoons$ AMP^{2-} + HP$_2$O$_7^{3-}$ + H$^+$	−31.8
HP$_2$O$_7^{3-}$ + H$_2$O $\rightleftharpoons$ 2HPO$_4^{2-}$ + H$^+$	−33.1
Phosphoarginine$^-$ + H$_2$O $\rightleftharpoons$ arginine$^+$ + HPO$_4^{2-}$	−38.1
Phosphocreatine^{2-} + H$_2$O $\rightleftharpoons$ creatine + HPO$_4^{2-}$	−42.7
Acetylphosphate^{2-} + H$_2$O $\rightleftharpoons$ acetate$^-$ + HPO$_4^{2-}$ + H$^+$	−43.1
1,3-bis-P-glycerate^{4-} + H$_2$O $\rightleftharpoons$ 3-P-glycerate^{2-} + HPO$_4^{2-}$	−49.4
Carbamoyl-P^{2-} + H$_2$O $\rightleftharpoons$ carbamate$^-$ + HPO$_4^{2-}$ + H$^+$	−51.5
Phosphoenolpyruvate^{3-} + H$_2$O $\rightleftharpoons$ pyruvate$^-$ + HPO$_4^{2-}$	−61.9
Isomerization	
Dihydroxyacetone P^{2-} $\rightleftharpoons$ glyceraldehyde 3-P^{2-}	+7.7
Citrate^{3-} $\rightleftharpoons$ isocitrate^{3-}	+6.7
Glycerate 3-phosphate^{3-} $\rightleftharpoons$ glycerate 2-phosphate^{3-}	+4.6
Fructose 6-phosphate^{2-} $\rightleftharpoons$ glucose 6-phosphate^{2-}	−1.7
Glucose 1-phosphate^{2-} $\rightleftharpoons$ glucose 6-phosphate^{2-}	−7.3
Oxidation (dehydrogenases)	
Lactate$^-$ + NAD$^+$ $\rightleftharpoons$ pyruvate$^-$ + NADH + H$^+$	+25.1
Ethanol + NAD$^+$ $\rightleftharpoons$ acetaldehyde + NADH + H$^+$	+22.6
Glyceraldehyde 3-P^{2-} + NAD$^+$ + HPO$_4^{2-}$ $\rightleftharpoons$ 1,3-bis-P-glycerate^{4-} + NADH + H$^+$	+6.3
Oxidation (molecular oxygen)	
Glucose + 6O$_2$ $\rightleftharpoons$ 6CO$_2$ + 6H$_2$O	−2870
Palmitic acid + 23O$_2$ $\rightleftharpoons$ 16CO$_2$ + 16H$_2$O	−9782

[a]P = "Phosphate" or "phospho-."

if you were to mix 1.0M CH$_3$COOH, 1.0M CH$_3$COO$^-$, and 10^{-7}M H$^+$ (pH 7.0), the reaction is exergonic and would proceed spontaneously. The low proton concentration favors the ionization by "pulling" the reaction from left to right. By contrast, we saw above that a high proton concentration ([H$^+$] = 1.0M; pH 0) depresses the ionization, so that the standard free energy change (ΔG°) is positive.

9.1.2B. Biochemical Actual Free Energy Change.
Free energy changes of nonstandard reaction mixtures—those having initial concentrations of reactants

and products that are not 1.0M each, but in which the pH is still 7.0—are called **biochemical actual free energy changes** and are designated $\Delta G'$. The quantity $\Delta G'$ is related to $\Delta G^{\circ\prime}$ much as ΔG is related to ΔG°:

$$\Delta G' = \Delta G^{\circ\prime} + RT \ln \frac{[\text{products}]}{[\text{reactants}]} \qquad (9.5)$$

Thus, $\Delta G'$ is a function of both $\Delta G^{\circ\prime}$ and the actual initial concentrations of reactants and products.

You may wonder how $\Delta G'$ differs from ΔG. The an-

Table 9.3. Effect of Temperature on Free Energy Changes[a,b]

Temperature (°C)	ΔH (J mol^{-1})	ΔS (J deg^{-1} mol^{-1})	$-T\Delta S$[c] (J mol^{-1})	ΔG[d] (J mol^{-1})
-10	-5619	-20.54	5406	-213
0	-6008	-21.99	6008	0
$+10$	-6397	-23.39	6623	$+226$

[a]Reprinted, with permission, from I. M. Klotz, *Energy Changes in Biochemical Reactions,* Academic Press, New York (1967).
[b]For the reaction $H_2O(l) \rightleftharpoons H_2O(s)$.
[c]Calculated for T in degrees Kelvin (see Eq. 9.1).
[d]$\Delta G = \Delta H - T\Delta S$ (Eq. 9.1).

swer is that the two quantities are identical ($\Delta G = \Delta G'$). There can be only *one actual free energy change for any given reaction,* regardless of how we designate it. Whether we base this actual free energy change on $\Delta G°$ (and designate it as ΔG) or base it on $\Delta G°'$ (and designate it as $\Delta G'$) is irrelevant. There can only be one set of actual conditions defined by the initial concentrations of reactants and products and by the pH.

Accordingly, the *mathematical* and *conceptual* definitions of ΔG apply to $\Delta G'$ as well, and you can calculate the actual free energy change by means of either Eq.(9.3) or Eq.(9.5). Keep in mind, though, that while the two actual free energy changes are identical, the corresponding standard free energy changes ($\Delta G°$ and $\Delta G°'$) are not. Standard free energy changes differ because we base them on *different standard states.* In order to be consistent, and in order to stress that free energy changes are computed for pH 7.0, we will henceforth use exclusively $\Delta G°'$ and $\Delta G'$.

9.1.3. The Effects of Variables on Free Energy Changes

Temperature and concentration represent two important variables that affect free energy changes of chemical reactions.

9.1.3A. The Effect of Temperature.
Temperature influences free energy changes because of its effect on equilibrium constants. Equilibrium constants vary with temperature as described by the **van't Hoff equation:**

$$\log \frac{K'_{eq_1}}{K'_{eq_2}} = \frac{\Delta H°}{2.303R} \left(\frac{1}{T_2} - \frac{1}{T_1} \right) \qquad (9.6)$$

where K'_{eq_1} and K'_{eq_2} are the equilibrium constants at the absolute temperatures T_1 and T_2, respectively, and $\Delta H°$ is the standard enthalpy change, assumed to be constant over the temperature range T_1–T_2. We can illustrate the effect of temperature by considering the conversion of liquid wa-

ter to ice at -10, 0, and 10°C (Table 9.3). At all three temperatures, the reaction has a negative entropy change; the system becomes less random as liquid water changes to solid ice. The reaction also has a negative enthalpy change at all three temperatures; it is *exothermic* and gives off heat. The free energy change, however, is negative, zero, or positive, and it determines whether the reaction proceeds as written.

At -10°C, the reaction has a negative free energy change and proceeds spontaneously. Water cooled to -10°C freezes to form ice. At 0°C, the reaction has a free energy change of zero, and the system is at equilibrium. A mixture of water and ice will keep as such, provided you maintain the temperature at 0°C. Lastly, at 10°C, the reaction has a positive free energy change and does not proceed as written. At this temperature, water cannot freeze spontaneously to form ice.

9.1.3B. The Effect of Concentration.
Table 9.4 illustrates the effect of concentration by listing free energy changes for the isomerization of glyceraldehyde 3-phosphate (G3P) and dihydroxyacetone phosphate (DHAP) at pH 7.0. This represents a key reaction in *glycolysis,* the major catabolic pathway of carbohydrates.

When reactant and product concentrations are 1.0*M* each (line 1 in Table 9.4), the free energy change consti-

Table 9.4 Effect of Concentration on Free Energy Changes[a]

Dihydroxyacetone phosphate (M)	Glyceraldehyde 3-phosphate (M)	Free energy change at pH 7.0 and 25°C (kJ mol^{-1})	
1.0	1.0	$+7.7$	($\Delta G°'$)
2.0×10^{-1}	9.0×10^{-3}	0	($\Delta G'$)
1.0×10^{-1}	1.0×10^{-4}	-9.5	($\Delta G'$)
1.0×10^{-4}	1.0×10^{-1}	$+24.8$	($\Delta G'$)

[a]For the reaction
Dihydroxyacetone phosphate $\rightleftharpoons$ glyceraldehyde 3-phosphate.
$\qquad$ (DHAP) $\qquad\qquad\qquad\qquad$ (G3P)

tutes $\Delta G^{\circ\prime}$. Its positive value indicates that the reaction is endergonic and does not proceed. The remaining free energy changes of Table 9.4 represent $\Delta G'$ values. Initial concentrations chosen for line 2 lead to a free energy change of zero; the system is *at equilibrium,* and no further net change takes place. You can calculate the ratio of [G3P]/[DHAP] at equilibrium from the data in Table 9.4:

$$(9.0 \times 10^{-3})/(2.0 \times 10^{-1}) = 0.045$$

or from Eq. (9.5) by setting $\Delta G' = 0$ and $\Delta G^{\circ\prime} = +7700$ J mol^{-1}:

$$0 = 7700 + (8.31)(298.2)(2.303) \log([G3P]/[DHAP])$$

$$\log([G3P]/[DHAP]) = -1.3492$$

$$[G3P]/[DHAP] = 0.045$$

Thus, to drive the isomerization toward G3P synthesis, the ratio [G3P]/[DHAP] can have a maximum value of 0.045. Ratios smaller than 0.045 result in exergonic reactions that proceed spontaneously (line 3). Ratios larger than 0.045 lead to endergonic reactions that cannot take place unless driven by added energy (line 4).

Our discussion of the effects of temperature, pH, and concentration on free energy changes leads to a most important conclusion:

> Any chemical reaction may be made to go in one direction or another by suitable changes in temperature and/or pH (if the reaction is pH dependent) and/or the concentrations of reactants and products.

In later chapters, we consider biochemical reactions in terms of their $\Delta G^{\circ\prime}$ values. Remember at all times that a reaction that is feasible on the basis of its $\Delta G^{\circ\prime}$ value *may or may not* be feasible on the basis of its $\Delta G'$ value. Feasibility under intracellular conditions depends on the biochemical *actual* ($\Delta G'$) and not the *standard* ($\Delta G^{\circ\prime}$) free energy change of the reaction. While temperature and pH are constant for both of these free energy changes, concentrations of reactants and products may differ widely. *What ultimately determines whether and in which direction a reaction proceeds inside the cell are the intracellular concentrations of reactants and products.*

9.2. ENERGY-RICH COMPOUNDS

The overall process whereby chemical energy of nutrients is released and used in metabolism involves a series of oxidation–reduction reactions termed **biological oxidation.**

In plants and animals, biological oxidation refers specifically to **aerobic cellular respiration** (Chapters 11 and 12). This comprises the oxidation of nutrients in the *citric acid cycle,* transport of the electrons thus released via the *electron transport system,* and synthesis of ATP in a process called *oxidative phosphorylation.*

9.2.1. Biological Oxidation

After scientists realized that they could treat biochemical reactions like ordinary chemical reactions, they concluded that biological oxidation must be accompanied by large decreases in free energy. The complete oxidation of glucose to CO_2 and H_2O, for example, has a $\Delta G^{\circ\prime}$ value of -2870 kJ mol^{-1}. Investigators then began to ask how these sizable amounts of energy are released and used by an organism.

Immediate release of the entire energy in the form of heat could be ruled out. First of all, producing such large quantities of heat would be devastating to most cells, as it would lead to denaturation of enzymes, other proteins, and nucleic acids. Denaturation of these vital biomolecules would most likely result in cell death. Second, living systems generally operate essentially in an *isothermal* fashion—they function at constant temperature. According to the second law of thermodynamics, one cannot derive useful work from heat in an isothermal system. Even if cells could tolerate the deleterious effects of a large release of heat, they could not derive usable energy from it.

The answer to the problem of energy utilization lies in compounds that serve as temporary storage forms for the energy released from nutrients. We call these compounds **energy-rich compounds** and can outline the entire process as follows:

- Catabolism of foodstuffs releases the chemical energy stored in nutrients in the form of free energy.
- Before the released free energy can be used in anabolism, it must be stored in energy-rich compounds.
- *ATP (5′*-adenosine triphosphate) represents the most prevalent energy-rich compound. Its synthesis from ADP (5′-adenosine diphosphate) and P_i requires the input of 30.5 kJ mol^{-1} and constitutes a strongly endergonic reaction.
- The free energy released from nutrients drives the synthesis of ATP and other energy-rich compounds. These reactions reshuffle the stored chemical energy of nutrients and convert it to stored chemical energy of energy-rich compounds. The process is characteristic of the catabolic phase of metabolism (Figure 9.1).

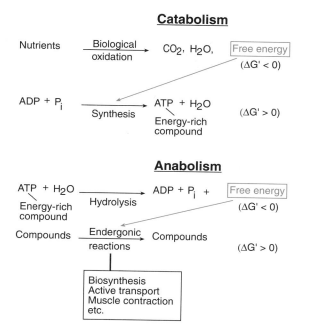

Figure 9.1. The link between free energy changes in metabolism and the role of ATP as an energy-rich compound. Oxidation of nutrients in catabolism releases energy, which drives ATP synthesis. Hydrolysis of ATP provides energy to drive endergonic reactions of anabolism.

- The cell or organism subsequently uses the energy stored in energy-rich compounds by a reversal of the above process. Hydrolysis or other cleavage of an energy-rich compound constitutes an exergonic reaction. The stored free energy released drives an endergonic reaction of metabolism. The process is characteristic of the anabolic phase of metabolism (Figure 9.1).
- The linking of exergonic and endergonic reactions that occurs in both catabolism and anabolism comprises *energetically coupled reactions,* discussed in Section 9.3. We often refer to the entire network of energy metabolism, based on ATP, as the **ATP cycle** (Figure 9.2).

Instead of simple hydrolysis, energy-rich compounds typically undergo an exergonic cleavage reaction, in which part of the molecule is *transferred* to another compound. For this reason, biochemists also describe energy-rich compounds as having **high chemical transfer potentials.** In the case of ATP, cleavage reactions typically involve transfer of either its phosphate group or its AMP moiety to another compound; ATP has a *high phosphate transfer potential.* In older terminology, energy-rich compounds were called *high-energy compounds.* In this book, we will use the term "energy-rich" exclusively.

9.2.2. Definition of Energy-Rich Compounds

Energy-rich compounds occupy a central position in the operation of metabolism. To understand their function clearly, we begin with a definition. *An energy-rich compound is a compound the hydrolysis of which has a highly negative biochemical standard free energy change.* This definition contains several parts.

First, note that we define energy-rich compounds by reference to their *hydrolysis* reaction. The free energy changes involved in other reactions in which these compounds participate are irrelevant. It is also immaterial whether an actual hydrolysis or some other cleavage takes place as the compound functions in metabolism. If the hydrolysis of a compound is highly exergonic, we deem the compound energy-rich. Some researchers define an energy-rich compounds more broadly as any compound that undergoes a reaction having a highly negative free energy change. We limit our discussion here to compounds undergoing hydrolysis, since that constitutes the commonly accepted definition.

Second, we need to clarify what we mean by a *highly negative biochemical standard free energy change.* The free energy change of interest is $\Delta G^{\circ}{}'$. Recall that other measures of free energy (ΔG°, $\Delta G'$) may differ significantly from $\Delta G^{\circ}{}'$. If hydrolysis of a compound has a highly negative $\Delta G^{\circ}{}'$, we consider the compound energy-rich.

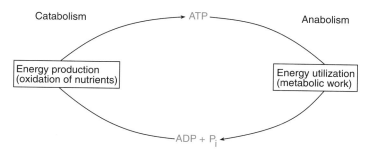

Figure 9.2. The ATP cycle.

The dividing line between moderately and highly negative free energy changes is arbitrary and depends on the values actually determined for various compounds in metabolism. Generally speaking, hydrolysis of an energy-rich compound has a biochemical standard free energy change equal to, or more negative than, about -30.5 kJ mol^{-1} (see Table 9.2).

Third, we have to stress that the definition of an energy-rich compound refers to the *free energy change of the reaction.* As you saw at the beginning of this chapter, a reaction's free energy change constitutes the difference between the sum of the free energies of *all of the products* and the sum of the free energies of *all of the reactants.* As an example, consider the hydrolysis of ATP:

$$ATP^{4-} + H_2O \rightleftharpoons ADP^{3-} + P_i^{2-} + H^+$$

The free energy change of this reaction equals the difference in free energy between $(ADP + P_i + H^+)$ and $(ATP + H_2O)$. The biochemical standard free energy change between $(ADP + P_i + H^+)$ and $(ATP + H_2O)$ is -30.5 kJ mol^{-1} (see Table 9.2).

We must distinguish clearly between the energy change of a *reaction* and the energy change for breakage of an individual bond. Biochemists refer to the bond involved in hydrolysis of an energy-rich compound, especially when dealing with a phosphorylated compound (like ADP or ATP), as an **energy-rich bond** (previously called a *high-energy bond*). We frequently designate such energy-rich bonds by a *squiggle* ($\sim$).

To a chemist, an energy-rich bond is one that requires a large amount of energy in order to be broken; the bond has a large *bond energy*. If breaking the bond between atoms A and B requires a large amount of energy, chemists consider the A—B bond to be energy-rich. *A biochemist understands something quite different by the term energy-rich bond!* To a biochemist, an energy-rich bond is a bond that, when broken by *hydrolysis,* leads to a highly negative free energy change *for the reaction.* The energy relates to hydrolysis and not to any other type of reaction that might lead to bond cleavage. Moreover, *the energy involved is that of the entire reaction and not that involved in bonding two atoms.* The biochemist considers the free energy content of all products and all reactants in a hydrolysis reaction; the chemist deals with the energy of the bond between two atoms.

9.2.3. Types of Energy-Rich Compounds

Having defined energy-rich compounds, let us see what types of compounds fall into this category and why their hydrolysis reactions are so exergonic. Three major types

of energy-rich compounds occur—**acid anhydrides, special esters,** and **derivatives of phosphamic acid** (Figure 9.3).

An ordinary acid anhydride has the following structure:

$$
\begin{array}{ccc}
O & & O \\
\parallel & & \parallel \\
R-C-O-&C-&R'
\end{array}
$$

Acid anhydrides of importance in biochemistry generally have one or both of the carbon atoms replaced by phosphorus atoms, so that the compound has the structure

$$
\begin{array}{ccccc}
O & O & & O & O \\
\parallel & \parallel & & \parallel & \parallel \\
-C-O-P- & & or & -P-O-P- \\
& | & & | & | \\
\mathbf{I} & & & \mathbf{II}
\end{array}
$$

We refer to compounds having structure **I** as *mixed anhydrides.* Acid anhydrides in which both carbons have been replaced by phosphorus atoms (**II**) include nucleoside di- and triphosphates like ADP, GDP, ATP, and GTP.

Special esters constitute a second type of energy-rich compounds. examples of such esters are sulfur esters (*acyl thioesters*) and *esters of enols.* The third type of energy-rich compounds consists of derivatives of *phosphamic acid,* such as *phosphocreatine.* These compounds have the general formula

$$
\begin{array}{c}
O \\
\parallel \\
R-NH-P-O^- \\
| \\
O^-
\end{array}
$$

9.2.4. Reasons for Negative $\Delta G^{\circ\prime}$ of Hydrolysis

Now we have to ask what makes the compounds shown in Figure 9.3 energy-rich. Why do their hydrolysis reactions proceed with highly negative free energy changes? The hydrolysis of ATP to ADP and P_i illustrates some of the factors that may be involved.

Consider the hydrolysis of ATP at pH 7–8. At that pH, adenine is uncharged (Figure 7.3). However, free phosphoric acid and the phosphate groups in ADP and ATP have each lost both their primary and their secondary proton by dissociation (Figure 9.4). Inspecting the structures of the reactants and products reveals three reasons for the highly negative free energy change of the hydrolysis reaction: resonance stabilization, electrical repulsion, and free energy of ionization.

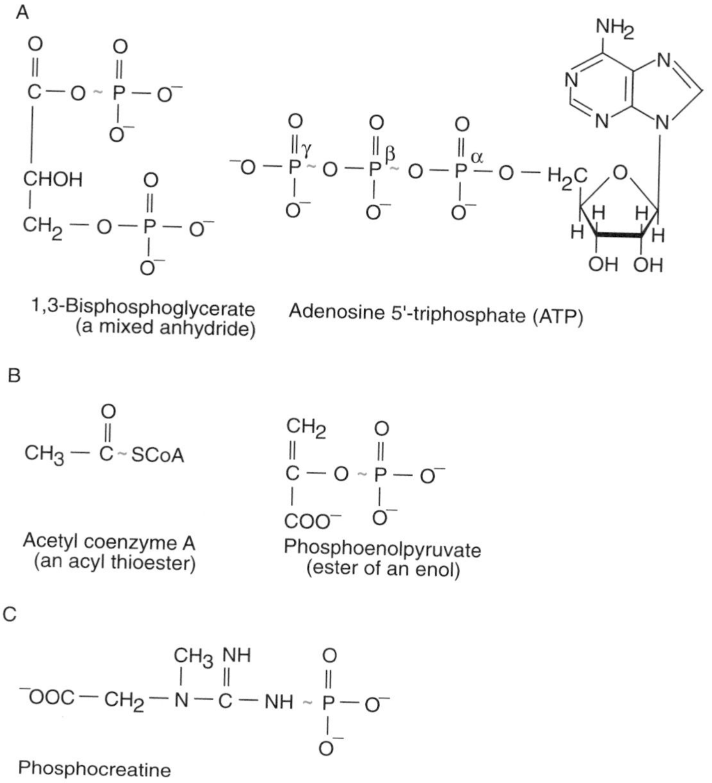

Figure 9.3. Examples of energy-rich compounds. (A) Acid anhydrides; (B) special esters; (C) derivatives of phosphamic acid; ~ designates an energy-rich bond.

9.2.4A. Resonance Stabilization.

ATP, ADP, and P_i occur in the form of many resonance structures that differ in the location of single and double bonds between the phosphorus atoms and the oxygen atoms to which they are linked. For example, HPO_4^{2-} has the following possible resonance structures:

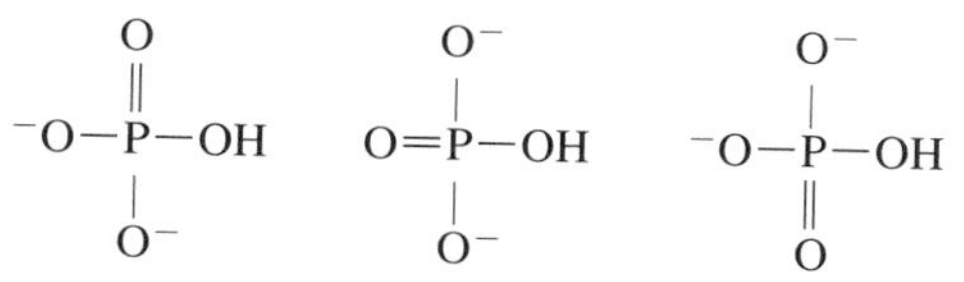

When you count all possible combinations of resonance structures, you find a larger number for the products than for the reactants. Because the stability of a compound increases with the number of resonance structures that it can assume, the products of ATP hydrolysis have greater stability than the reactants. In other words, ATP hydrolysis leads to **resonance stabilization**. A great tendency exists for the reaction to proceed as written, converting less stable reactants to more stable products. Thus, resonance stabilization contributes to making $\Delta G^{\circ\prime}$ highly negative.

9.2.4B. Electrical Repulsion.

ATP possesses four negative charges in close proximity, resulting in large internal **electrical repulsion**. The intramolecular strain is so great that the molecule tends toward breakage of some of its covalent bonds. In ADP, there occur only three such charges for a molecule of almost the same size as ATP. That still amounts to significant electrical repulsion (in fact, ADP is also an energy-rich compound) but also represents a sizable decrease in repulsion from that in ATP. Effectively, one-quarter of the charges have been eliminated. Thus, ADP, although still unstable, has greater stability than ATP. As in the case of resonance stabilization, electrical repulsion leads to the products having greater stability than the reactants. Therefore, electrical repulsion also enhances the tendency of the reactants to convert to products and contributes to making $\Delta G^{\circ\prime}$ highly negative.

9.2.4C. Free Energy of Ionization.

When you add up the charges for ATP and $(ADP + P_i)$ in the hydrolysis of ATP, you find that ATP has a net charge of -4 while the net charge of $(ADP + P_i)$ is -5. You conclude

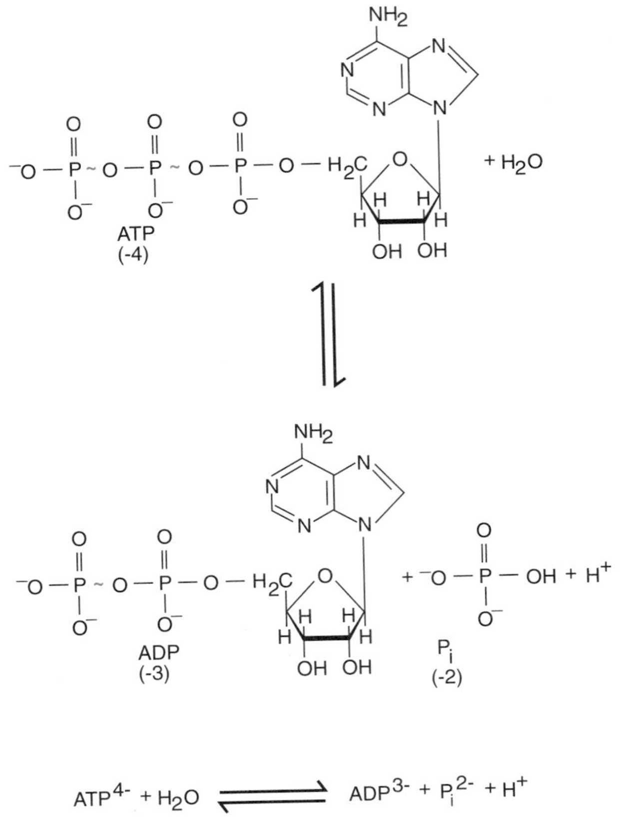

Figure 9.4. The hydrolysis of ATP to ADP and inorganic phosphate.

exergonic reaction *drives* an endergonic one, we refer to the two reactions as **coupled** or **energetically coupled.** Energy-rich compounds function via energetically coupled reactions. Coupling applies both to the synthesis of energy-rich compounds in catabolism and to their use in anabolism. To examine the nature of this coupling, consider the following two reactions:

$$\text{Glutamate}^- + \text{NH}_3 + \text{H}^+ \rightleftarrows \text{glutamine} + \text{H}_2\text{O}$$

$$\Delta G_1^{\circ\prime} = +15.0 \text{ kJ mol}^{-1} \tag{9.7}$$

$$\text{ATP}^{4-} + \text{H}_2\text{O} \rightleftarrows \text{ADP}^{3-} + \text{P}_i^{2-} + \text{H}^+$$

$$\Delta G_2^{\circ\prime} = -30.5 \text{ kJ mol}^{-1} \tag{9.8}$$

Here we have one reaction that, under biochemical standard conditions, is endergonic ($\Delta G^{\circ\prime} > 0$) and thermodynamically not feasible. The second reaction is exergonic ($\Delta G^{\circ\prime} < 0$) and can proceed readily. It appears that we should be able to accomplish the reaction having a positive $\Delta G^{\circ\prime}$ value by energetically coupling it to the one having a negative $\Delta G^{\circ\prime}$ value. Using the free energy released by the hydrolysis of ATP, it should be possible to "drive" the energy-requiring reaction of glutamine synthesis. We might attempt to do this by simply adding the two reactions together:

$$\text{Glutamate}^- + \text{NH}_3 + \text{H}^+ \rightleftarrows \text{glutamine} + \text{H}_2\text{O} \quad (9.7)$$

$$\text{ATP}^{4-} + \text{H}_2\text{O} \rightleftarrows \text{ADP}^{3-} + \text{P}_i^{2-} + \text{H}^+ \quad (9.8)$$

Overall reaction:

$$\text{Glutamate}^- + \text{NH}_3 + \text{ATP}^{4-} \rightleftarrows \text{glutamine} + \text{ADP}^{3-} + \text{P}_i^{2-}$$
$$\tag{9.9}$$

When we add two reactions, we obtain the overall free energy change by adding those of the component reactions. Hence, for the reaction given by Eq. (9.9),

$$\Delta G_{\text{overall}}^{\circ\prime} = \Delta G_1^{\circ\prime} + \Delta G_2^{\circ\prime} = 15.0 - 30.5 = -15.5 \text{ kJ mol}^{-1}$$

Because $\Delta G_{\text{overall}}^{\circ\prime}$ is negative, the reaction appears to be feasible. However, this reaction is merely the result of some "paper chemistry" that we engaged in and lacks physical reality. There exists no true chemical linkage between the two original reactions. For all we know, the two reactions may occur at different locations within the cell or at different times. How then can the energy released by one reaction actually be channeled into the other?

There exists only one way to bring about a proper linkage between the two reactions: we must *connect them chemically* by having a *product* of one reaction serve as a *reactant* for the other. Then, and only then, will the two

that another reaction, namely, an *ionization* step, must be hidden in the hydrolysis reaction. A proton must dissociate as ATP undergoes hydrolysis to ADP and P_i, and we have to write the reaction appropriately to indicate preservation of electrical neutrality. In actuality, a hydroxyl anion is removed from water, leaving an excess proton in solution. The OH^- from the water combines with the *phosphoryl group* (PO_3^{2-}), cleaved from ATP, to form HPO_4^{2-}.

Because of the ionization, the total free energy change of the hydrolysis reaction includes a contribution of **free energy of ionization.** Recall that we are evaluating ATP hydrolysis under *biochemical standard conditions,* that is, at pH 7.0. At that pH, the ionization step is exergonic and helps to make the overall free energy change highly negative.

9.3. COUPLED REACTIONS

9.3.1. Linking of Reactions

The essence of energy metabolism lies in linking endergonic and exergonic reactions (see Figure 9.1). When an

reactions be obligatorily linked both spacewise and time-wise. To illustrate this, consider coupling the following known reactions:

Glutamate$^-$ + ATP^{4-} ⇌ γ-glutamyl phosphate^{2-} + ADP^{3-}

$$\Delta G_1^{\circ\prime} = 0 \tag{9.10}$$

γ-Glutamyl phosphate^{2-} + NH$_3$ ⇌ glutamine + P$_i^{2-}$

$$\Delta G_2^{\circ\prime} = -15.5 \text{ kJ mol}^{-1} \tag{9.11}$$

Overall reaction:

Glutamate$^-$ + NH$_3$ + ATP^{4-} ⇌ glutamine + ADP^{3-} + P$_i^{2-}$

$$\Delta G_{\text{overall}}^{\circ\prime} = -15.5 \text{ kJ mol}^{-1} \tag{9.12}$$

The overall reaction is identical to that written earlier, but the two schemes differ radically. Simply adding the reactions makes no chemical sense, but linking them via a shared compound—γ-glutamyl phosphate—constitutes a plausible mechanism (Figure 9.5).

Note that both schemes have the same overall $\Delta G^{\circ\prime}$.

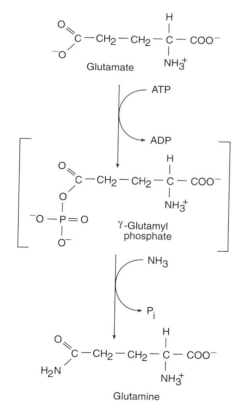

Figure 9.5. Synthesis of glutamine. The same enzyme, glutamine synthase, catalyzes both steps. The intermediate, γ-glutamyl phosphate, is unstable in aqueous solution but is protected from hydrolysis when bound to the enzyme's active site.

This is as it should be, since $\Delta G^{\circ\prime}$, like all thermodynamic functions, depends only on the final and initial states of a system and not on the path whereby the system proceeds from the initial to the final state. *γ-Glutamyl phosphate* constitutes a genuine link between the two reactions; it represents the product of one reaction and the reactant of the other reaction; it serves as a **common intermediate** of the two reactions. Both of the coupled reactions [Eqs. (9.10) and (9.11)] are catalyzed by the same enzyme, *glutamine synthase,* and the true common intermediate is actually not γ-glutamyl phosphate by itself, but rather the *enzyme-bound γ-glutamyl phosphate* complex.

Our coupling of reactions (9.10) and (9.11), under biochemical standard conditions, resulted in the first reaction having a free energy change of zero and the second reaction being strongly exergonic. The first reaction is at equilibrium, the second constitutes a spontaneous reaction, and the overall reaction proceeds readily. These specific aspects are *not* general coupling requirements; they just happen to apply in this particular example.

9.3.2. Parameters of Coupled Reactions

For two or more coupled reactions, we compute the equilibrium constant of the overall reaction by *multiplying* the equilibrium constants of the component reactions:

$$K_{\text{eq, overall}}' = K_{\text{eq}_1}' \times K_{\text{eq}_2}' \times \ldots \times K_{\text{eq}_n}' \tag{9.13}$$

Because the free energy is a logarithmic function of the equilibrium constant [and $\log(ab) = \log a + \log b$], it follows that

$$\begin{aligned}
\Delta G_{\text{overall}}^{\circ} &= -RT \ln K_{\text{eq, overall}}' \\
&= -RT \ln K_{\text{eq}_1}' - RT \ln K_{\text{eq}_2}' - \ldots - RT \ln K_{\text{eq}_n}' \\
&= \Delta G_1^{\circ} + \Delta G_2^{\circ} + \ldots + \Delta G_n^{\circ}
\end{aligned} \tag{9.14}$$

In other words, we obtain the free energy change for the overall reaction by *adding* the free energy changes for the individual reactions. For reactions at pH 7.0:

$$\Delta G_{\text{overall}}^{\circ\prime} = \Delta G_1^{\circ\prime} + \Delta G_2^{\circ\prime} + \ldots + \Delta_n^{\circ\prime} \tag{9.14a}$$

or

$$\Delta G_{\text{overall}}^{\circ\prime} = \Delta G_1' + \Delta G_2' + \ldots + \Delta G_n' \tag{9.14b}$$

and

$$K_{\text{bio, overall}}' = K_{\text{bio}_1}' \times K_{\text{bio}_2}' \times \ldots \times K_{\text{bio}_n}' \tag{9.13a}$$

9.3.3. Common Intermediate Principle

The preceding discussion leads to the following generalization:

> Any two energetically coupled reactions must always proceed via a common intermediate; there exists no other way of coupling reactions.

We refer to this concept as the **common intermediate principle.** The common intermediate may be a shared component, directly apparent by inspecting the two coupled reactions:

$$A + B \rightleftharpoons C + D$$

$$C \rightleftharpoons X + Y$$

Alternatively, the common intermediate may participate in the mechanism without its nature being immediately apparent. Enzyme-bound γ-glutamyl phosphate represents a common intermediate of this type.

Reactions involving energy-rich compounds constitute typical examples of energetically coupled reactions. Whenever an energy-rich compound is synthesized, the reaction must be coupled to some exergonic reaction by means of a common intermediate. Likewise, whenever an energy-rich compound is hydrolyzed or otherwise cleaved to drive some endergonic reaction, the two reactions must be coupled by means of a common intermediate.

In some energetically coupled reactions, ATP cleavage produces ADP and P_i; in others, it yields AMP and PP_i. Still other energetically coupled reactions proceed without ATP being involved at all. In some cases, a single enzyme catalyzes both energetically coupled reactions; in other cases, a different enzyme catalyzes each of the coupled reactions. Some coupled reactions require an energy-rich compound of some type; others proceed without the participation of a true energy-rich compound. You can see that many variations of energetically coupled reactions occur in biochemical systems.

However, regardless of the specifics of energy-rich compound involvement, or the specifics of enzyme involvement, *all energetically coupled reactions must proceed via a common intermediate.*

9.3.4. Coupling Requirements

In addition to the requirement for a common intermediate and an overall $\Delta G' \leq 0$, two other aspects of coupled reactions need to be addressed. First, note that it is customary to discuss coupled reactions in biochemistry textbooks in terms of $\Delta G^{\circ\prime}$ values, an approach that we have used as well. However, having done so, we must stress that this constitutes *an acceptable illustration* of the principle of coupled reactions but *not an acceptable description* of actual coupling in biochemical systems.

By using $\Delta G^{\circ\prime}$ values, *we have coupled reactions under, and only under, biochemical standard conditions.* In real biochemical systems, we have to deal with biochemical actual, not standard, free energy changes ($\Delta G'$ rather than $\Delta G^{\circ\prime}$). As you saw earlier, $\Delta G'$ values can differ significantly from corresponding $\Delta G^{\circ\prime}$ values. Only after obtaining $\Delta G'$ values can you proceed to determine whether, and how, any two reactions may be coupled. Moreover, recall that $\Delta G'$ can vary greatly, both in magnitude and in sign, as a result of changes in the concentrations of reactants and products. Hence, coupling of two reactions on the basis of their $\Delta G^{\circ\prime}$ values *is neither a necessary nor a sufficient requirement* for coupling these reactions under intracellular conditions.

The other aspect of coupled reactions that we have not yet mentioned we may call the "ultimate coupling requirement." Even after coupling two reactions on the basis of their $\Delta G'$ values, an additional condition has to be met. In order for the proposed mechanism to be thermodynamically feasible, it is necessary that, under biochemical actual conditions, *the free energy change ($\Delta G'$) be zero or less than zero (negative) for each step of the mechanism.* This requirement exists because any step having a positive free energy change is thermodynamically not feasible. For any step to proceed, it must have either $\Delta G' = 0$ (be at equilibrium) or $\Delta G' < 0$ (occur spontaneously). We can now summarize all of the requirements for the coupling of reactions:

1. Any two coupled reactions must proceed via a *common intermediate.*
2. Coupling must be based on *biochemical actual free energy changes ($\Delta G'$ values).*
3. The mechanism must be such that the *overall $\Delta G'$ is equal to or less than zero.*
4. Each step in the mechanism must be *thermodynamically feasible;* for each step in the mechanism, it is necessary that $\Delta G'$ be equal to or less than zero.

SUMMARY

Exergonic reactions have a negative free energy change and proceed spontaneously. Endergonic reactions have a positive free energy change and do not proceed spontaneously; they require an input of energy in order to proceed. The standard free energy change depends on the equilibrium constant of a reaction and on an arbitrarily selected reference state. The actual free energy change is a function of both the standard free energy change and the actual initial concentrations of reactants and products.

In biochemistry, we compute free energy changes for physiological conditions (pH 7) and refer to them as biochemical standard free energy changes ($\Delta G^{\circ\prime}$) and biochemical actual free energy changes ($\Delta G'$). The latter quantity determines whether or not a reaction will proceed as written under intracellular conditions. Free energy changes of reactions vary with temperature and concentration. Any chemical reaction may be made to go in one direction or another by suitable changes in pH (if the reaction is pH dependent), temperature, and the concentrations of reactants and products.

Energy-rich compounds are defined as compounds the hydrolysis of which has a high negative free energy change. We refer to the bond of the energy-rich compound undergoing hydrolysis as an energy-rich bond. Acid anhydrides, special esters, and derivatives of phosphamic acid represent typical energy-rich compounds. Hydrolysis of energy-rich compounds is strongly exergonic because of factors like resonance stabilization, changes in electrical repulsion, and free energy of ionization.

Free energy is produced in catabolism and stored in energy-rich compounds. The stored free energy is subsequently used to drive endergonic reactions characteristic of anabolism. Both synthesis and utilization of energy-rich compounds proceed via energetically coupled reactions in which an exergonic reaction drives an endergonic one. Coupling requires chemical linkage of the two reactions via a common intermediate, and the mechanism must be such that both the overall $\Delta G'$ and the $\Delta G'$ for each step are either equal to or less than zero.

SELECTED READINGS

Cantor, C. R., and Schimmel, P. R., *Biophysical Chemistry,* W. H. Freeman, San Francisco (1980).

Eisenberg, D., and Crothers, D., *Physical Chemistry with Applications to the Life Sciences,* Benjamin-Cummings (1979).

Fox, R. F., *Energy and the Evolution of Life,* W. H. Freeman, New York (1988).

Freifelder, D., *Principles of Physical Chemistry,* Jones & Bartlett, Boston (1985).

Garby, L., and Larsen, P. S., *Bioenergetics: Its Thermodynamic Foundations,* Cambridge University Press, Cambridge (1995).

Harris, D. A., *Bioenergetics at a Glance: An Illustrated Introduction,* Blackwell Scientific, Oxford (1994).

Kim, C. H., and Ozawa, T. (eds.), *Bioenergetics: Molecular Biology, Biochemistry and Pathology,* Plenum, New York (1990).

Makhatadze, G. I., and Privalov, P. L., Energetics of protein structure, *Adv. Protein Chem.* 47:308–425 (1995).

Nicholls, D. G., and Ferguson, S. J., *Bioenergetics,* 2nd ed., Academic Press, London (1992).

Stenesh, J., *Core Topics in Biochemistry,* Cogno Press, Kalamazoo, Michigan (1993).

REVIEW QUESTIONS

A. Define each of the following terms:

Gibbs free energy   Coupled reactions
Energy-rich bond   Resonance stabilization
K'_{bio} Common intermediate principle

B. Differentiate between the two terms in each of the following pairs:

$\Delta G^{\circ}/\Delta G^{\circ\prime}$ $\Delta G^{\circ}/\Delta G$
$\Delta G^{\circ\prime}/\Delta G'$ $\Delta G/\Delta G'$
Energy-rich bond/energy-   Enthalpy (heat content)/
  rich compound   entropy

C. (1) How is the ATP cycle linked to the reactions of catabolism and anabolism?
(2) Define energy-rich compounds. What types of compounds are energy-rich? Explain the factors that contribute toward making ATP an energy-rich compound.
(3) Explain what is meant by coupled reactions. List all of the coupling requirements.

PROBLEMS

9.1. Why is there only one actual free energy change for any given reaction, regardless of whether you base the calculations on $\Delta G°$ or $\Delta G°'$? In other words, explain why $\Delta G = \Delta G'$.

9.2. Free energy, like all other thermodynamic functions, always depends on the difference (designated delta, Δ) between a final and an initial state. Of what significance is this when comparing (a) *in vivo* and *in vitro* reactions and (b) enzyme-catalyzed and uncatalyzed reactions?

9.3. The binding of three noncompetitive inhibitors (A, B, C) at 25°C to an enzyme is described by the equation $E + I \rightleftarrows EI$ and is characterized by the following standard free energy changes ($\Delta G°$): -800, -4000, and $+2000$ J mol^{-1} for A, B, and C, respectively. Which of the three compounds is the strongest inhibitor under standard conditions? Why?

9.4.* Calculate $\Delta G'$ for the hydrolysis of ATP to ADP and P_i in an actively respiring cell at pH 7.0. Under those conditions, the intracellular $[P_i] = 1.00 \times 10^{-2}M$, the steady-state [ATP]/[ADP] ratio is 10.0, and the biochemical standard free energy change, $\Delta G°'$, for the hydrolysis of ATP is -30.5 kJ mol^{-1}. (Hint: Use [product]/[reactants] $=$ [ADP][P$_i$]/[ATP].)

9.5. A substrate binds to an enzyme according to the equation $E + S \rightleftarrows ES + H^+$. At equilibrium, the pH of this reaction is 5.0, the substrate concentration is $1.00 \times 10^{-2}M$, and 80.0% of the enzyme is in the form of the enzyme–substrate complex. What is the standard free energy change of the reaction?

9.6.* Given the following:

(1). Glucose 1-phosphate^{2-} $\rightleftarrows$
glucose 6-phosphate^{2-}
$K'_{bio_1} = 10.0$
(2). 2ADP^{3-} $\rightleftarrows$ ATP^{4-} + AMP^{2-}
$\Delta G_2°' = +2.03$ kJ mol^{-1}
(3). Glucose 6-phosphate^{2-} + ADP^{3-} + H$^+$ $\rightleftarrows$
glucose + ATP^{4-}
$\Delta G_3°' = +19.1$ kJ mol^{-1}

couple these three reactions such that glucose 1-phosphate would be synthesized from glucose and ADP. What is the overall reaction? What are the values for the overall free energy change ($\Delta G°'$) and the overall equilibrium constant (K'_{bio})?

9.7. Given the following:

$$A \rightleftarrows B \qquad K'_{eq} = 1.0 \times 10^{-5}$$
$$C \rightleftarrows D \qquad K'_{eq} = 1.0 \times 10^{-3}$$

calculate the equilibrium constant for the reaction $B + C \rightleftarrows A + D$.

9.8. Refer to Table 9.2 and indicate which of the following you would classify as energy-rich compounds: (a) phosphocreatine; (b) acetylcholine; (c) phosphoenolpyruvate; (d) ethyl acetate; (e) glucose.

9.9. The equilibrium constant for a reaction at 20°C has a value of 2.70×10^{-3}. What is the value of the equilibrium constant at 25°C if $\Delta H°$ is constant and equal to -6.28 kJ mol^{-1} over the temperature range of 20–25°C?

9.10.* Couple the following two reactions

(1) Glucose 6-phosphate^{2-} + H$_2$O $\rightleftarrows$
glucose + P$_i^{2-}$
$K'_{bio_1} = 1.00 \times 10^{-2}$
(2) ATP^{4-} + H$_2$O $\rightleftarrows$ AMP^{2-} + PP$_i^{3-}$ + H$^+$
$\Delta G_2°' = -31.8$ kJ mol^{-1}

so that ATP hydrolysis drives the synthesis of glucose 6-phosphate. Write a plausible two-step mechanism and then write the overall reaction. Calculate the overall free energy change ($\Delta G°'$) of the reaction. (Hint: K'_{bio} for ATP hydrolysis is [AMP][PP$_i$]/[ATP].)

9.11.* What is the biochemical actual free energy change ($\Delta G'$) for the overall reaction of the previous problem under the following steady-state conditions?

[AMP] $= 1.00 \times 10^{-2}M$ [ATP] $= 1.00 \times 10^{-1}M$
[PP$_i$] $= 1.00 \times 10^{-2}M$ [P$_i$] $= 1.00 \times 10^{-5}M$
[glucose] $= 1.00 \times 10^{-4}M$ [glucose 6-phosphate]
$= 1.00 \times 10^{-1}M$

Hint: Omit [H$^+$] from the [products]/[reactants] term.

9.12. Refer to Table 9.2 to calculate the equilibrium constant (K'_{bio}) for the reaction

Creatine + P$_i^{2-}$ $\rightleftarrows$ phosphocreatine^{2-} + H$_2$O

9.13.* What must be the minimum ratio of [creatine] to [phosphocreatine] in order to drive the reaction of the previous problem toward synthesis of phosphocreatine when $[P_i] = 1.00M$? (Hint: Omit $[H_2O]$ from the [products]/[reactants] term.)

9.14.* The hydrolysis of ATP to ADP and P_i has a free energy change ($\Delta G^{\circ\prime}$) of -30.5 kJ mol^{-1}. The catabolic conversion of compound A to compound D is an exergonic reaction ($\Delta G^{\circ\prime} = -91.5$ kJ mol^{-1}). Below are listed two theoretical mechanisms for coupling ATP synthesis to the catabolism of compound A. Both mechanisms have a 100% efficiency of energy trapping (three moles of ATP are synthesized per mole of compound A catabolized). Based on your general knowledge of chemistry, which of these two mechanisms represents a more plausible one and why?

$$
\begin{array}{cc}
(a) & (b) \\
A \longrightarrow D & A \longrightarrow B \longrightarrow C \longrightarrow D \\
\downarrow & \downarrow \quad \downarrow \quad \downarrow \\
3\ ATP & ATP \quad ATP \quad ATP
\end{array}
$$

9.15. In stage I of catabolism, polymeric molecules undergo degradation to numerous building blocks (see Figure 8.2). Since this amounts to an increase in randomness, the process is characterized by an increase in entropy. Hence, decide whether it is thermodynamically possible for stage I of catabolism to proceed such that $\Delta G = \Delta H$. [Hint: See Eq. (9.1).]

9.16. Consider the following two coupled reactions:

$$
\begin{array}{cc}
A \rightleftarrows B + C & K'_{eq_1} \\
C \rightleftarrows X + Y & K'_{eq_2} \\
A \rightleftarrows B + X + Y & K'_{eq,overall}
\end{array}
$$

Show that $K'_{eq,overall} = K'_{eq_1} \times K'_{eq_2}$ in the following two ways. (a) Write out the expression for $K'_{eq,overall}$. Then write out the expressions for K'_{eq_1} and K'_{eq_2}, multiply them, and compare the result with the expression for $K'_{eq,overall}$. (b) Solve the expressions of K'_{eq_1} and K'_{eq_2} for C. Because C is the common intermediate of the two reactions, its concentration can be expressed by either of the two component reactions. Equate the two expressions for C, and solve the resultant equation for $K'_{eq_1} \times K'_{eq_2}$.

9.17.* What is the lowest pH that would permit the ionization of acetic acid at 25°C when $[CH_3COOH] = 1.00 \times 10^{-3}M$ and $[CH_3COO^-] = 1.00M$? (Hint: Set $\Delta G' = 0$. The ionization of acetic acid has a $\Delta G^{\circ\prime}$ value of -12.8 kJ mol^{-1}.)

9.18.* A novel bacterium is suspected of being able to grow on citrate as its sole carbon and energy source. The first step in citrate utilization consists of its conversion to isocitrate (see Table 9.2). Energetically speaking, could this organism grow when the [citrate]/[isocitrate] ratio equals 1000?

9.19. Calculate the number of moles of ATP that, theoretically, could be synthesized from ADP for every 10 moles of maltose hydrolyzed to glucose. Assume a 100% efficiency of energy trapping and reactions occurring under biochemical standard conditions. Refer to Table 9.2.

9.20. A 0.10M solution of a weak monoprotic acid, HA, is 10% ionized at 25°C. Calculate (a) the pH of the solution, (b) the equilibrium (ionization) constant, (c) pK'_a, (d) and ΔG°.

9.21. What are the relative molar concentrations of citrate and isocitrate at equilibrium for the isomerization reaction listed in Table 9.2?

9.22.* According to Table 8.6, a normal man aged 19–24 requires a minimum intake of about 7500 kJ per day. Assume that a man who weighs 150 lb (68 kg) consumes only enough food to meet the minimum energy requirement and converts food energy into ATP at an efficiency of 40%. Assume further that $\Delta G'$ for the hydrolysis of ATP under intracellular conditions is -30.5 kJ mol^{-1}. The molecular weight of ATP is 507. Calculate the weight of ATP produced by the man from food per day. What percent of his body weight does this represent?

9.23. What is the difference in ΔG° at 25°C for two reactions for which the equilibrium constants differ by a factor of 10?

9.24. What is the value of the equilibrium constant at 25°C and pH 7.0 for the isomerization of fructose 6-phosphate to glucose 6-phosphate? (See Table 9.2.)

9.25. What is the biochemical actual free energy change ($\Delta G'$) for the conversion of phosphoarginine to arginine (see Table 9.2) when the following initial concentrations are used? (Hint: Omit $[H_2O]$ from the [products]/[reactants] term.)

$$[\text{Phosphoarginine}] = 3.00 \times 10^{-2}M$$

$$[\text{Arginine}] = 1.00 \times 10^{-3}M$$

$$[HPO_4^{2-}] = 2.00 \times 10^{-5}M$$

9.26.* Use the data in Table 9.2 to calculate $\Delta G^{\circ\prime}$ for the following reactions:

(a) Glucose 6-phosphate^{2-} + $6O_2 \rightleftarrows HPO_4^{2-}$ + $6CO_2$ + $5H_2O$

(b) Maltose + $2HPO_4^{2-} \rightleftarrows$ 2 glucose 6-phosphate^{2-} + H_2O

(c) Lactate$^-$ + acetaldehyde $\rightleftarrows$ pyruvate$^-$ + ethanol

(d) Glucose 6-phosphate^{2-} + fructose $\rightleftarrows$ sucrose + P_i^{2-}

Carbohydrate Metabolism

10

We begin our study of metabolism with carbohydrates because of their central role in the generation, use, and storage of metabolic energy. Additionally, carbohydrates are of major importance in metabolism because their degradations and interconversions provide the carbon skeletons for the biosynthesis of most other metabolites, from small coenzymes to large structural molecules. After a look at some general aspects of carbohydrate metabolism, we will discuss specific metabolic pathways.

We can divide carbohydrate metabolism into five major parts—*digestion, transport, storage, degradation, and biosynthesis* (Figure 10.1). Starch and glycogen represent the major dietary polysaccharides of animals; their digestion produces **glucose.** As you will see in this chapter, other dietary carbohydrates are either converted to glucose or to intermediates of glucose metabolism. Thus, carbohydrate metabolism becomes effectively glucose metabolism; most major pathways either start or end with glucose (Figure 10.2).

Monosaccharides, the end products of oligo- and polysaccharide digestion, are carried to the liver and distributed from there throughout the body. Blood transports carbohydrates, primarily in the form of glucose, from the liver to muscle tissue and, in the form of *lactate* (formed from glucose), from muscle tissue to the liver.

Carbohydrates and lipids constitute the storage forms of energy in animals. Carbohydrates are stored as glycogen, principally in muscles and the liver. Major degradative pathways of carbohydrates include the breakdown of glycogen to glucose, **glycogenolysis,** and two catabolic pathways for glucose. One, **glycolysis,** leads to production of usable energy; the other, the **pentose phosphate pathway,** generates reducing power and pentoses.

Biosynthesis of carbohydrates includes three major pathways: synthesis of glycogen, **glycogenesis;** synthesis of glucose, **gluconeogenesis;** and fixation of carbon dioxide, **photosynthesis.** In this chapter, we cover all of these catabolic and anabolic pathways except photosynthesis, which we discuss separately in Chapter 15.

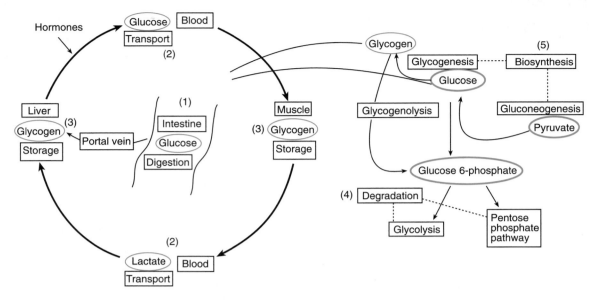

Figure 10.1. Diagram showing the major aspects of carbohydrate metabolism in animals: (1) digestion, (2) transport, (3) storage, (4) degradation, and (5) biosynthesis. Varying portions of the degradative and biosynthetic pathways occur in muscle tissue, the liver, and the blood. An additional biosynthetic pathway, photosynthesis, occurs in plants and other photosynthetic organisms.

10.1. TRANSPORT AND STORAGE OF CARBOHYDRATES

Blood transports carbohydrates primarily in the form of glucose, or *blood sugar.* Normal blood sugar levels fall in the range of 65–100 mg % (mg/100 ml or mg/dl of blood). Glucose transport is linked to that of lactate by reactions of the *Cori cycle* (Section 10.3). Three hormones control blood sugar levels:

- *Insulin,* a polypeptide (Figure 2.7) secreted by the pancreas, *lowers the level of blood sugar* by stimulating glucose transport into cells and by stimulating glycogen synthesis from glucose in both muscles and the liver.

- *Epinephrine (adrenaline),* a derivative of tyrosine (Figure 10.3) secreted by the adrenal gland, *raises the level of blood sugar* by stimulating degradation of glycogen to glucose in muscles and the liver.

- *Glucagon,* a polypeptide of 29 amino acids secreted by the pancreas, *raises the level of blood sugar* by stimulating degradation of glycogen to glucose in the liver.

Clinicians use measurements of blood sugar to construct *glucose tolerance curves* (Figure 10.4), which serve as screening tests for **diabetes.** For this test, an individual in a fasting state first provides a blood sample to establish the baseline level of blood sugar. The person then drinks a concentrated glucose solution. Samples of blood are withdrawn periodically and analyzed for glucose. In a normal individual, the blood sugar concentration first rises as excess glucose enters the bloodstream, and subsequently drops to its normal level as the glucose is metabolized.

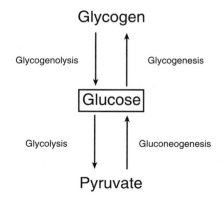

Figure 10.2. Major catabolic and anabolic pathways of glucose metabolism.

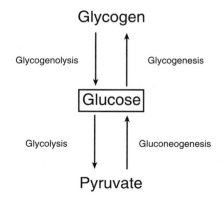

Figure 10.3. Epinephrine.

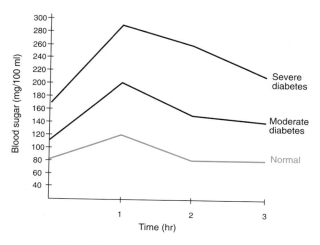

Figure 10.4. Typical glucose tolerance curves.

Typically, this occurs within about two hours. For a diabetic, the curve has a different shape. A diabetic suffers from excessive and uncontrolled levels of blood sugar, so the initial baseline level is higher than that of a healthy person. Ingesting a large amount of glucose exacerbates the problem of impaired glucose metabolism. The blood sugar concentration first rises, as in any individual, but it takes more than two hours to return to the baseline level.

There are two types of diabetes, *juvenile-onset* (insulin-dependent, Type I) and *adult-onset* (insulin-independent, Type II) diabetes. Juvenile-onset diabetes occurs in childhood and is due to an insulin deficiency that may result from insufficient synthesis, accelerated breakdown, or inhibition of insulin. Adult-onset diabetes occurs in middle-aged individuals who have insulin near or even above normal levels. The defect results from a decreased response of cells to insulin due to a scarcity of active insulin receptors in the cell membrane. Glucose tolerance curves are abnormal for both types of diabetics.

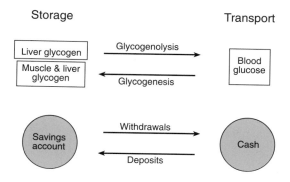

Figure 10.5. The relationship between transport and storage of carbohydrates resembles that between cash and savings. Degradation of muscle glycogen yields lactate rather than glucose (see Figure 10.26).

Carbohydrate transport and storage represent related processes (Figure 10.5). When the blood glucose concentration goes up, as after a meal, some of the glucose is metabolized to provide energy, and some is converted to glycogen and stored in muscles and the liver. Conversely, when blood glucose concentration decreases and energy is needed, some of the glycogen stored in the liver is converted to glucose. Muscle glycogen is catabolized differently (see Figure 10.26).

The interdependence of carbohydrate transport and storage resembles that of cash and savings. When you accumulate a reasonable amount of cash, you may deposit some of it in a savings account. When you need cash to pay for expenses, you withdraw the necessary funds from the savings account.

10.2. GLYCOLYSIS—INDIVIDUAL REACTIONS

The major pathway for generating usable energy from carbohydrates consists of a set of 10 reactions called *glycolysis* ("sugar splitting," from the Greek *glykos,* sweet, and *lysis,* dissolution). The reaction sequence, also known as the *Embden–Meyerhof pathway,* takes place in the cytosol, the fluid portion of the cytoplasm.

It is fitting that we begin our detailed study of metabolism with this pathway. First, glycolysis occurs almost universally in animals, plants, and microorganisms. Second, the reaction sequence plays a central role in generating both energy and metabolic intermediates for other pathways. Third, glycolysis was the first metabolic pathway to be elucidated in detail. Lastly, the regulation of glycolysis is particularly well understood. In fact, glycolysis probably represents the most completely understood metabolic pathway.

The study of glycolysis had its start in 1897 with the discovery by Eduard Buchner (awarded the Nobel Prize in 1907) that a cell-free extract of yeast could carry out fermentation. This groundbreaking work was soon followed by two other important findings. Between 1905 and 1910, Arthur Harden and William Young showed that fermentation requires inorganic phosphate for incorporation into a sugar phosphate, later shown to be fructose 1,6-bisphosphate. Additionally, Harden and Young succeeded in separating a yeast cell-free extract by dialysis into two fractions, both of which are required for fermentation. The nondialyzable fraction is heat-sensitive and was called *zymase;* the dialyzable fraction is heat-stable and was named *cozymase.* The work of other researchers established subsequently that zymase consists of a mixture of enzymes and that cozymase consists of a mixture of cofactors—including ATP, ADP, and NAD$^+$—and metal ions.

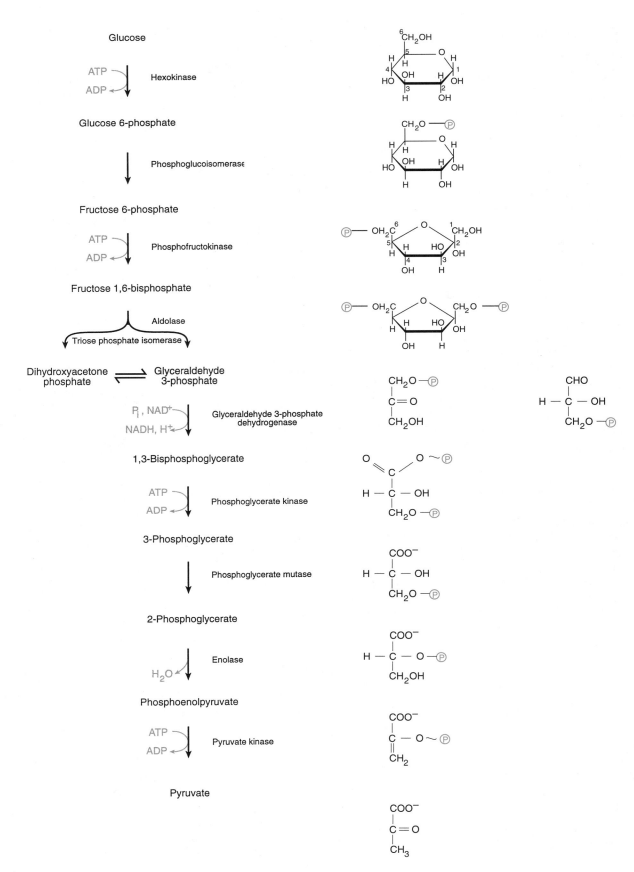

Figure 10.6. The reactions of glycolysis.

These initial investigations were followed by widespread research that dealt both with the alcoholic fermentation of yeast and the related topic of glucose catabolism in muscles. By 1940, the combined efforts of many scientists culminated in elucidation of the complete pathway of glycolysis. The field of biochemistry had been provided with its first major enzymatic pathway. Investigators who made notable contributions in this effort include Gustav Embden, Otto Meyerhof (awarded the Nobel Prize in 1922), Jacob Parnas, Carl and Gerti Cori, Carl Neuberg, Robert Robison, and Otto Warburg.

Glycolysis is an anaerobic degradative process leading to the conversion of *one* molecule of glucose to *two* molecules of pyruvate and capture of a limited amount of energy in the form of ATP. The glycolytic pathway consists of two stages (Figure 10.6). *Stage I* represents an *energy-consuming stage* that accomplishes the conversion of one molecule of glucose (six carbons) to two molecules of glyceraldehyde 3-phosphate (three carbons). This stage includes two *ATP-requiring* reactions; the energy-rich ATP "primes" glucose for glycolysis and also drives a second *phosphorylation* reaction (transfer of a *phosphoryl group*). *Stage II* represents an *energy-generating stage* that accomplishes the conversion of glyceraldehyde 3-phosphate (three carbons) to pyruvate (three carbons). This stage includes two *ATP-yielding* reactions; synthesis of ATP is *coupled* to cleavage of energy-rich bonds in two glycolytic intermediates.

10.2.1. Stage I

10.2.1A. Hexokinase. Glycolysis commences
with the **hexokinase** reaction, in which glucose is phosphorylated to *glucose 6-phosphate* by means of ATP:

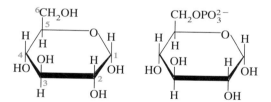

Glucose + ATP → glucose 6-phosphate + ADP + H⁺

A **kinase** catalyzes the transfer of a phosphoryl group from ATP to a metabolite or from an energy-rich compound to ADP. Kinases usually require Mg^{2+}, which is chelated by the phosphoryl groups in ATP (Figure 10.7). The metal ion helps to orient ATP at the enzyme's active site and provides electrostatic shielding for the negative charges of phosphoryl groups. In the absence of such shielding, electron pairs of attacking nucleophiles would be repelled.

In the hexokinase reaction, ATP provides both the en-

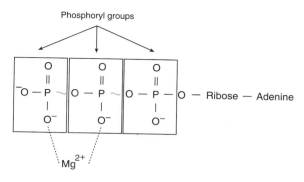

Figure 10.7. Chelation of magnesium ions by phosphoryl groups of ATP.

ergy and the phosphoryl group for producing glucose 6-phosphate. The reaction is highly exergonic (Table 10.1), making it an irreversible step.

Glucose phosphorylation has two advantages. First, it converts the initial metabolite to a more reactive form so that subsequent reactions proceed readily; glucose 6-phosphate is chemically more reactive than glucose. Second, phosphorylation traps glucose inside the cell, since free glucose moves across the cell membrane but negatively charged glucose 6-phosphate does not. Phosphorylation ensures that glucose will not be lost from the cell. Instead, it is catabolized for the generation of energy or used to form glycogen.

All of the remaining glycolytic intermediates, except pyruvate, also are phosphorylated compounds that carry a negative charge at physiological pH. These intermediates do not move across the cell membrane since cell membranes are generally impermeable to charged molecules.

Table 10.1. The Energetics of Glycolysis[a]

Enzyme	$\Delta G^{\circ\prime}$ (kJ mol^{-1})	$\Delta G^{\prime}$ (kJ mol^{-1})
Stage I		
1. Hexokinase	−16.7	−33.5
2. Phosphoglucoisomerase	+1.7	−2.5
3. Phosphofructokinase	−14.2	−22.2
4. Aldolase	+23.8	−1.3
5. Triose-phosphate isomerase	+7.5	+2.5
Total	+2.1	−57.0
Stage II		
6. Glyceraldehyde 3-phosphate dehydrogenase	+6.3	−1.7
7. Phosphoglycerate kinase	−18.8	+1.3
8. Phosphoglyceromutase	+4.6	+0.8
9. Enolase	+1.7	−3.3
10. Pyruvate kinase	−31.4	−16.7
Total	−37.6	−19.6

[a]$\Delta G^{\prime}$ values have been calculated on the basis of the approximate intracellular concentrations of glycolytic intermediates in rabbit skeletal muscle.

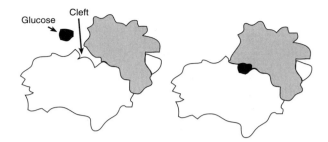

Figure 10.8.   Model of hexokinase, showing closure of the cleft upon binding glucose.

Glucose (uncharged) and lactate and pyruvate (both charged) can move across cell membranes because of the existence of specific transport mechanisms.

Hexokinase was discovered by Otto Meyerhof in 1927. The yeast enzyme is a dimer (MW = 55,000/subunit), but mammalian hexokinase is a monomer (MW = 100,000). Hexokinase catalyzes transfer of a phosphoryl group from ATP to a number of different hexoses. The mammalian enzyme is inhibited by glucose 6-phosphate, the reaction product. X-ray diffraction studies have revealed that all hexokinases have a cleft to which the substrate becomes bound (Figure 10.8). Upon binding of substrate, the enzyme undergoes a large conformational change that leads to closing of the cleft. Glucose phosphorylation takes place with the cleft in its closed state. This type of enzyme–substrate interaction illustrates *induced fit* or a *flexible active site.*

In addition to hexokinase, liver tissue contains a second kinase, *glucokinase,* that acts specifically on glucose but has a large Michaelis constant (K_m) for it. Because of the large K_m, the enzyme becomes effective only at high glucose concentrations, at which time it provides glucose 6-phosphate for glycogen synthesis. Thus glycogen synthesis from glucose occurs only when there exists an abundant supply of glucose.

10.2.1B. Phosphoglucoisomerase. The
second reaction of glycolysis, catalyzed by **phosphoglucoisomerase,** results in isomerization of glucose 6-phosphate, an aldose, to *fructose 6-phosphate,* a ketose:

Glucose 6-phosphate $\rightleftharpoons$ fructose 6-phosphate

The reaction proceeds via formation of an enediol [—C(OH)=C(OH)—] intermediate. Mammalian phosphoglucoisomerase is a dimer (MW = 61,000/subunit) and requires Mg^{2+} as cofactor.

10.2.1C. Phosphofructokinase. **Phosphofructokinase** catalyzes a second phosphorylation by means of ATP:

Fructose 6-phosphate + ATP^{4-} → fructose 1, 6-bisphosphate + ADP^{3-} + H^+

As in the hexokinase reaction, ATP provides both the energy to drive the phosphorylation and the phosphoryl group transferred. The phosphofructokinase reaction constitutes the *committed step* (see Section 8.2) of glycolysis. *Fructose 1,6-bisphosphate,* formed in this reaction, has no other metabolic role except serving as an intermediate in glycolysis. This fact and the highly exergonic nature of the reaction ensure that glycolysis proceeds in the direction of glucose → pyruvate.

Conversion of fructose 6-phosphate to fructose 1,6-bisphosphate constitutes the most important regulatory site of glycolysis. Phosphofructokinase is an *allosteric enzyme* that requires Mg^{2+} as cofactor and that has various allosteric effectors. Activators of the enzyme include AMP, ADP, fructose 6-phosphate, fructose 1,6-bisphosphate, *fructose 2,6-bisphosphate* (discussed in Section 10.6), and P_i. Inhibitors include ATP, citrate, long-chain fatty acids, and NADH (the reduced form of nicotinamide adenine dinucleotide). The mammalian enzyme is a tetramer (MW = 78,000/subunit).

10.2.1D. Aldolase. The enzyme catalyzing the fourth reaction of stage I is called **aldolase** because the reverse reaction constitutes an *aldol condensation:*

Fructose 1,6-bisphosphate $\rightleftharpoons$ dihydroxyacetone phosphate +

$$
\begin{array}{c}
O \quad H \\
\diagdown\!\!/ \\
{}^4C \\
| \\
H-{}^5C-OH \\
| \\
{}^6CH_2OPO_3^{2-}
\end{array}
$$

glyceraldehyde 3-phosphate

Aldolase brings about cleavage of fructose 1,6-bisphosphate (six carbons) to two three-carbon compounds, *dihydroxyacetone phosphate* and *glyceraldehyde 3-phosphate*. Under biochemical standard conditions, the aldolase reaction is strongly endergonic (see Table 10.1) and will not proceed as written, but under intracellular conditions the reaction is exergonic and proceeds spontaneously. Intracellular concentrations of reactants and products are such that the free energy change ($\Delta G'$) is negative. This illustrates the important point made earlier that free energy changes of reactions may be greatly affected by changes in the concentrations of reactants and products.

Two types of aldolases exist. In **class I aldolases,** found in animals and plants, the mechanism of the reaction involves formation of a **Schiff base** (Figure 10.9). A Schiff base represents a condensation product between a primary amine and either an aldehyde or a ketone, forming a carbon–nitrogen double bond (an *imine* group). In the aldolase reaction, the imine group forms between the ε-amino group of a lysine residue in the enzyme and the carbonyl group of the open-chain form of fructose 1,6-bisphosphate. Mammalian aldolase is a tetramer (MW = 40,000/subunit); the predominant form in muscles consists of two types of subunits ($\alpha_2\beta_2$).

Class II aldolases, present in bacteria and protista, are dimers that require Zn^{2+} as cofactor. These enzymes do not form a Schiff base intermediate.

10.2.1E. Triose-Phosphate Isomerase.

The last reaction of stage I comprises an isomerization between dihydroxyacetone phosphate and glyceraldehyde 3-phosphate, produced in the previous step. These two compounds are structural isomers that readily interconvert by means of **triose-phosphate isomerase:**

$$
\begin{array}{ccc}
{}^1CH_2OH & & O \quad H \\
| & & \diagdown\!\!/ \\
{}^2C{=}O & & {}^1C \\
| & & | \\
{}^3CH_2OPO_3^{2-} & & H-{}^2C-OH \\
& & | \\
& & {}^3CH_2OPO_3^{2-}
\end{array}
$$

Dihydroxyacetone phosphate $\rightleftarrows$ glyceraldehyde 3-phosphate

Under intracellular conditions, the reaction is slightly endergonic. It is driven by being coupled to the next reaction, an exergonic step.

10.2.2. Stage II

Of the two products formed in the aldolase reaction, *only glyceraldehyde 3-phosphate enters stage II* of glycolysis. Dihydroxyacetone phosphate isomerizes to glyceraldehyde 3-phosphate, which then enters stage II. Thus, *one molecule of glucose,* when processed through stage I, ultimately yields *two molecules of glyceraldehyde 3-phosphate* that enter the reaction sequence of stage II.

10.2.2A. Glyceraldehyde 3-Phosphate Dehydrogenase.

The first reaction of stage II is an *oxidation–reduction* step catalyzed by **glyceraldehyde 3-phosphate dehydrogenase:**

$$
\begin{array}{ccc}
O \quad H & & O \quad OPO_3^{2-} \\
\diagdown\!\!/ & & \diagdown\!\!/ \\
C & & C \\
| & & | \\
H-C-OH & & H-C-OH \\
| & & | \\
CH_2OPO_3^{2-} & & CH_2OPO_3^{2-}
\end{array}
$$

Glyceraldehyde + NAD$^+$ + P$_i^{2-}$ $\rightleftarrows$ 1,3-bisphos- + NADH + H$^+$
3-phosphate $\qquad\qquad\qquad\qquad$ phoglycerate

Like other biological oxidation–reduction reactions, this particular one requires involvement of a coenzyme. The coenzyme of glyceraldehyde 3-phosphate dehydrogenase is *nicotinamide adenine dinucleotide,* or *NAD$^+$* (see Figure 11.3). As glyceraldehyde 3-phosphate (an aldehyde) undergoes oxidation to 1,3-bisphosphoglycerate (an acid), the oxidized form of the coenzyme (NAD$^+$) converts to the reduced form (NADH):

$$NAD^+ + H^- \rightleftarrows NADH$$

Reduction of NAD$^+$ requires transfer of a hydride ion, H$^-$ (also designated H:), which, in this case, is abstracted from the aldehyde carbon atom of glyceraldehyde 3-phosphate. A hydride ion comprises a proton and two electrons (H$^+$ + 2$\bar{e}$). Oxidation of a metabolite by a dehydrogenase typically involves removal of two hydrogen atoms (2H·, equivalent to 2H$^+$ + 2$\bar{e}$) from the metabolite. Of these, a hydride ion (equivalent to H$^+$ + 2$\bar{e}$) reduces NAD$^+$, leaving one proton (H$^+$) to appear as a product of the reaction. Accordingly, we usually de-

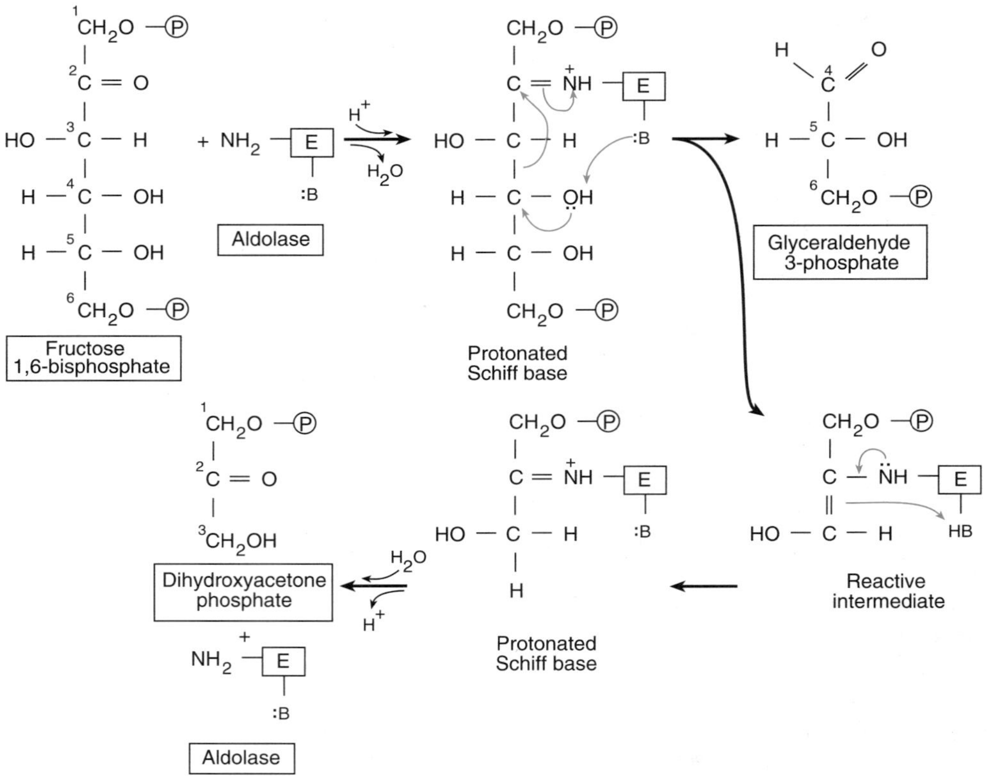

Figure 10.9. The aldolase reaction. Formation of a Schiff base is followed by release of the first product, glyeraldehyde 3-phosphate. A second Schiff base forms, followed by release of the second product, dihydroxyacetone phosphate (E, enzyme; B, a basic group).

pict the interconversion of NAD^+ and NADH as follows:

$$NAD^+ + H^- + H^+ \rightleftharpoons NADH + H^+$$

or simply as

$$NAD^+ \rightleftharpoons NADH + H^+$$

1,3-Bisphosphoglycerate represents an *energy-rich compound* because of its mixed acid anhydride structure. Only the linkage between the phosphoryl group and C(1) constitutes an *energy-rich bond;* the linkage between the phosphoryl group and C(3) is an ordinary ester bond. In addition to being an intermediate in glycolysis, 1,3-bisphosphoglycerate also serves as the source of *2,3-bisphosphoglycerate* (via the enzyme *bisphosphoglycerate mutase*), an allosteric effector of hemoglobin.

Glyceraldehyde 3-phosphate dehydrogenase contains an active sulfhydryl group that participates in the reaction (Figure 10.10). Mammalian glyceraldehyde 3-phosphate dehydrogenase is a tetramer of four identical subunits (MW = 37,000/subunit), and each subunit has

one binding site for NAD^+. Arsenate ($HAsO_4^{2-}$) can substitute for phosphate in the enzyme mechanism. When that occurs, glycolysis can proceed, but ATP synthesis in stage II is inhibited. Decreased ATP synthesis in the presence of arsenate accounts for the toxicity of this compound.

10.2.2B. Phosphoglycerate Kinase.
Transfer of the phosphoryl group, linked via an energy-rich bond in 1,3-bisphosphoglycerate, to ADP constitutes the second reaction of stage II. Cleavage of the energy-rich bond is energetically coupled to phosphorylation of ADP. **Phosphoglycerate kinase** catalyzes this reaction, and 1,3-bisphosphoglycerate provides both the energy and the phosphoryl group for phosphorylation of ADP to ATP:

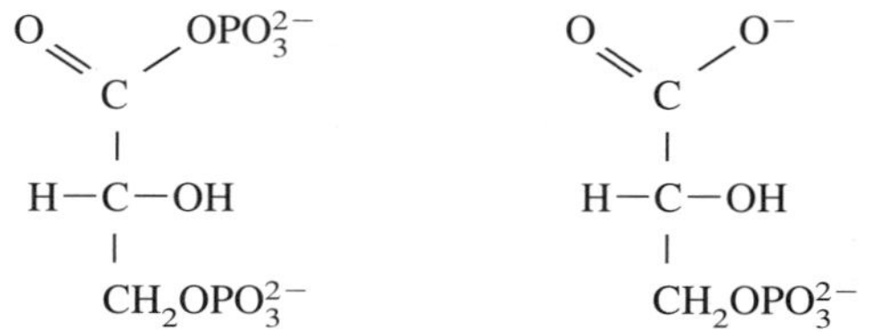

1, 3-Bisphospho- + ADP^{3-} $\rightleftharpoons$ 3-phospho- + ATP^{4-}
    glycerate   glycerate

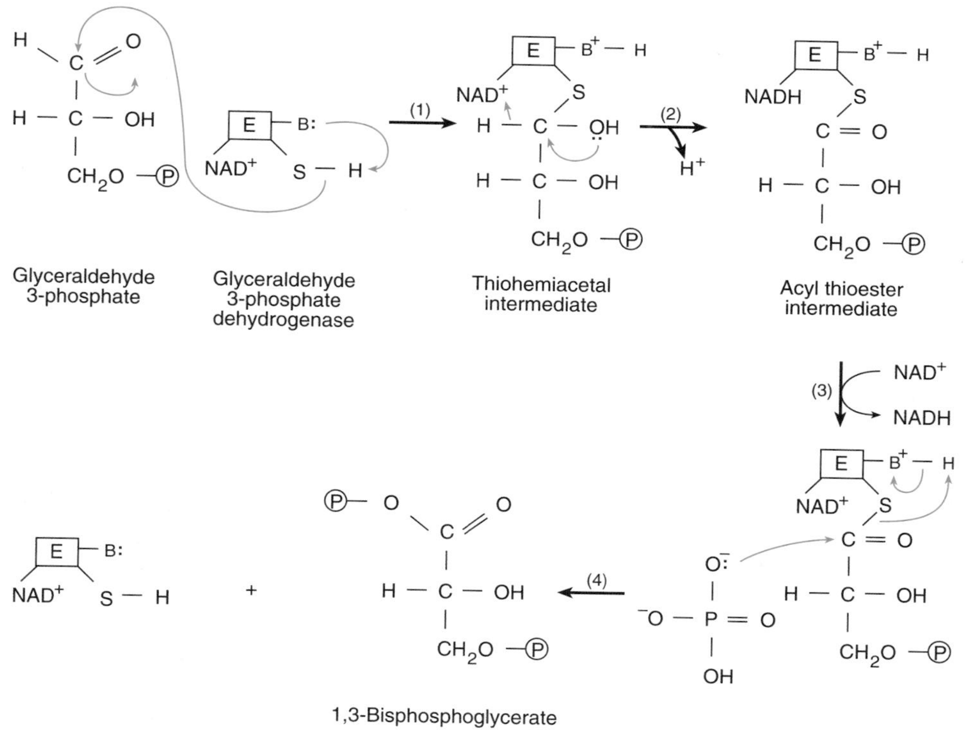

Figure 10.10. The glyceraldehyde 3-phosphate dehydrogenase reaction. (1) A thiohemiacetal forms between glyceraldehyde 3-phosphate and a reactive SH group of the enzyme (B: is a basic group); (2) NAD+ oxidation yields a thioester plus NADH and H+; (3) NAD+ replaces NADH on the enzyme; (4) phosphate attack produces 1,3-bisphosphoglycerate and restores the enzyme to its initial state.

ATP synthesis, catalyzed by phosphoglycerate kinase, illustrates what we term *substrate-level phosphorylation*—phosphorylation of ADP to ATP coupled to cleavage of an energy-rich bond in a metabolite. Substrate-level phosphorylation constitutes one mechanism for ATP synthesis; you will encounter two other mechanisms—*oxidative phosphorylation* and *photosynthetic phosphorylation*—in later chapters. Mammalian phosphoglycerate kinase is a monomer (MW = 64,000) and requires Mg^{2+} as cofactor.

10.2.2C. Phosphoglyceromutase. An isomerization reaction, catalyzed by **phosphoglyceromutase**, is the next reaction of stage II. This reaction proceeds via the intermediate 2,3-bisphosphoglycerate:

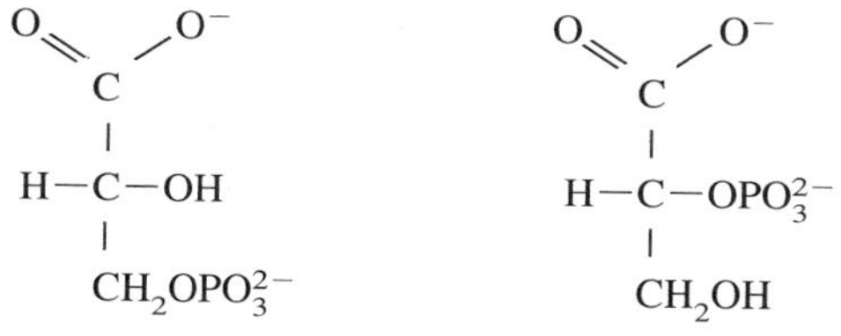

3-Phosphoglycerate ⇌ 2-phosphoglycerate

A **mutase** catalyzes the transfer of a functional group from one position to another on the same molecule. The mammalian enzyme is a dimer (MW = 27,000/subunit) and requires Mg^{2+} as cofactor.

10.2.2D. Enolase. The next reaction, catalyzed by **enolase**, forms 2-phosphoenolpyruvate (phosphoenolpyruvate for short) by dehydration of 2-phosphoglycerate. Chemically, this constitutes an α,β-elimination reaction in which H_2O forms from substituents of the α and β carbons:

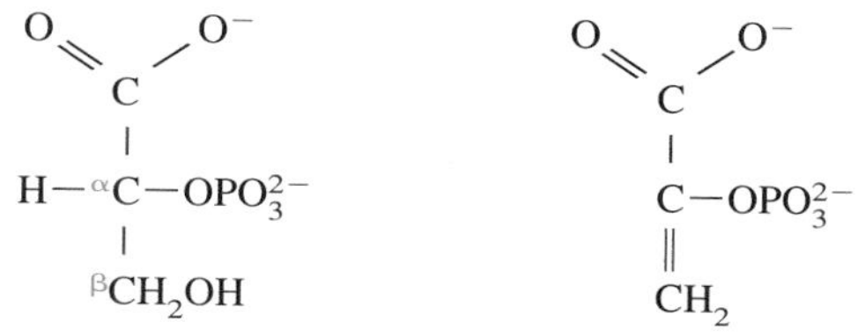

2-Phosphoglycerate ⇌ phosphoenolpyruvate + H_2O

Mammalian enolase is a dimer (MW = 41,000/subunit) and requires Mg^{2+} as cofactor. Fluoride (F^-) is a potent inhibitor of the enzyme, presumably owing to formation of ionic complexes between enzyme-bound Mg^{2+} and fluorophosphate ions (FPO_3^{3-}).

Phosphoenolpyruvate, aptly abbreviated as *PEP,* is the second energy-rich compound formed in stage II. Its energy-rich character comes from being the ester of an enol rather than of an ordinary alcohol. Cleavage of the energy-rich bond linking the phosphoryl group to C(2) is strongly exergonic because the free energy change includes that due to ester bond hydrolysis and that due to the highly favored shift from the enol to the keto form.

10.2.2E. Pyruvate Kinase. The last reaction of stage II, catalyzed by **pyruvate kinase,** accomplishes the transfer of the phosphoryl group from phosphoenolpyruvate to ADP:

Phosphoenolpyruvate + ADP^{3-} + H^+ → pyruvate + ATP^{4-}

This reaction, like that catalyzed by phosphoglycerate kinase, constitutes a *substrate-level phosphorylation.* Phosphoenolpyruvate provides both the energy and the phosphoryl group for synthesizing ATP.

Mammalian pyruvate kinase is a tetramer of four identical subunits (MW = 57,000/subunit). Several isozymes are known. One type of isozyme, designated L_4, occurs in the liver; another, designated M_4, is found in muscles. The L_4 isozyme is allosterically inhibited by ATP, acetyl coenzyme A, and fatty acids; it is activated by fructose 1,6-bisphosphate. Pyruvate kinase also undergoes covalent modification (under hormonal control), consisting of phosphorylation and dephosphorylation. Dephosphorylated pyruvate kinase represents the active form. Pyruvate kinase requires K^+ and either Mg^{2+} or Mn^{2+} for activity.

Summing up the 10 reactions of stages I and II yields the following overall reaction for glycolysis:

Glucose + $2NAD^+$ + $2ADP^{3-}$ + $2P_i^{2-}$ → 2 pyruvate$^-$ + $2ATP^{4-}$ + 2NADH + $2H^+$

10.3. GLYCOLYSIS—END PRODUCTS, ENERGETICS, AND CONTROL

10.3.1. Aerobic and Anaerobic Conditions

The glycolytic sequence from glucose to pyruvate constitutes an anaerobic set of reactions; none of the steps requires oxygen. However, glycolysis can occur when the system as a whole is either aerobic or anaerobic (hence the terms aerobic and anaerobic glycolysis). The sequence glucose → pyruvate is unchanged whether oxygen is present or absent, but the metabolic fate of pyruvate depends on whether the system as a whole is aerobic or anaerobic (Figure 10.11).

10.3.1A. Acetyl Coenzyme A and Lactate.
Under aerobic conditions, the *pyruvate dehydrogenase complex* (discussed in Section 11.2) catalyzes the conversion of pyruvate to *acetyl coenzyme A* (see Figure 11.7), which enters the *citric acid cycle.* The cycle, in conjunction with the *electron transport system,* leads to complete oxidation of the acetyl group of acetyl coenzyme A to CO_2 and H_2O. Thus, starting with glucose, and using the combined metabolic systems of glycolysis, the citric acid cycle, and the electron transport system, the cell accomplishes *complete oxidation of carbohydrate to CO_2 and H_2O.*

Recall that the first reaction of stage II results in reduction of NAD^+ to NADH. It is essential to regenerate NAD^+ from NADH so that subsequent substrates can be oxidized. If NADH is not reoxidized—or more significantly, if NAD^+ is not regenerated—glycolysis stops owing to lack of NAD^+ for glyceraldehyde 3-phosphate dehydrogenase. Under aerobic conditions, the electron transport system accomplishes the reoxidation of NADH by means of molecular oxygen (more in Section 12.3).

Under anaerobic conditions, pyruvate does not form acetyl coenzyme A, and NADH cannot be reoxidized via the electron transport system. Instead, regeneration of NAD^+ is accomplished by coupling the oxidation of NADH to NAD^+ with the reduction of pyruvate to *lactate* (CH_3-CHOH-COO$^-$) in a reaction catalyzed by *lactate dehydrogenase (LDH;* a tetramer; MW = 33,500/subunit). Therefore, under anaerobic conditions, glucose catabolism produces lactate rather than pyruvate:

Pyruvate + NADH + H^+ ⇄ NAD^+ + lactate
$$\Delta G^{\circ\prime} = -25.1 \text{ kJ mol}^{-1}$$

An aerobic organism cannot survive in a strictly anaerobic environment—the complete absence of oxygen—but it may be able to tolerate a *temporary oxygen shortage.* In humans, strenuous exercise rapidly depletes oxygen by operation of the electron transport system. Under such relatively anaerobic conditions, pyruvate is converted to lactate. The conversion occurs in muscle tissues, and lactate accumulates there first. From the muscles, lactate diffuses into the bloodstream (see Figure 10.1), which transports it to the liver, where it is reconverted to glucose.

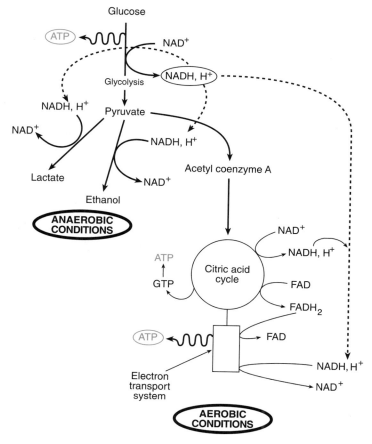

Figure 10.11. Glycolysis under aerobic and anaerobic conditions.

10.3.1B. The Cori Cycle. We refer to the cyclic set of reactions just described as the **Cori cycle** (Figure 10.12), after Carl and Gerti Cori, who first described it. The cycle involves the interplay of muscle, liver, and blood and begins with conversion of glucose to lactate in muscle tissue through glycolysis. The blood then transports lactate to the liver, where it is reoxidized to pyruvate. Pyruvate is converted to glucose via *gluconeogenesis* (discussed in Section 10.6), using liver ATP. Lastly, the blood transports glucose back to muscle tissue.

The Cori cycle has high activity during periods of strenuous exercise. The extra liver ATP, used during gluconeogenesis to process accumulated lactate, must be regenerated by *oxidative phosphorylation* (covered in Section 12.4). Increased oxidative phosphorylation requires extra oxygen, more than that needed for normal operation of the electron transport system. The additional oxygen requirement accounts for the shortness of breath, the panting, you experience with strenuous exercise. We refer to the oxygen required to regenerate used-up liver ATP as *oxygen debt*. It may take a long time before your oxygen consumption rate returns to normal.

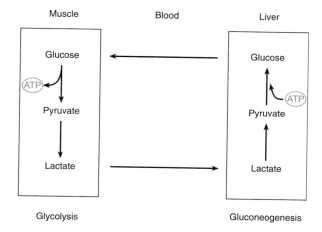

Figure 10.12. The Cori cycle. Muscle glycolysis converts glucose to lactate, which is carried by the blood to the liver, where it is reoxidized to pyruvate. Gluconeogenesis converts pyruvate to glucose, using liver ATP as a source of energy, and the blood transports the glucose back to muscle tissue.

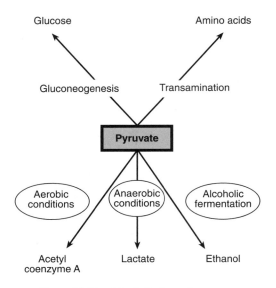

Figure 10.13.   Metabolic fates of pyruvate.

The Cori cycle is linked to both glycogen break-down, *glycogenolysis,* and glycogen synthesis, *glycogenesis.* In muscles, glycogenolysis yields glucose that can be converted to lactate via glycolysis. In the liver, glucose formed by gluconeogenesis can be used for glycogenesis.

10.3.2. Metabolic Fates of Pyruvate

Glycolysis is best defined as the *sequence of reactions from glucose to pyruvate.* Some authors include the pyruvate-to-lactate step in their definition, but the lactate dehydrogenase reaction occurs only under certain conditions. The sequence from glucose to pyruvate, on the other hand, is common to all glycolytic systems.

Pyruvate has several metabolic fates (Figure 10.13). As you saw, conversion of pyruvate to acetyl coenzyme A or to lactate depends on whether the system as a whole is relatively aerobic or anaerobic. However, it also depends on the type of tissue involved and its content of lactate dehydrogenase isozymes.

Consider cardiac muscle and skeletal muscle, which

contain, respectively, the H_4 and M_4 isozymes of lactate dehydrogenase (Table 10.2). Kinetic properties of these isozymes are such that cardiac muscle tissue favors conversion of pyruvate to acetyl coenzyme A whereas skeletal muscle tissue favors conversion of pyruvate to lactate. The heart, the most critical organ of animals, is programmed to derive maximum energy from glucose by linking glycolysis to the aerobic citric acid cycle/electron transport system, which yields a large amount of ATP. Skeletal muscle, on the other hand, can function adequately by deriving a smaller amount of energy from glucose by linking glycolysis to the anaerobic production of lactate.

Pyruvate also serves as the initial substrate for *gluconeogenesis.* In yeast, pyruvate is a substrate for *alcoholic fermentation,* the reaction sequence that converts glucose to ethanol. During alcoholic fermentation, *pyruvate decarboxylase* catalyzes the conversion of pyruvate to acetaldehyde (CH_3-CHO). The acetaldehyde is then reduced to ethanol by means of NADH produced in the glyceraldehyde 3-phosphate dehydrogenase reaction. In the process, NADH is oxidized to NAD^+:

$$\text{Pyruvate} + H^+ \xrightarrow{\text{pyruvate decarboxylase}} CO_2 + \text{acetaldehyde}$$

$$\text{Acetaldehyde} + \text{NADH} + H^+ \xrightarrow{\text{alcohol dehydrogenase}} \text{ethanol} + NAD^+$$

The *alcohol dehydrogenase* reaction also constitutes the first step of alcohol metabolism in the liver, except that under those conditions it proceeds in the reverse direction, from ethanol to acetaldehyde. Some of the effects associated with alcohol intake are directly related to aspects of this reaction, namely, the lowering of pH (production of H^+) and the formation of an aldehyde (acetaldehyde).

10.3.3. Energetics

Table 10.1 lists the free energy changes for the 10 reactions of glycolysis. You can see that in all cases there exist significant differences between biochemical standard

Table 10.2. Properties of Lactate Dehydrogenase Isozymes

Property	H_4 isozyme	M_4 isozyme
K_m for pyruvate	High	Low
k_{cat} (for pyruvate → lactate)	Low	High
Inhibition by pyruvate (of lactate → pyruvate)	Strong	Weak
Fate of pyruvate (Pyr)	Pyr → acetyl coenzyme A	Pyr → Lactate
Metabolism	Aerobic	Anaerobic

and actual free energy changes. Remember that only bio-chemical actual free energy changes truly describe the tendency of reactions to proceed under intracellular conditions. On that basis, three reactions (those catalyzed by hexokinase, phosphofructokinase, and pyruvate kinase) are strongly exergonic and, therefore, irreversible. These reactions make the entire glycolytic sequence from glucose to pyruvate a "downhill" (exergonic) process, a unidirectional pathway with an overall $\Delta G'$ of -76.6 kJ mol^{-1}.

Stage I of glycolysis requires the input of two molecules of ATP (Figure 10.14). Stage II yields two molecules of ATP. At first glance, it may appear that the *net* production of ATP equals zero, but this is incorrect. Every molecule of glucose ultimately gives rise to *two* molecules of glyceraldehyde 3-phosphate. Processing of these in stage II results in formation of *two* molecules of pyruvate and a total production of *four* molecules of ATP. Thus, there occurs a *net gain* of *two* molecules of ATP per molecule of glucose converted to pyruvate.

Under anaerobic conditions, lactate dehydrogenase catalyzes the oxidation of the NADH formed in glycolysis to NAD$^+$. This reaction does not yield any additional energy. Hence, *the maximum energy yield from glycolysis under anaerobic conditions is two molecules of ATP per molecule of glucose* (for the sequence glucose $\rightarrow$ 2 lactate).

The energy yield of glycolysis under anaerobic conditions is low because glucose undergoes only *limited oxidation* in forming pyruvate. As you will see later (Section 12.6), when glycolysis becomes linked to the citric acid cycle under aerobic conditions, glucose undergoes *complete oxidation* to CO_2 and H_2O. The more extensive the oxidation of a metabolite is, the larger the number of hydrogens abstracted from it and the greater the energy yield. You can compare the oxidation states of pyruvate and glucose by computing their C:H:O ratios—1:2:1 for glucose ($C_6H_{12}O_6$) and 1:1.33:1 for pyruvic acid ($C_3H_4O_3$). Thus, pyruvate is only slightly more oxidized (less reduced) than glucose. Because pyruvate production from glucose captures only a small amount of energy in the form of ATP, you might ask whether glycolysis constitutes a wasteful pathway. The answer is that, on the contrary, glycolysis represents a well-engineered process that succeeds in extracting a limited amount of energy from glucose without oxidizing it appreciably.

You can calculate the *efficiency of energy conservation* attained by anaerobic glycolysis as follows. Conversion of glucose $\rightarrow$ 2 lactate has a theoretical $\Delta G^{\circ\prime}$ of -196.6 kJ mol^{-1} and results in formation of 2 ATP. Since synthesis of ATP requires 30.5 kJ mol^{-1}, a total of 61.0 kJ mol^{-1} of glucose catabolism is conserved in the form of ATP synthesis. This represents an efficiency of

$$\frac{61.0 \text{ kJ mol}^{-1}}{196.6 \text{ kJ mol}^{-1}} \times 100 = 31\%$$

Keep in mind, though, that this represents an approximate calculation, based on $\Delta G^{\circ\prime}$ values. Calculating actual efficiencies must be based on pertinent intracellular concentrations of reactants and products and use $\Delta G'$ values. Thus, the true efficiency of energy conservation in any given cell may be higher or lower than 31%.

10.3.4. Pasteur and Crabtree Effects

Scientists have recognized two general control mechanisms of glycolysis and related metabolic pathways for many years. The mechanisms are known, after their discoverers, as the *Pasteur effect* and the *Crabtree effect*. We can understand both effects qualitatively by considering the complete catabolism of glucose, which can be divided into two parts (Figure 10.15). The first part comprises the conversion of glucose to pyruvate via glycolysis. These reactions require no oxygen and yield only a *small amount of ATP*. The second part consists of the conversion of pyruvate to acetyl coenzyme A, which enters the citric acid cycle. Operation of the citric acid cycle in conjunction with the electron transport system requires aerobic conditions and yields a *large amount of ATP*.

Now consider the **Pasteur effect.** Louis Pasteur observed more than 100 years ago (1861) that when yeast

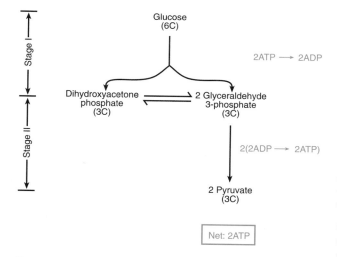

Figure 10.14. The energetics of glycolysis. Stage I requires the input of 2 ATP per molecule of glucose; stage II generates 2 ATP per molecule of glyceraldehyde 3-phosphate. Since one molecule of glucose yields two molecules of glyceraldehyde 3-phosphate, the net yield is 2 ATP per molecule of glucose.

cells were exposed to aerobic conditions, their glucose consumption and ethanol production decreased sharply. The more aerobic the system was, the greater the decrease in glycolysis. The Pasteur effect refers to the *inhibition of glycolysis by oxygen.*

Under aerobic conditions, the citric acid cycle/electron transport system part of glucose catabolism functions well and produces large amounts of energy in the form of ATP. Accordingly, no need exists to process every available glucose molecule through glycolysis in order to derive the maximum amount of energy possible from glucose catabolism. The greater is the activity of the citric acid cycle/electron transport system, the fewer glucose molecules need to be processed via glycolysis. Hence, the more aerobic the system, the greater is the inhibition of glycolysis. We now know the molecular basis of the Pasteur effect. Increased aerobic conditions result in increased production of ATP, and ATP strongly inhibits phosphofructokinase, thereby slowing down glycolysis.

In some ways, the Crabtree effect constitutes the opposite of the Pasteur effect. The **Crabtree effect** refers to the *inhibition of oxygen consumption produced by increasing concentrations of glucose.* When the glucose concentration is high, large amounts of it are catabolized via glycolysis. This results in appreciable production of energy in the form of ATP despite the low energy yield of the pathway. Under these conditions, no need exists to process every pyruvate molecule formed in glycolysis by means of the citric acid cycle/electron transport system to produce still much larger amounts of

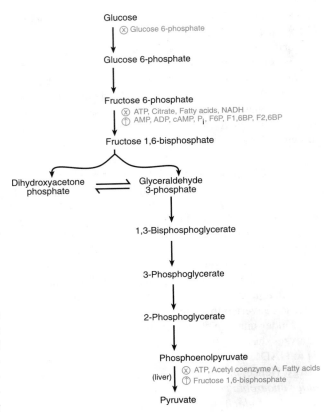

Figure 10.16.   Regulation of glycolysis. ⊗ Inhibitor, ⬆ activator. F6P, F1,6BP, and F2,6BP designate fructose 6-phosphate, fructose 1,6-bisphosphate, and fructose 2,6-bisphosphate, respectively.

ATP. The greater the glucose concentration, the lower is the activity of the citric acid cycle/electron transport system part of the combined pathway. Hence, increasing glucose concentrations inhibit oxygen consumption. We do not yet know the molecular basis of the Crabtree effect.

Control of glycolysis itself occurs at the three highly exergonic, irreversible steps involving hexokinase, phosphofructokinase, and pyruvate kinase (Figure 10.16). Control occurs largely via allosteric enzyme effectors, including adenine nucleotides, NADH, and phosphorylated monosaccharides. Low cellular ATP concentration and high AMP and ADP concentrations stimulate glycolysis and increase production of needed ATP. By contrast, high cellular levels of ATP and NADH depress glycolysis and prevent synthesis of unnecessary ATP.

10.3.5. Catabolism of Other Carbohydrates

Glycolysis provides a catabolic pathway not just for glucose but also for other carbohydrates. Important

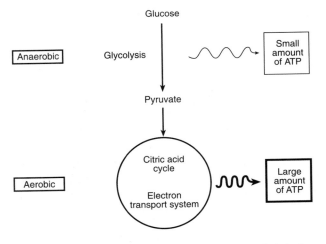

Figure 10.15.   Basis of the Pasteur and Crabtree effects. Carbohydrate metabolism consists of an anaerobic segment having a low energy yield (glycolysis) and an aerobic segment having a high energy yield (citric acid cycle/electron transport system). Extensive activity in one segment depresses activity in the other.

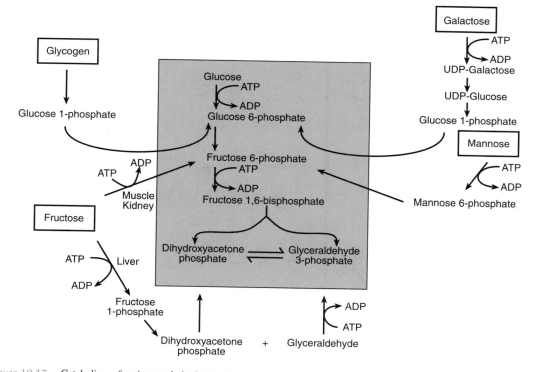

Figure 10.17. Catabolism of various carbohydrates. Many form glycolytic intermediates and enter the sequence at those points.

nonglucose carbohydrates include glycogen, disaccharides, and monosaccharides. We will discuss glycogen catabolism in Section 10.5. Disaccharides are hydrolyzed to their component monosaccharides in the intestinal tract. Many monosaccharides form glycolytic intermediates (Figure 10.17). After entering glycolysis at specific points in the pathway, these intermediates are catabolized by the remaining portion of the pathway.

10.3.5A. Fructose.
Fructose enters glycolysis via two routes. In muscles and kidneys, it undergoes phosphorylation to fructose 6-phosphate in a reaction catalyzed by hexokinase:

$$\text{Fructose} + \text{ATP}^{4-} \rightarrow \text{fructose 6-phosphate}^{2-} + \text{ADP}^{3-} + \text{H}^+$$

In the liver, fructose is phosphorylated at C(1) to form fructose 1-phosphate in a reaction catalyzed by *fructokinase:*

$$\text{Fructose} + \text{ATP}^{4-} \rightarrow \text{fructose 1-phosphate}^{2-} + \text{ADP}^{3-} + \text{H}^+$$

Aldolase catalyzes the cleavage of fructose 1-phosphate to dihydroxyacetone phosphate and glyceraldehyde:

Fructose 1-phosphate $\rightleftarrows$ dihydroxyacetone phosphate +

glyceraldehyde

Dihydroxyacetone phosphate can enter glycolysis directly. Glyceraldehyde must first be phosphorylated to glyceraldehyde 3-phosphate in a reaction catalyzed by *triose kinase:*

$$\text{Glyceraldehyde} + \text{ATP}^{4-} \rightarrow \text{glyceraldehyde 3-phosphate}^{2-} + \text{ADP}^{3-} + \text{H}^+$$

Each of the two routes by which fructose enters glycolysis requires the input of 2 ATP in stage I. Hence, con-

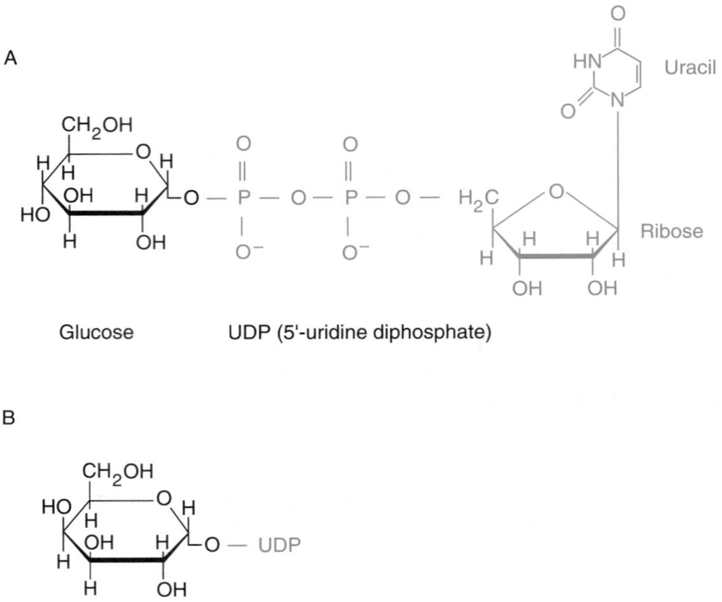

A

Glucose   UDP (5'-uridine diphosphate)

B

Figure 10.18.   (A) UDP-glucose; (B) UDP-galactose.

version of fructose to pyruvate via the combined reactions of stages I and II yields 2 ATP per molecule of fructose for either route. This means that, in terms of energy yield, fructose is equivalent to glucose. However, fructose catabolism in the liver bypasses the phosphofructokinase step with its multiple regulatory factors (see Figure 10.16). Consequently, diets rich in fructose or sucrose can lead to overproduction of pyruvate. Pyruvate, in turn, forms acetyl coenzyme A, a precursor for cholesterol and fatty acid biosynthesis. You can see that in terms of metabolic consequences, fructose is not equivalent to glucose, and high-fructose diets may lead to undesirable nutritional effects for humans.

10.3.5B. Mannose. Hexokinase catalyzes the phosphorylation of mannose to mannose 6-phosphate. *Phosphomannoisomerase* catalyzes the isomerization of mannose 6-phosphate to fructose 6-phosphate:

$$\text{Mannose} + \text{ATP}^{4-} \rightarrow \text{mannose 6-phosphate}^{2-} + \text{ADP}^{3-} + \text{H}^+$$

$$\text{Mannose 6-phosphate}^{2-} \rightleftarrows \text{fructose 6-phosphate}^{2-}$$

10.3.5C. Galactose. The first step in galactose catabolism also involves a phosphorylation, catalyzed by *galactokinase:*

$$\text{Galactose} + \text{ATP}^{4-} \rightarrow \text{galactose 1-phosphate}^{2-} + \text{ADP}^{3-} + \text{H}^+$$

Subsequent steps involve *nucleoside diphosphates* that function as carriers of hexose groups. In these steps, galactose 1-phosphate first reacts with *uridine diphos-*

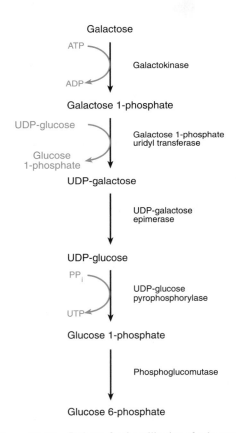

Figure 10.19.   Pathway for the utilization of galactose.

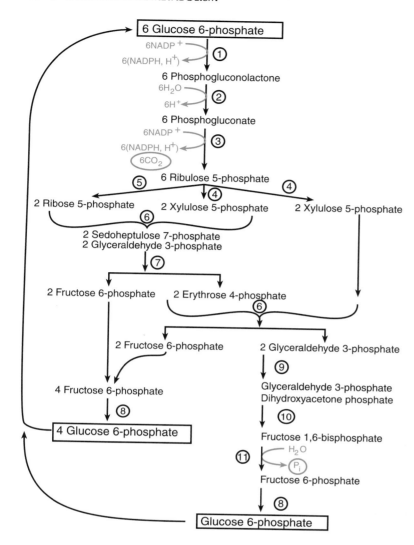

Figure 10.20. The pentose phosphate pathway. Six molecules of glucose 6-phosphate enter the pathway, and five are regenerated. The cycle accomplishes the equivalent of degrading one glucose 6-phosphate to 6 CO_2 and P_i and produces 12 NADPH and 12 H^+. Numbers designate enzymes. ① Glucose 6-phosphate dehydrogenase; ② 6-phosphogluconolactonase; ③ 6-phosphogluconate dehydrogenase; ④ phosphopentose epimerase; ⑤ phosphopentose isomerase; ⑥ transketolase; ⑦ transaldolase; ⑧ phosphoglucoisomerase; ⑨ triose-phosphate isomerase; ⑩ aldolase; ⑪ fructose 1,6-bisphosphatase.

phate glucose or *UDP-glucose* (Figure 10.18) to form *UDP-galactose*. UDP-galactose then forms UDP-glucose and, ultimately, glucose 6-phosphate (Figure 10.19). The pathway is of particular interest in humans because of the occurrence of a genetic disease, **galactosemia**, characterized by inability to convert galactose to glucose metabolites. A deficiency of galactokinase leads to a mild disorder, but a deficiency of *galactose 1-phosphate uridyl transferase* results in severe symptoms that may include lack of growth, liver failure, and mental retardation.

10.4. PENTOSE PHOSPHATE PATHWAY

The *pentose phosphate pathway*—a second major catabolic pathway of carbohydrates—represents an alternate mechanism for glucose degradation. The pathway occurs in the cytoplasm of animal, plant, and bacterial cells. Figure 10.20 shows an overview of the reaction sequence.

The pentose phosphate pathway comprises a cyclic set of reactions that effects the oxidation of glucose 6-phosphate to CO_2 and P_i and leads to the production of large amounts of *reduced nicotinamide adenine dinucleotide phosphate* or *NADPH* (see Figure 11.3).

10.4.1. Oxidative Phase

We divide the reactions of the pentose phosphate pathway into two phases. The first of these, the *oxidative phase,* accomplishes the decarboxylation of glucose (six carbons) to *ribulose* (five carbons) and generates reducing power in the form of NADPH. This phase (Figure 10.21) consists of three reactions:

1   Oxidation of glucose 6-phosphate by means of *glucose 6-phosphate dehydrogenase* to *6-phosphogluconolactone,* an intramolecular ester formed between C(1) and C(5).

2   Hydrolysis of the lactone to *6-phosphogluconate* by means of *6-phosphogluconolactonase.*

3   Oxidative decarboxylation of 6-phosphogluconate, catalyzed by *6-phosphogluconate dehydrogenase,* to *ribulose 5-phosphate.*

Note that in order for glucose to be metabolized via the pentose phosphate pathway, it must first be phosphorylated to glucose 6-phosphate. The phosphorylation requires an input of ATP and can be catalyzed by hexokinase or glucokinase.

10.4.2. Nonoxidative Phase

The remaining reactions of the pentose phosphate pathway consist of interconversions of carbohydrates by reshuffling of their carbon skeletons. Two key enzymes of this phase, *transaldolase* and *transketolase,* are similar in their substrate specificities. Both transfer a carbon fragment from a ketose, serving as donor, to an aldose, serving as acceptor. The enzymes differ in the size of fragment transferred (Figure 10.22).

Transaldolase catalyzes the transfer of a *three-carbon* fragment (a *dihydroxyacetone group*) from a 2-keto sugar to C(1) of an aldose, thereby forming a new ketose and a new aldose. The mechanism of action resembles that of aldolase in glycolysis.

Transketolase catalyzes the transfer of a *two-carbon* fragment (a *glycolaldehyde group*) from a 2-keto sugar to C(1) of an aldose, thereby producing a new ketose and a new aldose. Transketolase requires Mg^{2+} and *thiamine pyrophosphate (TPP)* as cofactors. (See Figure 11.9 for the structure of TPP.) Figure 10.23 shows the reshuffling of carbon skeletons in the pentose phosphate pathway in schematic fashion.

In the operation of the pentose phosphate pathway, *six* molecules of glucose 6-phosphate enter the cycle. Each glucose 6-phosphate loses one carbon as CO_2. Ulti-

mately, five molecules of glucose 6-phosphate are regenerated:

Overall reaction:

$$6 \text{ Glucose 6-phosphate}^{2-} + 7H_2O + 12NADP^+ \rightarrow 6CO_2 + 5 \text{ glucose 6-phosphate}^{2-} + 12NADPH + 12H^+ + P_i^{2-}$$

Consequently, what the pathway accomplishes is *equivalent* to the complete oxidation of *one* molecule of glucose 6-phosphate to six molecules of CO_2 and P_i:

Net reaction:

$$\text{Glucose 6-phosphate}^{2-} + 7H_2O + 12NADP^+ \rightarrow 6CO_2 + 12NADPH + 12H^+ + P_i^{2-}$$

10.4.3. Functions of the Pathway

The pentose phosphate pathway has three main functions:

1   It generates reducing power in the form of NADPH. Note that NADH and NADPH have different roles and *cannot be interchanged* in metabolism. NADH functions in the electron transport system to generate energy in the form of ATP. NADPH serves as a reducing agent for metabolic systems requiring reduction, such as photosynthesis, fatty acid biosynthesis, and conversion of ribonucleotides to deoxyribonucleotides.

2   It is a source of carbohydrates of different structures (three-, four-, five-, six-, and seven-carbon skeletons).

3   It provides a means for interconversions of different monosaccharides.

Pentose phosphate pathway activity is low in muscle but high in adipose tissue, where the cycle generates re-

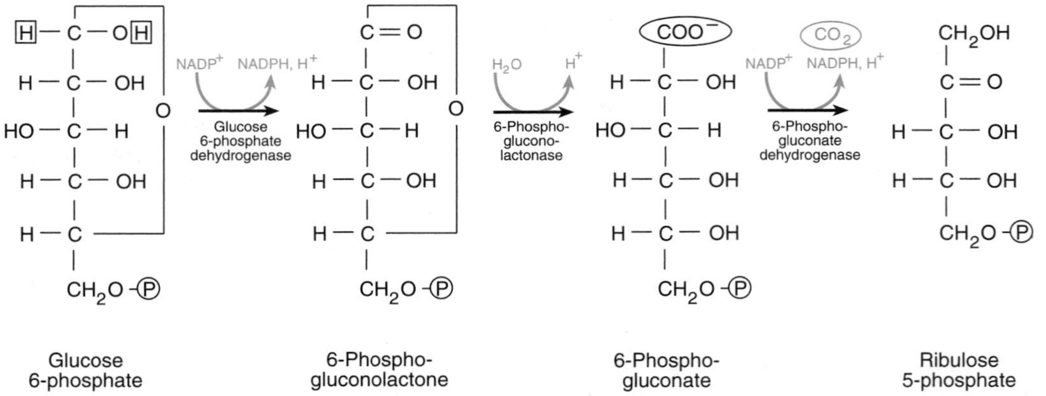

Figure 10.21.   Oxidative phase of the pentose phosphate pathway.

ducing power for fatty acid biosynthesis. Pathway activity is also high in vertebrate red blood cells, where the cycle ensures a supply of reduced glutathione (GSH) formed by action of *glutathione reductase:*

$$GSSG + NADPH + H^+ \xrightarrow{\text{glutathione reductase}} 2GSH + NADP^+$$

Red blood cells require reduced glutathione in large amounts to (a) maintain the structural integrity of proteins by shielding their —SH groups; (b) protect membrane lipids against oxidation by peroxides; and (c) maintain the iron of hemoglobin in its divalent state (Fe^{2+}). Decreased

levels of NADPH may result in alteration of proteins, peroxidation of lipids, and formation of methemoglobin (Fe^{3+}), all changes that weaken red blood cell membranes and render the cells sensitive to rupture (*hemolysis*).

Some individuals have a genetic defect characterized by a glucose 6-phosphate dehydrogenase deficiency. They have low concentrations of NADPH in their red blood cells, are prone to severe cases of anemia, and are sensitive to certain seemingly harmless drugs, including the antimalarial drug *primaquine.* The drugs oxidize some of the available NADPH, thereby putting added strain on cells that have a low concentration of NADPH to start with. Because of the decreased concentration of NADPH, affected

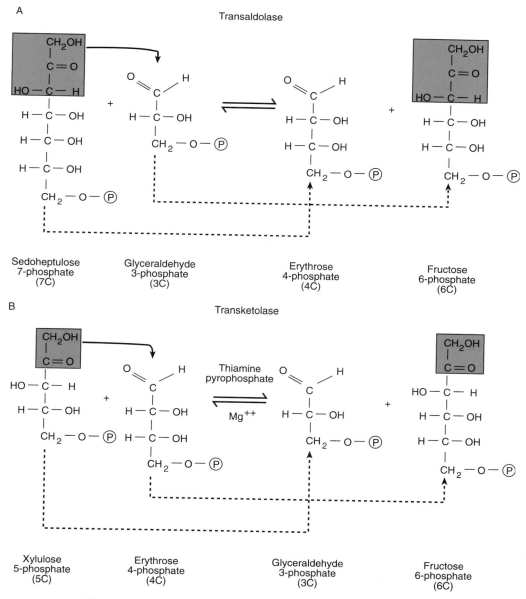

Figure 10.22.   Illustrative reactions of transaldolase (A) and transketolase (B).

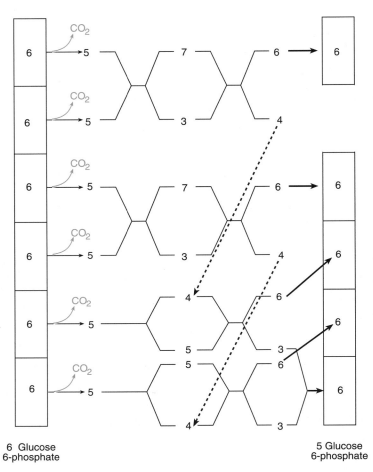

Figure 10.23.   Schematic diagram of the rearrangements of carbon skeletons in the pentose phosphate pathway. Six molecules of glucose 6-phosphate enter the pathway on the left and regenerate five molecules on the right. Numbers indicate the number of carbon atoms per molecule.

cells rupture easily, and administration of such "harmless" drugs may result in *hemolytic anemia*—an extensive anemia that includes massive destruction of red blood cells.

10.5. GLYCOGEN DEGRADATION AND SYNTHESIS

10.5.1. Glycogenolysis

Glycogen, the storage form of carbohydrates in animals, must be broken down to glucose before it can be used for generation of energy. The same holds for starch, the storage carbohydrate of plants. Degradation of both polysaccharides involves stepwise reactions in which the chains are shortened by removal of one glucose residue at a time. **Glycogen phosphorylase,** or **phosphorylase** for short, catalyzes the degradation of glycogen (*glycogenolysis*); a similar enzyme, *starch phosphorylase,* catalyzes the degradation of starch.

10.5.1A. Glycogen Phosphorylase. In the phosphorylase reaction, glycogen undergoes cleavage at its *nonreducing end,* and the terminal glucose residue is released as *glucose 1-phosphate* (Figure 10.24). Phosphorylation of glucose occurs without expenditure of ATP. In this respect, glycogenolysis differs from glycolysis and the pentose phosphate pathway, both of which begin with an ATP-dependent phosphorylation of glucose to glucose 6-phosphate.

We refer to the type of reaction catalyzed by phosphorylase as **phosphorolysis.** Generally speaking, phosphorolysis represents a reaction in which a covalent bond is cleaved and the elements of phosphoric acid are added across the bond (see also Figure 10.10). If phosphoric acid is in the form of H_3PO_4, then one of the products combines with H^+ and the other product combines with $H_2PO_4^-$. In the phosphorylase reaction, the reacting form of phosphoric acid is HPO_4^{2-}, and the products combine with H^+ and PO_4^{3-}, respectively. Cleavage by phosphoric acid in phosphorolysis is analogous to cleavage by water in hydrolysis.

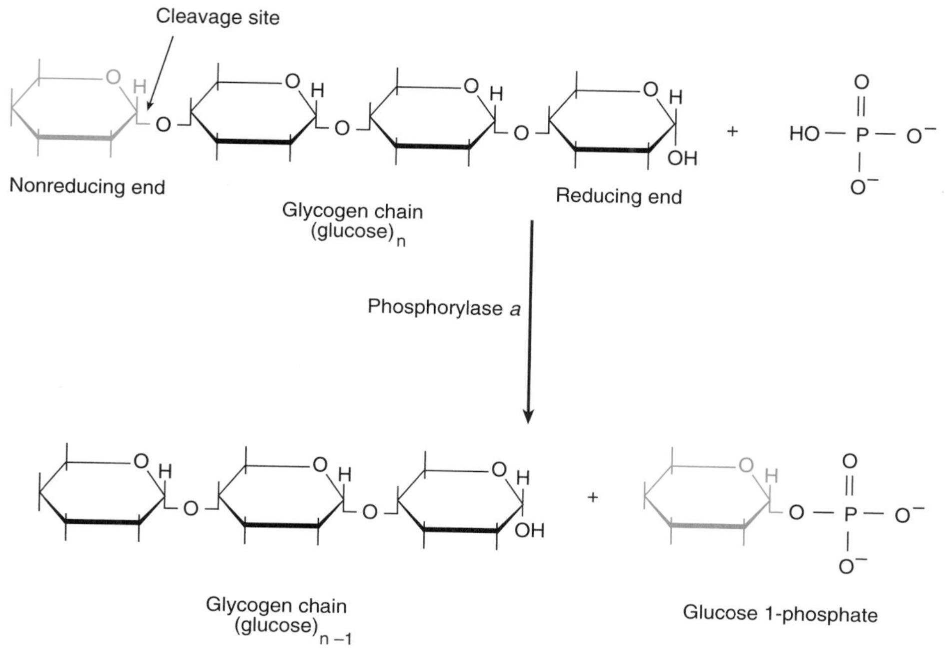

Cleavage site

Nonreducing end

Glycogen chain
(glucose)$_n$

Reducing end

Phosphorylase *a*

Glycogen chain
(glucose)$_{n-1}$

Glucose 1-phosphate

Figure 10.24.   Action of glycogen phosphorylase. The enzyme catalyzes a phosphorolysis reaction whereby one glucose residue at a time is removed from the nonreducing end of glycogen in the form of glucose 1-phosphate.

Following the phosphorylase reaction, **phosphoglucomutase** catalyzes the isomerization of glucose 1-phosphate to glucose 6-phosphate, an intermediate in glycolysis:

Glucose 1-phosphate $\rightleftarrows$ glucose 6-phosphate

Phosphorylase causes digestion of glycogen chains by catalyzing the breaking of $\alpha(1 \rightarrow 4)$ glycosidic bonds between glucose residues. The enzyme cannot catalyze the breaking of $\alpha(1 \rightarrow 6)$ glycosidic bonds at branch points in the molecule, and its action stops within four residues of a branch point. Complete degradation of glycogen requires the action of an additional enzyme called **debranching enzyme.** Debranching enzyme has two enzymatic activities, a transfer function and a hydrolytic function, making it a *bifunctional enzyme.*

When phosphorylase action stops, debranching enzyme catalyzes a transfer of three residues to the nonreducing end of some other branch or of the core chain (Figure 10.25). Following transfer, debranching enzyme catalyzes the hydrolysis of the glucose residue left at the original branch point. The extended branch, or core chain, can now be attacked further by phosphorylase.

10.5.1B. Interaction of Hormones with Cells. Glycogenolysis represents a key link between stored energy of carbohydrates (glycogen) and their workable currency (glucose). Not surprisingly, a process of

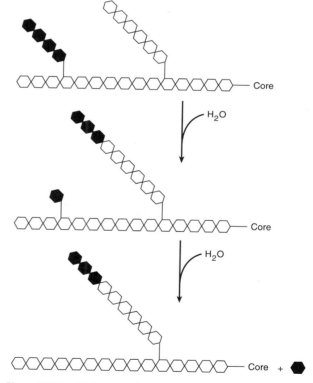

Figure 10.25.   Mechanism of action of debranching enzyme. Three glucose residues of a "limit branch" are transferred to some other branch, followed by hydrolysis of the glucose residue at the original branch point.

such pivotal importance is subject to complex control. The control mechanism of glycogenolysis involves the interaction of *hormones* with the cell membrane and an intracellular *enzyme cascade.* The control mechanism illustrates the chain of events that is triggered when a hormone binds to the cell membrane of a target cell and produces a specific response within the cell.

Hormone interactions with target cells involve three important components—*receptors, G proteins,* and *second messengers.* **G proteins** (so called because they bind guanine nucleotides) serve as membrane-bound signal transducers between hormone receptors in the cell membrane and some intracellular signaling system; they bind GDP in their inactive state and GTP in their active state.

When a hormone binds to a receptor on the outer side of the cell membrane, the receptor undergoes a conformational change that allows it to interact with an inactive G protein. The interaction of receptor and G protein leads to activation of the G protein and causes dissociation of its bound GDP. The dissociating GDP is replaced by GTP. The active G protein, with its bound GTP, interacts with and activates an intracellular signaling system. Subsequent to this interaction, a specific substance is released inside the cell and initiates intracellular changes. We term the substance functioning in this capacity a *second messenger,* and we term the hormone binding to the cell membrane a first messenger.

Active G proteins possess *GTPase* (GTP hydrolase) activity. GTPase catalyzes the hydrolysis of GTP, bound to the active G protein, and thereby converts it back to GDP. Because of this enzymatic activity, the activation of a G protein is short-lived. Nevertheless, the receptor–G protein interaction results in *amplification* of the hormonal signal because each hormone–receptor complex interacts with and activates many G proteins. Moreover, each activated G protein activates many intracellular signaling systems before it becomes inactivated.

10.5.1C. Enzyme Cascade of Glycogenolysis.
Receptors for the control system of glycogenolysis are located on the outside of the cell membrane and bind the hormones *epinephrine* and *glucagon* (Figure 10.26). When either of these two hormones becomes bound to the receptor, a G protein is activated that in turn activates a molecule of **adenylate cyclase.** Adenylate cyclase is bound to the interior side of the cell membrane and catalyzes the formation of **cyclic AMP (3′,5′-cyclic adenylic acid; cAMP)** from ATP:

$$\text{ATP}^{4-} + \text{H}_2\text{O} \longrightarrow \text{cAMP}^- + \text{PP}_i^{3-}$$

Cyclic AMP constitutes the second messenger of this control system, transmitting the signal from the first messenger, the hormone, to a cellular enzyme and thereby initiating the intracellular changes. Cyclic AMP functions as an allosteric effector of **protein kinase,** the first enzyme in the intracellular enzyme cascade. Protein kinase activates phosphorylase kinase, which in turn activates phosphorylase. Investigators estimate that the total cascade mechanism of glycogenolysis provides an amplification of about 25×10^6 for the incoming signal. In other words, during the time that one hormone molecule acts on a membrane receptor, some 25×10^6 molecules of active glycogen phosphorylase form.

Protein kinase, the enzyme activated by cAMP, is a tetramer of two catalytic (C) and two regulatory (R) subunits. When cAMP binds to R subunits, C subunits dissociate. Catalytic subunits are active only after they dissociate from the tetramer. Protein kinase activates **phosphorylase kinase,** another tetramer but composed of four nonidentical subunits. Two of the larger subunits contain serine residues that become phosphorylated when the enzyme is activated. The smallest subunit consists of a regulatory protein called *calmodulin* (*cal*cium-*modul*ating prote*in*) that binds calcium. Calmodulin, a small protein of 148 amino acids, has four high-affinity binding sites for Ca^{2+}. Calmodulin's amino acid sequence is highly conserved, and the protein is widespread among eukaryotes. Calmodulin participates in the regulation of numerous cellular processes by controlling the level of Ca^{2+}.

Binding of calcium by phosphorylase kinase leads to partial activation of the enzyme. The enzyme has greatest activity when it is both phosphorylated and has calcium bound to it. The fourth subunit of phosphorylase kinase contains the active site that catalyzes the phosphorylation and activation of glycogen phosphorylase, the last enzyme in the cascade. Glycogen phosphorylase exists as a dimer of two identical subunits. Its active form has a phosphate group esterified at a serine residue in each subunit. Both phosphorylase kinase and phosphorylase can be dephosphorylated, and inactivated, by specific *phosphatases.* We refer to enzymes, such as protein kinase or phosphatase, that catalyze interconversions of two forms of another enzyme as **converter enzymes.**

10.5.1D. "Fight or Flight" Response.
Glycogen structure, the specificity of glycogen phosphorylase, and hormonal control of glycogenolysis play key roles in an important physiological response called the **"fight or flight" response.** This response is elicited in vertebrate animals when they are subjected to stress. Various stress stimuli lead to secretion of epinephrine, the principal hormone governing the response. Epinephrine helps the animal prepare for the emergency and cope with the imposed stress by triggering a number of changes, including an increase in heart rate and blood pressure and an increase in the generation of ATP.

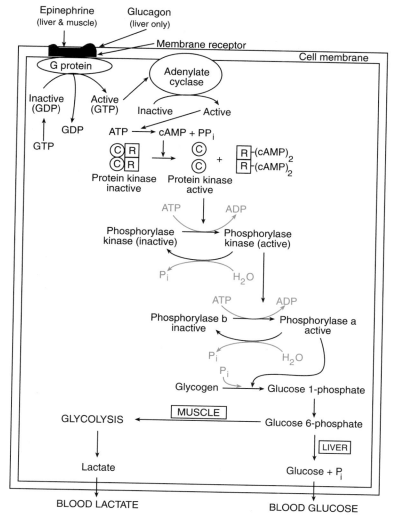

Figure 10.26. The hormonally controlled enzyme cascade regulating the activity of glycogen phosphorylase.

Having $\alpha(1 \rightarrow 6)$ branch points means that glycogen molecules possess a larger number of nonreducing ends than if the same mass were to exist in straight-chain molecules. Because glycogen phosphorylase acts only at non-reducing ends, the branched structure of glycogen leads to production of larger amounts of glucose 1-phosphate *per unit time* than would be produced by phosphorylase from a comparable mass of straight-chain molecules. Increased production of glucose 1-phosphate presents an advantage for the "fight or flight" response because it allows for rapid generation of ATP over and above that resulting from the enzyme cascade.

Formation of glucose 1-phosphate in muscle stimulates glycolysis and results in generation of ATP (Figure 10.26). Simultaneously, production of glucose 1-phosphate in the liver leads to an increase in the level of blood sugar that further fuels muscles with glucose for generation of

additional ATP. The ATP formed helps muscles cope with the stress that triggered the release of epinephrine.

10.5.2. Glycogenesis

10.5.2A. Glucose Activation. *Glycogenesis,* or the synthesis of glycogen from glucose, occurs when the concentration of glucose exceeds that needed for generation of energy by glycolysis and when glycogen stores must be replenished. Glycogenesis involves a glycolytic enzyme and several other enzymes that are unique to this anabolic pathway. The process begins with activation of glucose to glucose 6-phosphate, catalyzed by the glycolytic enzyme hexokinase, or by glucokinase when the glucose concentration is high:

$$\text{Glucose} + \text{ATP}^{4-} \rightarrow \text{glucose 6-phosphate}^{2-} + \text{ADP}^{3-} + \text{H}^{+}$$

Next, phosphoglucomutase catalyzes the isomerization of glucose 6-phosphate to glucose 1-phosphate:

$$\text{Glucose 6-phosphate}^{2-} \rightleftarrows \text{glucose 1-phosphate}^{2-}$$

Glucose 1-phosphate, the product of glycogen breakdown, is also the building block for glycogen synthesis. In a reaction catalyzed by *UDP-glucose pyrophosphorylase,* glucose 1-phosphate is converted to uridine diphosphate glucose, or *UDP-glucose* (Figure 10.18):

$$\text{Glucose 1-phosphate}^{2-} + \text{UTP}^{4-} + \text{H}^+ \rightarrow$$
$$\text{UDP-glucose}^{2-} + \text{PP}_i^{3-}$$

In this reaction, the phosphoryl oxygen of glucose 1-phosphate attacks the α-phosphorus atom of UTP. UDP-glucose and similar *nucleotide-linked sugars* represent compounds in which the nucleoside diphosphate functions as a carrier of an activated *glycosyl group.* Nucleotide-linked sugars constitute energy-rich compounds that play important roles in carbohydrate metabolism.

The reaction catalyzed by UDP-glucose pyrophosphorylase is freely reversible ($\Delta G^{\circ\prime} \approx 0$). You might expect this since both the reactants and the products have a total of two energy-rich bonds each (two in UTP, one each in UDP-glucose and PP_i). However, whenever pyrophosphate forms in metabolism, it generally undergoes hydrolysis to inorganic phosphate ($\text{PP}_i \rightarrow 2\text{P}_i$) in a reaction catalyzed by *pyrophosphatase.* This reaction is strongly exergonic because PP_i constitutes an energy-rich compound of the acid anhydride type. Hydrolysis of PP_i in the UDP-glucose pyrophosphorylase reaction ensures that formation of UDP-glucose is irreversible.

10.5.2B. Glycogen Synthase. **Glycogen synthase** catalyzes the polymerization of activated glucose residues. The enzyme occurs tightly bound to intracellular glycogen granules. During polymerization, one glucose residue at a time adds to the nonreducing end of the growing polysaccharide chain (Figure 10.27):

$$(\text{Glucose})_n + \text{UDP-glucose}^{2-} \rightarrow (\text{glucose})_{n+1} + \text{UDP}^{3-} + \text{H}^+$$

The reaction is exergonic ($\Delta G^{\circ\prime} = -13.4$ kJ mol^{-1}), but the enzyme requires a *primer* for initiation of polymerization. Glycogen molecules present in the cell can serve in this capacity and become extended by the action of glycogen synthase. It appears that the best primers consist of short-chain glycogen molecules. The enzyme has a small K_m for short chains but larger K_m values for longer chains.

Complete glycogen synthesis requires a mechanism for the synthesis of primers. Primer synthesis occurs via two steps and involves a protein named *glycogenin* (MW = 37,000). The first step seems to require a specific glucosyltransferase that catalyzes transfer of a glucose residue from UDP-glucose to glycogenin. The glucose becomes attached via a glycosidic bond involving the 1′-OH of glucose (the

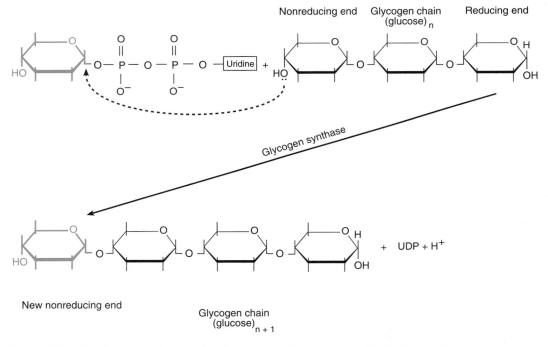

Figure 10.27.   The glycogen synthase reaction. One glucose residue at a time is added to the nonreducing end of glycogen.

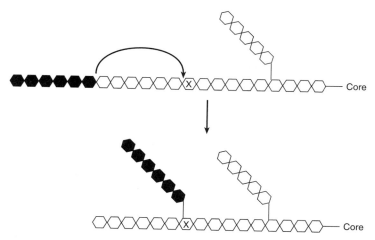

Figure 10.28. Mechanism of action of branching enzyme. A segment of 6 or 7 glucose residues is removed from a chain containing at least 11 residues and is reanchored as a branch at least 4 residues removed from another branch.

reducing end) and the phenolic hydroxyl group of a tyrosine residue in glycogenin. The second step, catalyzed by glycogenin itself, consists of extending the bound glucose with up to seven additional glucose residues, using UDP-glucose as a substrate. Subsequent linear extension of the primer, attached to glycogenin, is catalyzed by glycogen synthase. Thus, glycogenin serves both as an enzyme and as a scaffold for glycogen synthesis. UDP, released during synthesis of the primer, is reconverted to UTP by action of *nucleoside diphosphate kinase:*

$$UDP + ATP \rightleftarrows UTP + ADP$$

Action of glycogen synthase produces straight chains in which glucose residues are linked by $\alpha(1 \rightarrow 4)$ glycosidic bonds. To complete the synthesis of glycogen and form branches attached via $\alpha(1 \rightarrow 6)$ glycosidic bonds requires the action of **branching enzyme** (*1,4 → 1,6 transglycosylase*). Branching enzyme catalyzes transfer of a segment containing 6 or 7 residues—from a chain containing at least 11 residues—to some place on the same chain or on a different chain. The new branch point must be at least four residues removed from an adjacent branch point (Figure 10.28). Branching proceeds readily and does not involve large free energy changes because the energy required for breaking an $\alpha(1 \rightarrow 4)$ bond is very similar, but opposite in sign, to that required for forming an $\alpha(1 \rightarrow 6)$ bond.

10.5.2C. Regulation of Glycogenesis. Con-
trol of glycogenesis is complex. One regulatory effect results from the concentration of glycogen itself. Glycogen exerts feedback inhibition on its own formation by inhibiting glycogen synthase; consequently, the K_m of the enzyme increases with increasing glycogen chain length.

A second regulatory mechanism involves covalent modification of glycogen synthase. Glycogen synthase is an oligomeric protein consisting of four identical subunits (MW = 85,000 each). Each subunit can be phosphorylated and dephosphorylated at the OH of a serine residue (Figure 10.29). Phosphorylation and dephosphorylation of glycogen synthase are catalyzed by, respectively, the same kinase and the same phosphatase that catalyze phosphorylation and dephosphorylation of glycogen phospho-

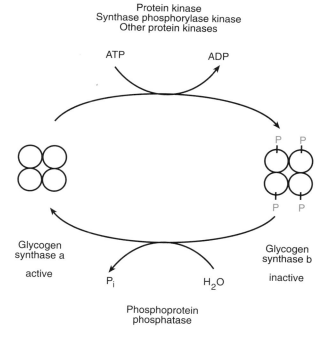

Figure 10.29. Covalent modification of glycogen synthase. The enzyme is a tetramer, composed of four identical subunits.

rylase. Glucagon and epinephrine, which regulate glycogen phosphorylase, also control activation of the kinase and the phosphatase that act on glycogen synthase. In both instances, activation involves a similar enzyme cascade (Figure 10.30); cyclic AMP serves as a second messenger for both glycogen synthesis and glycogen degradation.

Dephosphorylated glycogen synthase, designated "*a*," constitutes the *active* form of the enzyme. Glycogen synthase *a* is independent of the concentration of glucose 6-phosphate. *Phosphorylated* glycogen synthase, designated "*b*," constitutes the *inactive* form of the enzyme. Several phosphorylated varieties of glycogen synthase occur because at least six other protein kinases, in addition to that also acting on glycogen phosphorylase, can deactivate glycogen synthase by phosphorylation. The various phosphorylations involve at least nine different serine residues per subunit. A number of secondary messengers of hormone action, including cAMP, Ca^{2+}, and diacylglycerol, regulate the activity of several kinases that function in glycogenolysis and glycogenesis.

The "*b*" form of glycogen synthase is active only in the presence of glucose 6-phosphate. At high concentrations of glucose 6-phosphate, the inactive enzyme becomes active; glucose 6-phosphate functions as an allosteric effector of glycogen synthase *b*. The allosteric effect of glucose 6-phosphate constitutes yet another regulatory mechanism of the enzyme and makes good metabolic sense. Activation of glycogen synthase by high concentrations of glucose 6-phosphate ensures that excess glucose 6-phosphate is converted to glycogen and stored, rather than degraded via glycolysis to generate unnecessary energy.

Insulin also has an effect on glycogen synthesis. Insulin stimulates phosphatase, thereby increasing the dephosphorylation and activation of glycogen synthase. Accordingly, insulin opposes the action of glucagon and epinephrine, promotes glycogen synthesis from glucose, and leads to a lower level of blood sugar.

From our discussion, you can see that glycogenolysis and glycogenesis are *reciprocally regulated* (Figure 10.31); when one process is activated, the other is inhibit-

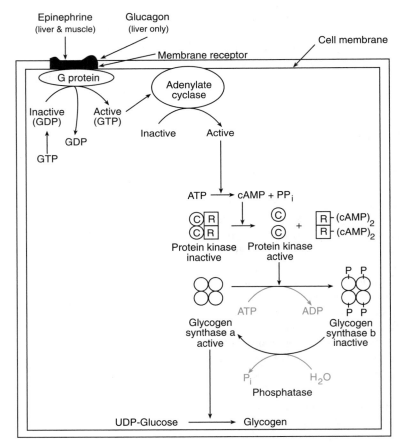

Figure 10.30.   The hormonally controlled enzyme cascade regulating the activity of glycogen synthase. Initial reactions are identical to those of the glycogen phosphorylase cascade (Figure 10.26).

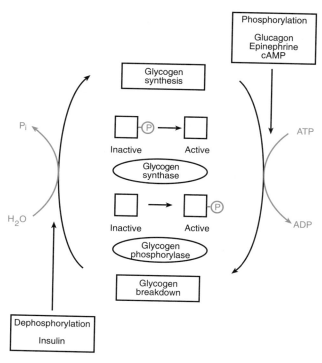

Figure 10.31. Reciprocal regulation of glycogen synthesis and degradation. Activation of one pathway leads to inhibition of the other.

ed. Phosphorylation leads to activation of glycogen phosphorylase but inactivation of glycogen synthase. Dephosphorylation has the opposite effects. Reciprocal regulation allows for effective control of carbohydrate metabolism.

While both glycogen phosphorylase and glycogen synthase are regulated by similar hormonally controlled enzyme cascades, note that the glycogen breakdown cascade has one more cycle than the glycogen synthesis cascade. This means that glycogen degradation has greater sensitivity than glycogen synthesis; degradation can be amplified to a greater extent than synthesis. Researchers have estimated that glycogen degradation can proceed about 300 times faster than glycogen synthesis. The different sensitivity of the two cascades is metabolically useful. It is critical that the cell possess the capacity to rapidly generate glucose from glycogen stores and convert it to usable energy. It is far less critical that the cell be capable of rapidly storing excess glucose in the form of glycogen.

Scientists have identified a number of hereditary disorders of glycogen metabolism, called **glycogen storage diseases** (see Table 8.3). Some of these lead to glycogen that has an abnormal structure, such as glycogen having short outer chains or no outer chains or glycogen having long and unbranched chains. Other glycogen storage diseases, while resulting from specific enzymatic deficien-

cies, lead to glycogen that has a normal molecular structure. Abnormalities in glycogen metabolism are usually most pronounced in the liver, but heart and muscle glycogen metabolism can also be defective.

10.6. GLUCONEOGENESIS

Gluconeogenesis refers to the synthesis of glucose from noncarbohydrate precursors such as lactate, pyruvate, glycerol, and amino acids (Figure 10.32). Gluconeogenesis is crucial for survival of humans and other animals because these organisms possess a number of tissues, including the central nervous system, kidney medulla, red blood cells, and testes, that use primarily glucose as fuel to generate energy. An adequate level of blood sugar must be maintained so that these tissues receive the necessary amounts of glucose. Generally, dietary intake of carbohydrates and carbohydrate stores are insufficient to meet the organism's need for glucose.

Much of the dietary supply of glucose becomes rapidly exhausted by catabolism. Strenuous exercise, a carbohydrate-deficient diet, and a fast longer than one day all lead to accelerated depletion of dietary glucose and require glucose synthesis at even greater than normal rates.

Stored carbohydrate is also insufficient to meet the organism's need for glucose. In humans, body fluids contain about 20 g of glucose, and carbohydrate stored as glycogen amounts to about 190 g. By contrast, metabolism requires about 160 g of glucose per day, of which approximately 120 g/day is needed for operating the brain. Based on these quantities, the readily available glucose in a human can only meet the energy demands for approximately one day.

Thus, despite dietary intake of glucose and the presence of carbohydrate stores, glucose must be synthesized at all times to provide a continuous supply to those tissues using it as primary energy source.

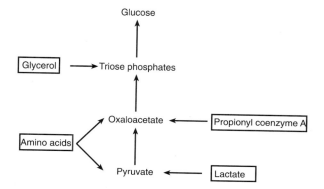

Figure 10.32. Major precursors for the biosynthesis of glucose. Some arrows represent multiple-step reactions.

10.6.1. Reversal of Glycolysis

In a sense, gluconeogenesis constitutes a reversal of glycolysis: glucose is synthesized from pyruvate. Recall, however, that an anabolic pathway *cannot* be the exact reverse of a catabolic sequence (Section 8.2); hence, gluconeogenesis cannot be an exact reversal of glycolysis. Glycolysis proceeds unidirectionally from glucose to pyruvate because of three *strongly exergonic* and irreversible steps that involve the conversions of (a) glucose to glucose 6-phosphate, (b) fructose 6-phosphate to fructose 1,6-bisphosphate, and (c) phosphoenolpyruvate to pyruvate. In order to accomplish a reversal of glycolysis, each of these three steps, which are *strongly endergonic* in the reverse direction, must be replaced by a suitable pathway alteration termed a **bypass**. The three bypasses of gluconeogenesis occur in the following order (Figure 10.33):

Bypass I: Pyruvate to phosphoenolpyruvate

Bypass II: Fructose 1,6-bisphosphate to fructose 6-phosphate

Bypass III: Glucose 6-phosphate to glucose

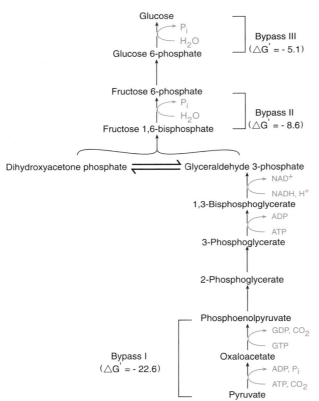

Figure 10.33. The pathway of gluconeogenesis. Free energy changes ($\Delta G'$) of the bypasses are in kilojoules per mole, determined for liver.

The remaining steps of glycolysis are readily reversible reactions ($\Delta G' \approx 0$) that can be traversed in either direction. These steps form part of both glycolysis and gluconeogenesis. Biochemists usually consider oxaloacetate to be the true starting material of gluconeogenesis. Although gluconeogenesis commences with pyruvate, the capacity to form oxaloacetate constitutes the actual step permitting a reversal of glycolysis.

10.6.1A. Bypass I.

Bypass I achieves the conversion of pyruvate to phosphoenolpyruvate. The energy barrier is overcome by changing the reaction to a *two-step mechanism, requiring two nonglycolytic enzymes and a nonglycolytic intermediate (oxaloacetate). Both the reactants and the products of the glycolytic step are altered.*

Pyruvate carboxylase catalyzes the first step, involving a carboxylation of pyruvate to form oxaloacetate ($^-OOC\text{-}CH_2\text{-}CO\text{-}COO^-$), an intermediate of the citric acid cycle:

$$Pyruvate^- + CO_2 + ATP^{4-} + H_2O \rightarrow oxaloacetate^{2-} + ADP^{3-} + P_i^{2-} + 2H^+$$

Pyruvate carboxylase, a tetramer composed of four identical subunits (MW $\approx$ 120,000 each), occurs in the mitochondrial matrix. Each subunit binds Mg^{2+} and carries *biotin* (see Section 11.5) as a prosthetic group. Acetyl coenzyme A is a powerful allosteric activator of pyruvate carboxylase; without bound acetyl coenzyme A, the enzyme has essentially no activity.

Phosphoenolpyruvate carboxykinase catalyzes the second step:

$$Oxaloacetate^{2-} + GTP^{4-} \rightarrow phosphoenolpyruvate^{3-} + GDP^{3-} + CO_2$$

The overall reaction is:

$$Pyruvate^- + ATP^{4-} + GTP^{4-} + H_2O$$
$$\downarrow$$
$$phosphoenolpyruvate^{3-} + ADP^{3-} + GDP^{3-} + P_i^{2-} + 2H^+$$

The subcellular location of phosphoenolpyruvate carboxykinase varies among different organisms. In mouse and rat livers, the enzyme occurs almost exclusively in the cytosol; in pigeon and rabbit livers, it occurs inside the mitochondria; and in guinea pigs and humans, it is more or less evenly distributed between cytosolic and mitochondrial compartments. The intracellular location of phosphoenolpyruvate carboxykinase necessitates specific transport systems for either oxaloacetate or phosphoenolpyruvate. If the enzyme is located in the cytosol, oxaloacetate must leave the mitochondria for conversion to phosphoenolpyruvate in the cytosol. If the enzyme is located in the mitochondria, phosphoenolpyruvate formed inside the mitochondria must move out into the cytosol for participation in gluconeogenesis. All subsequent reactions of gluconeogenesis, beginning with phosphoenolpyruvate, occur in the cytosol. Other sources of oxaloacetate for gluconeogenesis include degradation of amino acids and degradation of fatty acids with an odd number of carbons atoms.

10.6.1B. Bypass II.
Bypass II occurs at the conversion of fructose 1,6-bisphosphate to fructose 6-phosphate. Strict reversal of the glycolytic reaction that converts fructose 6-phosphate to fructose 1,6-bisphosphate, with phosphate provided by ATP, requires synthesis of ATP from ADP, a step that is strongly endergonic. The difficult ATP synthesis is avoided by substituting a simple hydrolysis reaction that involves a *different enzyme. Both the reactants and the products of the glycolytic step are altered.* The reaction of bypass II is catalyzed by fructose 1,6-bisphosphatase:

$$\text{Fructose 1,6-bisphosphate}^{4-} + H_2O \rightarrow$$
$$\text{fructose 6-phosphate}^{2-} + P_i^{2-}$$

10.6.1C. Bypass III.
Bypass III occurs at the conversion of glucose 6-phosphate to glucose and is chemically analogous to bypass II. Here, too, reversal of the corresponding glycolytic step requires synthesis of ATP from ADP. Once again, the difficult ATP synthesis is avoided by substituting a simple hydrolysis reaction that involves a *different enzyme. Both the reactants and the products of the glycolytic step are altered.* The reaction is catalyzed by glucose 6-phosphatase, an enzyme bound to the endoplasmic reticulum:

$$\text{Glucose 6-phosphate}^{2-} + H_2O \rightarrow \text{glucose} + P_i^{2-}$$

10.6.1D. Overall Reaction.
Bypass I requires energy in the form of both ATP and GTP. Because GTP is equivalent to ATP in terms of energy-rich bonds, the energy requirement of bypass I totals 2 ATP (two energy-rich bonds). Synthesis of 1,3-bisphosphoglycerate from 3-phosphoglycerate requires an additional ATP, making a total requirement of 3 ATP. However, synthesis of one molecule of glucose requires two molecules of pyruvate, forming two molecules of 1,3-bisphosphoglycerate. Accordingly, the total energy cost of gluconeogen-

esis is 6 ATP per molecule of glucose synthesized and we can write the overall reaction as

$$2 \text{ Pyruvate}^- + 2NADH + 4ATP^{4-} + 2GTP^{4-} + 6H_2O$$
$$\downarrow$$
$$\text{glucose} + 2NAD^+ + 4ADP^{3-} + 2GDP^{3-} + 6P_i^{2-} + 2H^+$$

Actually, the true energy cost exceeds the six energy-rich bonds indicated by this equation. Conversion of two molecules of 1,3-bisphosphoglycerate to two molecules of glyceraldehyde 3-phosphate requires input of 2 NADH. If not thus expended, NADH could be oxidized to NAD^+ via the electron transport system, resulting in a gain of 4 or 5 ATP depending on the type of shuttle required (see Section 12.6). Loss of 4 or 5 ATP raises the *effective energy cost* of gluconeogensis to 10 or 11 ATP per molecule of glucose synthesized.

10.6.2. Glucose–Alanine Cycle

Two important cycles function to provide the liver, the central organ for gluconeogenesis, with appropriate substrates for glucose synthesis: the *Cori cycle,* discussed in Section 10.3, and the **glucose–alanine cycle.**

In the glucose–alanine cycle, glucose is converted to pyruvate in muscle tissue, and *transamination* (Section 14.2) converts pyruvate to alanine. Blood then transports the alanine to the liver, where transamination reconverts it to pyruvate, which then serves as substrate for gluconeogenesis. Blood returns the glucose produced to the muscles (Figure 10.34).

A major difference between the glucose–alanine and Cori cycles lies in the nature of the three-carbon com-

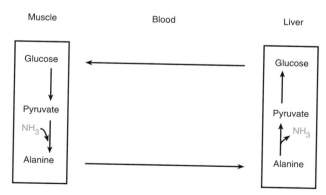

Figure 10.34. The glucose–alanine cycle. Muscle glycolysis converts glucose to pyruvate. Gluconeogenesis converts pyruvate to glucose in the liver. Interconversions of alanine and pyruvate in both tissues involve transaminations.

pound transported by the blood from muscle tissue to the liver. Blood transports lactate in the Cori cycle but transports alanine in the glucose–alanine cycle.

10.6.3. Effect of Alcohol

Ingesting alcohol inhibits gluconeogenesis and leads to lower blood sugar levels. A potentially dangerous low level of blood sugar, termed *hypoglycemia,* can develop. The effect results from the first reaction of alcohol catabolism—the oxidation of ethanol to acetaldehyde, catalyzed by *alcohol dehydrogenase:*

$$\text{Ethanol} + NAD^+ \rightarrow \text{acetaldehyde} + NADH + H^+$$

This reaction produces NADH and leads to an increase in the NADH/NAD$^+$ ratio of the cell. Increased concentrations of NADH drive the *lactate dehydrogenase* and *malate dehydrogenase* (Section 11.3) reactions toward formation of lactate and malate, respectively:

$$\text{Pyruvate}^- + NADH + H^+ \rightarrow \text{lactate}^- + NAD^+$$

$$\text{Oxaloacetate}^{2-} + NADH + H^+ \rightarrow \text{malate}^{2-} + NAD^+$$

As lactate and malate form, concentrations of pyruvate and oxaloacetate decrease. Low levels of pyruvate and oxaloacetate result in a strong depression of gluconeogenesis and a severe lowering of the blood sugar level.

10.6.4. Regulation of Gluconeogenesis

If neither glycolysis nor gluconeogenesis were regulated, the combined pathways would constitute a giant **futile cycle** that achieves nothing except dissipation of the free energy of hydrolysis of ATP and GTP (Figure 10.35). Fortunately, this is not the case.

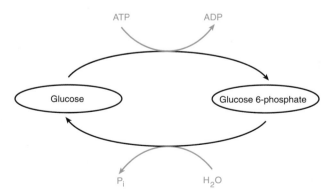

Figure 10.35.   Hypothetical example of a futile cycle.

Like glycogenolysis and glycogenesis, glycolysis and gluconeogenesis are reciprocally regulated. Control points occur at the three steps where the forward and reverse reactions can be independently regulated—in other words, where bypasses function in gluconeogenesis. Regulatory effects in animals occur at all three steps for glycolysis, but only at two steps for gluconeogenesis (Figure 10.36). Major control of gluconeogenesis, much as that of glycolysis, involves allosteric enzyme effectors.

10.6.4A. Fructose 2,6-Bisphosphate.
Fructose 2,6-bisphosphate, an isomer of fructose 1,6-bisphosphate, is a key allosteric effector in the regulation of glycolysis and gluconeogenesis; it functions as an *activator of phosphofructokinase in glycolysis* and as an *inhibitor of fructose 1,6-bisphosphatase in gluconeogenesis.* The concentration of fructose 2,6-bisphosphate is regulated by controlling both its rate of synthesis and its rate of degradation. *Phosphofructokinase-2* (the glycolytic enzyme is termed phosphofructokinase-1) catalyzes the synthesis of fructose 2,6-bisphosphate, and *fructose bisphosphatase-2* catalyzes its degradation.

The enzymatic activities of both phosphofructokinase-2 and fructose bisphosphatase-2 occur on the same *bifunctional* protein. Both activities are subject to multiple allosteric control and to covalent modification by phosphorylation. Phosphorylation inactivates phosphofructokinase-2 and activates fructose bisphosphatase-2. Phosphorylation/dephosphorylation is controlled by cAMP-dependent protein kinase and a phosphoprotein phosphatase. These multiple factors combine to provide a finely tuned regulation of the concentration of fructose 2,6-bisphosphate.

10.6.4B. Carbohydrate Catabolism and Anabolism.
We can summarize the overall regulation of glycolysis, glycogenolysis, glycogenesis, and gluconeogenesis as follows. A well-fed organism has a good dietary supply of glucose and does not require additional production of blood sugar by metabolic processes. Under these conditions, the liver functions to conserve metabolic fuels so that both carbohydrate and lipid stores become augmented. Some glucose serves directly for biosynthesis of glycogen (glycogenesis), and the rest undergoes catabolism (glycolysis) to acetyl coenzyme A, which serves as the starting material for fatty acid synthesis.

An organism that has fasted has a poor dietary supply of glucose and requires continuous production of blood sugar by metabolic processes. Under these conditions, the liver functions to draw on stored metabolic fuels by stimulating glycogen breakdown to glucose (glyco-

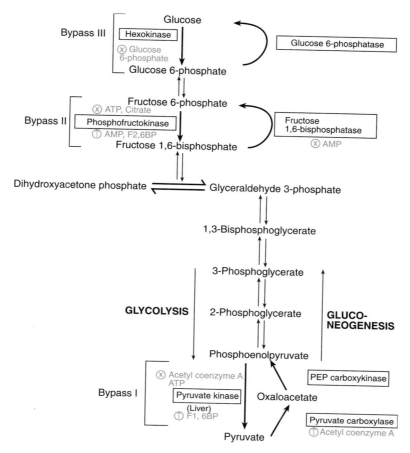

Figure 10.36. Reciprocal regulation of glycolysis and gluconeogenesis. ⊗ Inhibitor, ⊕ activator. Only major inhibitors and activators of phosphofructokinase are shown. F2,6BP and F1,6BP designate fructose 2,6-bisphosphate and fructose 1,6-bisphosphate, respectively.

genolysis). The liver also functions to enhance glucose biosynthesis (gluconeogenesis), using primarily protein degradation products as precursors of pyruvate and oxalo-acetate.

10.7. BIOSYNTHESIS OF OTHER CARBOHYDRATES

10.7.1. Polysaccharides

Most polysaccharides are synthesized from **nucleotide-linked sugars.** Both muscle and liver glycogen form from UDP-glucose, and ADP-glucose serves as precursor for the biosynthesis of glycogen in bacteria and for the biosynthesis of starch (amylose and amylopectin) in plants. For cellulose synthesis, different plant species use either UDP-glucose, ADP-glucose, or CDP-glucose as the activated form of the glycosyl residue. Chitin, a

homopolysaccharide of N-acetylglucosamine, forms from UDP-N-acetylglucosamine. Biosynthesis of hyaluronic acid, a polymer of alternating units of glucuronic acid and N-acetylglucosamine, requires the action of two enzymes whose substrates are UDP-glucuronate and UDP-N-acetylglucosamine, respectively.

Synthesis of some bacterial **dextrans** constitutes an exception to the involvement of nucleotide-linked sugars. Dextrans are storage polysaccharides in yeasts and bacteria. They consist of D-glucose residues linked largely by $\alpha(1 \rightarrow 6)$ glycosidic bonds, with occasional branches formed by $\alpha(1 \rightarrow 2)$, $\alpha(1 \rightarrow 3)$, or $\alpha(1 \rightarrow 4)$ glycosidic bonds. Some bacterial species growing in the human mouth synthesize dextran—a major component of dental plaque—by transglycosylation, using sucrose as a substrate:

$$n \text{ Sucrose} \rightarrow (\text{glucose})_n + n \text{ fructose}$$

Dextran

Because dextran formation utilizes sucrose, consumption of execssive amounts of sucrose creates a serious concern for dental hygiene.

10.7.2. Oligosaccharides

Biosynthesis of sucrose in plants also proceeds via nucleotide-linked sugars:

$$\text{UDP-glucose}^{2-} + \text{fructose 6-phosphate}^{2-} \rightarrow$$
$$\text{sucrose 6-phosphate}^{2-} + \text{UDP}^{3-} + \text{H}^+$$

$$\text{H}_2\text{O} \searrow \\ \text{P}_i \nearrow$$

Sucrose

Lactose forms in mammary glands by action of lactose synthase, an enzyme consisting of two subunits. One subunit, called galactosyl transferase, catalyzes the following reaction when present by itself:

$$\text{UDP-galactose}^{2-} + N\text{-acetylglucosamine} \rightarrow$$
$$N\text{-acetyllactosamine} + \text{UDP}^{3-} + \text{H}^+$$

The second subunit, called α-lactalbumin, is a mammary gland protein that associates with preexisting galactosyl transferase and alters its specificity. After interaction with α-lactalbumin, the enzyme can use glucose as an acceptor of the galactosyl group instead of N-acetylglucosamine, thereby leading to lactose synthesis:

$$\text{UDP-galactose}^{2-} + \text{glucose} \rightarrow \text{lactose} + \text{UDP}^{3-} + \text{H}^+$$

In animals, the dimeric form of the enzyme occurs only in the mammary gland. Having the capacity to change galactosyl transferase activity to that of lactose synthase permits the gland to synthesize large quantities of lactose, required for the production of milk.

Biosynthesis of glycoproteins proceeds via two main pathways. O-linked oligosaccharides form by action of specific glycosyl transferases that catalyze transfer of a monosaccharide unit from a nucleotide-linked sugar to the nonreducing end of an oligosaccharide or to a functional group on a protein. These reactions occur in the Golgi apparatus.

N-linked oligosaccharides form as lipid-linked precursors. They are joined to dolichol, a long-chain polyisoprenoid compound that serves as a carbohydrate carrier. Initial synthesis and some processing take place in the endoplasmic reticulum. Glycoproteins are then transported to the Golgi apparatus for further processing.

10.7.3. Monosaccharides

Glucose can serve as a precursor for all other monosaccharides found in biological systems. Conversely, various monosaccharides can be converted to glucose derivatives. Interconversions of monosaccharides frequently involve isomerization, epimerization, and phosphorylation, and many of the reactions require nucleotide-linked sugars.

Amino sugars, building blocks of glycolipids and glycoproteins, derive from glucosamine. Glucosamine forms by transfer of an amino group from glutamine to fructose 6-phosphate:

$$\text{Fructose 6-phosphate}^{2-} + \text{glutamine}$$
$$\downarrow \text{transamidase}$$
$$\text{glucosamine 6-phosphate}^{2-} + \text{glutamate}^- + \text{H}^+$$

Glucosamine 6-phosphate leads to UDP-N-acetylglucosamine, which is converted to N-acetylneuraminic acid (sialic acid).

SUMMARY

Carbohydrate metabolism consists essentially of the metabolism of glucose. Most major pathways begin or end with glucose, and other carbohydrates are converted to glucose or to intermediates of glucose metabolism. Blood transports carbohydrates in the form of glucose (blood sugar) or lactate. The hormones insulin, epinephrine, and glucagon regulate the level of blood sugar.

Glycolysis comprises the catabolism of glucose to pyruvate and occurs in two stages. Stage I results in degradation of one molecule of glucose (six carbons) to two molecules of glyceraldehyde 3-phosphate (three carbons) and requires input of energy in the form of ATP. Stage II leads to conversion of glyceraldehyde 3-phosphate to pyruvate (three carbons) and is accompanied by production of NADH and generation of energy in the form of ATP. ATP synthesis proceeds via substrate-level phosphorylation. Under aerobic conditions, pyruvate leads to acetyl coenzyme A, which enters the citric acid cycle. Under anaerobic conditions, pyruvate leads to lactate. Conversion of glucose to pyruvate or lactate yields two molecules of ATP per molecule of glucose catabolized. Combination of glycolysis and the citric acid cycle/electron transport system results in complete oxidation of carbohydrates to CO_2 and H_2O.

In the pentose phosphate pathway, glucose 6-phosphate undergoes oxidation to CO_2 and P_i. The cycle produces reducing power in the form of NADPH, serves to synthesize a number of carbohydrates, and allows for interconversions of monosaccharides.

Glycogen catabolism (glycogenolysis) involves removal of one glucose residue at a time from the nonreducing end of glycogen. Glucose is removed in the form of glucose 1-phosphate in a reaction catalyzed by glycogen phosphorylase. Activation of phosphorylase requires a hormonally controlled enzyme cascade. Glycogen biosynthesis (glycogenesis) is catalyzed by glycogen synthase, which uses UDP-glucose as substrate. Activation of glycogen synthase also requires a hormonally controlled enzyme cascade. Glycogenolysis and glycogenesis are reciprocally regulated; activation of one process is coupled to inhibition of the other.

Synthesis of glucose from noncarbohydrate precursors (gluconeogenesis) consists of a series of reactions that proceed from pyruvate to oxaloacetate to glucose. Some readily reversible reactions of glycolysis form part of gluconeogenesis as well. Three strongly exergonic glycolytic reactions cannot be simply reversed and are bypassed by means of different reactions. Glycolysis and gluconeogenesis are reciprocally regulated. Fructose 2,6-bisphosphate serves as an important allosteric effector for both pathways.

SELECTED READINGS

Beylot, M., Soloviev, M. V., David, F., Landau, B. R., and Brunengraber, H., Tracing hepatic gluconeogenesis relative to citric acid cycle activity *in vitro* and *in vivo*, *J. Biol. Chem.* 270:1509–1514 (1995).

DiDonato, L., Des Rosiers, C., Montgomery, J. A., David, F., Garneau, M., and Brunengraber, H., Rates of gluconeogenesis and citric acid cycle in perfused livers, assessed from the mass spectrometric assay of the carbon-13-labeling pattern of glutamate, *J. Biol. Chem.* 268:4170–4180 (1993).

Hanson, R. W., and Patel, Y. M., Phosphoenolpyruvate carboxykinase (GTP): The gene and the enzyme, *Adv. Enzymol. Relat. Areas Mol. Biol.* 69:203–281 (1994).

Hardie, D. G., *Biochemical Messengers,* Chapman & Hall, London (1991).

Kyriakis, J. M., and Avruch, J., Sounding the alarm: Protein kinase cascades activated by stress and inflammation, *J. Biol. Chem.* 271:24313–24316 (1996).

Lebioda, L., and Stec, B., Crystal structure of enolase indicates that enolase and pyruvate kinase evolved from a common ancestor, *Nature (London)* 333:683–686 (1988).

Leschine, S. B., Cellulose degradation in anaerobic environments, *Annu. Rev. Microbiol.* 49:399–426 (1995).

Lienhard, G. E., Slot, J. W., and Mueckler, M. M., How cells absorb glucose, *Sci. Am.* 266(1):86–91 (1992).

Pilkis, S. J., El-Maghrabi, M. R., and Claus, T. H., Hormonal regulation of hepatic gluconeogenesis and glycolysis, *Annu. Rev. Biochem.* 57:755–784 (1988).

Pilkis, S. J., Claus, T. H., Kurland, I. J., and Lange, A. J., 6-Phosphofructo-2-kinase/fructose-2,6-bisphosphatase: A metabolic signaling enzyme, *Annu. Rev. Biochem.* 64:799–835 (1995).

Smythe, C., and Cohen, P., The discovery of glycogenin and the priming mechanism for glycogen biosynthesis, *Eur. J. Biochem.* 200:625–631 (1991).

Strader, C. D., Fong, T. M., Tota, M. R., and Underwood, D., Structure and function of G protein-coupled receptors, *Annu. Rev. Biochem.* 63:101–132 (1994).

Taylor, S. S., Buechler, J. A., and Yonemoto, W., cAMP-dependent protein kinase: Framework for a diverse family of regulatory enzymes, *Annu. Rev. Biochem.* 59:971–1005 (1990).

Wood, T., *The Pentose Phosphate Pathway,* Academic Press, Orlando (1985).

REVIEW QUESTIONS

A.   Define each of the following terms:

Glycolysis	Gluconeogenesis
Schiff base	Phosphorolysis
Converter enzyme	G protein
Galactosemia	Bypass
Mutase	Cyclic AMP
"Fight or flight" response	Glycogen storage disease
Futile cycle	Nucleotide-linked sugar
Adenylate cyclase	Enolase

B.   Differentiate between the two terms in each of the following pairs:

Transaldolase/transketolase	Hexokinase/pyruvate kinase
Pasteur effect/Crabtree effect	
Cori cycle/glucose-alanine cycle	Glycogenesis/glycogenolysis
Phosphofructokinase/ phosphogluco- isomerase	Bypass I/bypass III
	Phosphoglycero- mutase/phospho- glucomutase
Branching enzyme/ debranching enzyme	Class I aldolase/class II aldolase

C.   (1) Write balanced equations for the 10 individual reactions of glycolysis and name each enzyme. What is the overall reaction for (a) stage I and (b) stage II?

(2) Discuss the conditions under which carbohydrates are (a) transported as glucose, (b) transported as lactate, (c) subjected to glycogenesis, and (d) subjected to glycogenolysis. What hormones control the level of glucose in the blood and what are their effects?

(3) Outline the operation of the pentose phosphate pathway. Distinguish between the oxidative and nonoxidative phases, and describe the pathway's functions.

(4) Discuss the mechanism of action of the following enzymes: (a) aldolase; (b) glucose 6-phosphate dehydrogenase; (c) phosphorylase; (d) debranching enzyme.

(5) Why can gluconeogenesis not be an exact reversal of glycolysis? How do the two pathways differ and how are they regulated? What are the energy requirements of gluconeogenesis?

(6) What are G proteins and how do they function?

PROBLEMS

10.1.   Would strenuous exercise (partially anaerobic conditions) tend to exacerbate or diminish the effect of ethanol on gluconeogensis? Explain your answer.

10.2.   Given that

$$ATP^{4-} + H_2O \rightleftarrows ADP^{3-} + P_i^{2-} + H^+$$
$$\Delta G^{\circ\prime} = -30.5 \text{ kJ mol}^{-1}$$

calculate $\Delta G^{\circ\prime}$ for the following reactions (see Table 10.1):

(a) Phosphoenolpyruvate^{3-} + H$_2$O $\rightleftarrows$ pyruvate$^-$ + P$_i^{2-}$

(b) Glucose + P$_i^{2-}$ $\rightleftarrows$ glucose 6-phosphate^{2-} + H$_2$O

10.3.*   Calculate the *net* yield of ATP molecules per molecule of metabolite (underlined) for the following cases. The reactions occur under anaerobic conditions.

(a)   Glycolysis of <u>fructose</u> to lactate in liver

(b)   Glycolysis of <u>fructose</u> to lactate in kidney

(c)   Glycolysis of <u>mannose</u> to pyruvate

(d)   Glycolysis of <u>3-phosphoglycerate</u> to pyruvate

10.4.*   Calculate the *net* yield of ATP molecules per molecule of sucrose for anaerobic glycolysis of sucrose to lactate, given the following:

$$\text{Sucrose} + H_2O \rightarrow \text{glucose} + \text{fructose}$$

$$\text{Fructose} + ATP^{4-} \rightarrow \text{fructose 6-phosphate}^{2-} + ADP^{3-} + H^+$$

10.5.   To evaluate the energy savings achieved by phosphorylase, assume that, in the absence of phosphorylase, cleavage of each glucose residue from glycogen required hydrolysis of one molecule of ATP. On that basis, calculate the number of ATP molecules required to degrade 1.0 g of glycogen to glucose. Assume that the molecular weight of any glucose residue is 162.

10.6.   Explain why a high K_m for pyruvate and a low k_{cat} for conversion of pyruvate to lactate (Table 10.2) for the H$_4$ isozyme of lactate dehydrogenase indicate that conversion of pyruvate to acetyl coenzyme A is favored over conversion of pyruvate to lactate.

10.7.   What is the ratio of 1,3-bisphosphoglycerate to 3-phosphoglycerate under biochemical standard conditions when the [ATP]/[ADP] ratio is 10.0?

10.8.*   A classmate has proposed that the molecular basis

of the Crabtree effect involves an allosteric effect of glucose on a specific enzyme. What effect would you look for in the following enzymes to see if the suggested hypothesis has any merit? (a) Lactate dehydrogenase (LDH); (b) cytochrome oxidase (a key enzyme of the electron transport system); (c) pyruvate carboxylase (converts pyruvate to oxaloacetate, an intermediate of the citric acid cycle)

10.9. The maximum velocity of muscle glycogen phosphorylase is large, whereas that of the liver enzyme is small. How can you explain this difference in maximum velocity?

10.10. The extreme toxicity of methanol arises not so much from methanol itself but rather from its metabolic conversion to formaldehyde by the action of alcohol dehydrogenase. Part of the treatment for methanol poisoning entails administering large doses of ethanol. Why is this treatment effective?

10.11. A student uses the alcoholic fermentation of yeast to prepare a wine that has an alcohol content of 5.0% w/v (5.0 g of ethanol/100 ml of wine). What is the minimum molar concentration of glucose the fermentation mixture must contain? (MW of ethanol = 46)

10.12.* Four samples of glucose are selectively labeled with ^{14}C at one carbon atom: (a) at C(1); (b) at C(3); (c) at C(4); and (d) at C(6). Subjecting the four samples to anaerobic glycolysis results in conversion of the glucose to lactate. Which carbon atoms of lactate will become labeled in each case?

10.13.* The glucose samples from the previous problem are converted to glucose 6-phosphate by action of hexokinase, and the glucose 6-phosphate is passed once through the pentose phosphate pathway. Under these conditions, which samples would yield ^{14}C-labeled CO_2?

10.14. Consider two hypothetical carbohydrates, a 10-carbon 2-ketose and an 8-carbon aldose. What carbohydrates would you obtain when you treat the two compounds with (a) transaldolase and (b) transketolase?

10.15. Energetically speaking, how many molecules of glucose could be synthesized via gluconeogenesis from the energy released by hydrolysis of 90 molecules of ATP to ADP, followed by hydrolysis of ADP to AMP?

10.16. How many molecules of ATP would be required if the starting material for the pentose phosphate pathway consisted of six molecules of glucose rather than of glucose 6-phosphate?

10.17. In the absence of glycogen breakdown, the hexokinase reaction constitutes the rate-determining step of glycolysis. Would this be true in the presence of glycogen breakdown as well? Explain.

10.18. Distinguish between the rate-determining step (hexokinase) and the committed step (phosphofructokinase) of glycolysis.

10.19. When you add phosphoglucomutase to an equilibrium mixture of glucose 1-phosphate and glucose 6-phosphate, no net reaction takes place. If you were to use ^{32}P-labeled glucose 1-phosphate in this experiment, would some of the label exchange into glucose 6-phosphate? If so, why?

10.20. Why must the NADH produced in glycolysis be oxidized to regenerate NAD^+, regardless of whether the system as a whole is aerobic or anaerobic?

The Citric Acid Cycle

11

The citric acid cycle, together with the electron transport system, constitutes stage III of catabolism (see Figure 8.2), also called *cellular respiration*. Because the citric acid cycle functions in both catabolism (Figure 11.1) and anabolism (Figure 11.2), we call it an **amphibolic pathway.** The citric acid cycle plays a pivotal role in cellular respiration, has multiple interconnections with other pathways, and provides for interconversions of numerous metabolites.

We use the term **citric acid cycle** because citrate is the first compound produced in the cyclic set of reactions. The pathway is also known as the **Krebs cycle** in honor of Hans Krebs, the British biochemist who proposed it in 1937. (Krebs was awarded the Nobel Prize in 1953.) Lastly, we refer to the reaction sequence as the **tricarboxylic acid (TCA) cycle** because the first few intermediates are tricarboxylic acids. The citric acid cycle has three major roles:

1. It is the central oxidative pathway by which all nutrients—carbohydrates, lipids, and proteins—are catabolized in aerobic organisms and tissues.
2. It is an important source of intermediates for a large number of anabolic pathways leading to the biosynthesis of a variety of biomolecules.
3. It is the main source of metabolic energy, which derives from the oxidation–reduction steps of the cycle and their linkage to the electron transport system.

Krebs' proposal of the citric acid cycle followed a number of important findings dealing with cellular respiration. In 1935, Albert Szent-Györgyi showed that adding catalytic amounts of succinate, fumarate, malate, or oxaloacetate to minced muscle tissue greatly stimulated cellular respiration. Oxygen uptake and carbon dioxide production increased beyond the quantities required for oxidation of the added compound. In addition, Szent-Györgyi established that the compounds were interconverted according to the sequence succinate → fumarate → malate → oxaloacetate. Shortly thereafter, Carl Martius and Franz Knoop demonstrated the sequence citrate → *cis*-aconitate → isocitrate → α-keto-

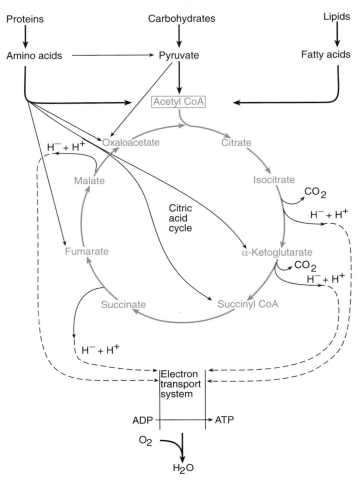

Figure 11.1.   The citric acid cycle in catabolism.

glutarate. Since scientists already knew that α-ketoglutarate could be decarboxylated to succinate, they concluded that a pathway extends from citrate to oxaloacetate. Based on this information and on his own extensive studies, Krebs proposed a cyclic set of reactions consisting of the above sequences that was closed by the conversion of oxaloacetate to citrate.

The contribution of Krebs ranks as one of the most important milestones of metabolic biochemistry. Ironically, when he first submitted his manuscript for publication, it was rejected for lack of publishing space.

The citric acid cycle represents one of three catalytic cycles Krebs proposed. In 1932, Krebs and Kurt Henseleit elucidated the *urea cycle* (Section 14.3), the first known metabolic cycle. In 1957, Krebs and H. R. Kornberg proposed a modifed form of the citric acid cycle, called the *glyoxylate cycle* (Section 11.6).

Although Krebs established the existence of the citric acid cycle, the precise mechanism of citrate formation remained unknown for several years. How oxaloacetate leads to citrate became clear only after Nathan Kaplan and Fritz Lipmann discovered *coenzyme A* (1945) and Severo Ochoa and Feodor Lynen showed that acetyl coenzyme A condenses with oxaloacetate to form citrate (1951).

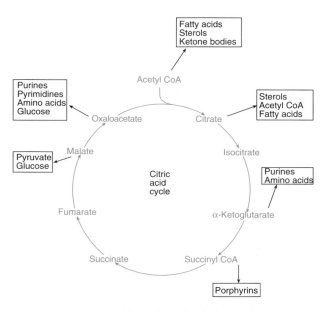

Figure 11.2. The citric acid cycle in anabolism.

11.1. COENZYMES OF THE CYCLE

A key feature of the citric acid cycle consists of four oxidation–reduction reactions, catalyzed by specific dehydrogenases that use either nicotinamide adenine dinucleotide (NAD⁺) or flavin adenine dinucleotide (FAD) as coenzyme. As the metabolites undergo oxidation, the coenzymes undergo reduction. In these reactions, the coenzyme functions like a second substrate. Protons and electrons derived from the metabolites reduce NAD⁺ and FAD to NADH and $FADH_2$. The reduced coenzymes are subsequently oxidized by the electron transport system with the concomitant synthesis of ATP in a process called oxidative phosphorylation (Section 12.4). In this way, protons and electrons removed from metabolites become channeled into the electron transport system to generate metabolic energy in the form of ATP.

11.1.1. Pyridine-Linked Dehydrogenases

We use the term **pyridine-linked dehydrogenases** to describe enzymes that use either *nicotinamide adenine dinucleotide (NAD⁺)* or the related compound *nicotinamide adenine dinucleotide phosphate (NADP⁺)* as coenzyme. Some pyridine-linked dehydrogenases use NAD⁺, and others use NADP⁺. The term derives from the fact that a pyridine nucleus (a nitrogen-containing six-membered ring) forms part of the coenzyme molecule (Figure 11.3). You have already encountered a number of such enzymes in Chapter 10.

Because the structures of NAD⁺ and NADP⁺ bear

some similarity to nucleic acid building blocks, the coenzymes' names include the word "nucleotide." NAD⁺ and NADP⁺ are loosely bound to their respective dehydrogenases and easily dissociate from them. They do *not* constitute prosthetic groups.

We call NAD⁺ and NADP⁺ *pyridine nucleotide coenzymes.* The pyridine ring forms part of the coenzyme's **nicotinamide** moiety. Nicotinamide is a derivative of *niacin (nicotinic acid),* a water-soluble vitamin and member of the vitamin B complex. Like other water-soluble vitamins, niacin is required for assembly of a specific coenzyme molecule, in this case NAD⁺ and NADP⁺.

Niacin is unique in that it can be synthesized in many mammals from the amino acid tryptophan. The extent to which the dietary requirement for niacin can be decreased by dietary tryptophan varies according to the species. The deficiency disease associated with niacin is *pellagra.*

Oxidation–reduction reactions catalyzed by dehydrogenases typically involve removal of two hydrogen atoms from a substrate, as in the oxidation of ethanol to acetaldehyde catalyzed by alcohol dehydrogenase:

$$CH_3\text{-}CH_2OH + NAD^+ \longrightarrow CH_3\text{-}CHO + NADH + H^+$$

Two hydrogen atoms (2H·) are formally equivalent to two protons and two electrons (2H⁺ + 2e⁻). However, oxidation–reduction reactions catalyzed by dehydrogenases re-

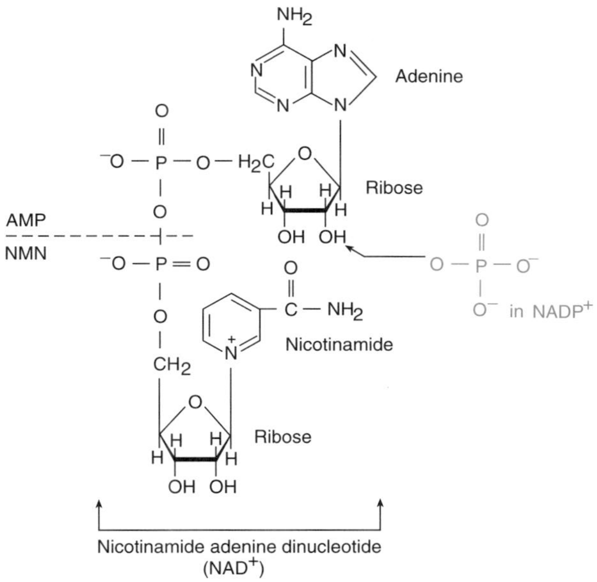

Figure 11.3. Pyridine nucleotide coenzymes. Nicotinamide adenine dinucleotide (NAD⁺) consists of adenosine 5′-monophosphate (AMP) linked to nicotinamide mononucleotide (NMN). The structure of NADP⁺ is identical to that of NAD⁺ except for an additional phosphate group at C(2′) of the ribose in AMP.

quire transferring these protons and electrons in the form of hydride ions (H^- or H:) and protons. Because a hydride ion consists of a proton and two electrons ($H^+ + 2e^-$), we have the following equivalence:

$$2H\cdot = H^- + H^+ = 2H^+ + 2e^-$$

As the substrate of a dehydrogenase is oxidized by removal of hydrogen, the enzyme's coenzyme is reduced. In pyridine-linked dehydrogenases, niacin forms the "business end" of the coenzyme; the nicotinamide portion of NAD^+ and $NADP^+$ undergoes reduction (or oxidation) as a metabolite undergoes oxidation (or reduction). In this process, the pyridine nucleus gains (or loses) a hydride ion, and a proton appears in (or disappears from) the medium:

NAD$^+$ NADH
(NADP$^+$)   (NADPH)

We can follow the progress of pyridine-linked dehydrogenase reactions by means of spectrophotometric measurements, since NADH and NADPH, but not NAD^+ and $NADP^+$, absorb strongly at 340 nm (Figure 11.4). X-ray diffraction studies have shown that the overall molecular structures of pyridine-linked dehydrogenases differ, but their NAD^+ binding sites have similar architectures. Each binding site consists of two parts, one specific for the nicotinamide portion, and the other for the adenine moiety.

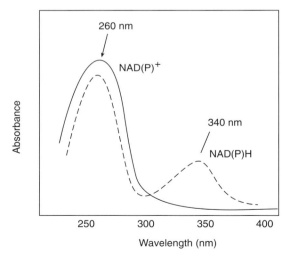

Figure 11.4. Absorption spectra of NAD(P)$^+$ and NAD(P)H.

11.1.2. Flavoproteins

Pyridine-linked dehydrogenases, using NAD^+ as coenzyme, catalyze three of the oxidation–reduction reactions of the citric acid cycle. The fourth is catalyzed by a dehydrogenase using FAD as coenzyme. We refer to *flavin adenine dinucleotide (FAD)* and the related compound *flavin mononucleotide (FMN)* as *flavin coenzymes* (Figure 11.5). Unlike NAD^+ and $NADP^+$, FAD and FMN are tightly bound to their specific dehydrogenases and constitute *prosthetic groups.* Accordingly, we call the complete enzymes **flavoproteins.** Some dehydrogenases use FAD, and others use FMN. Like NAD^+ and $NADP^+$, FAD and FMN have some structural similiarity to nucleic acid building blocks.

Flavin coenzymes contain *vitamin B_2 (riboflavin)* as a structural component. Riboflavin, like niacin, constitutes the "business end" of the coenzyme. Complete reduction of the *isoalloxazine* ring system of FAD and FMN, as distinct from the reduction of NAD^+ or $NADP^+$, requires both a hydride ion and a proton. Hence, oxidation of a metabolite (MH_2) by a flavoprotein does *not* yield a proton as a reaction product:

$$MH_2 + FMN \rightleftharpoons M + FMNH_2$$
$$\text{(FAD)} \qquad\qquad\qquad \text{(FADH}_2)$$

Reduction of flavins differs from that of pyridine nucleotide coenzymes in yet another aspect. Whereas NAD^+ and $NADP^+$ undergo only two-electron reductions, FMN and FAD can undergo either one- or two-electron reductions. Both coenzymes can participate in two sequential one-electron transfers or in a simultaneous two-electron transfer (Figure 11.6). When flavins undergo a one-electron reduction, a relatively stable free radical forms. The redox versatility of flavins comes into play in the electron transport system, in which FMN serves as a link between electron carriers requiring two- and one-electron transfers.

In addition to their roles in the citric acid cycle, pyridine nucleotide coenzymes and flavin coenzymes also mediate a variety of other biological oxidation–reduction reactions.

11.1.3. Coenzyme A

In addition to NAD^+ and FAD, three other coenzymes—coenzyme A, lipoic acid, and thiamine pyrophosphate—participate in the citric acid cycle.

Next to ATP, **coenzyme A** and its derivative **acetyl coenzyme A** (Figure 11.7) are probably the most important low-molecular-weight biomolecules of metabolism. Coenzyme A (the *A* stands for *acetylation*) activates and carries acetyl and other acyl groups. It functions in reactions that require transfers of these groups.

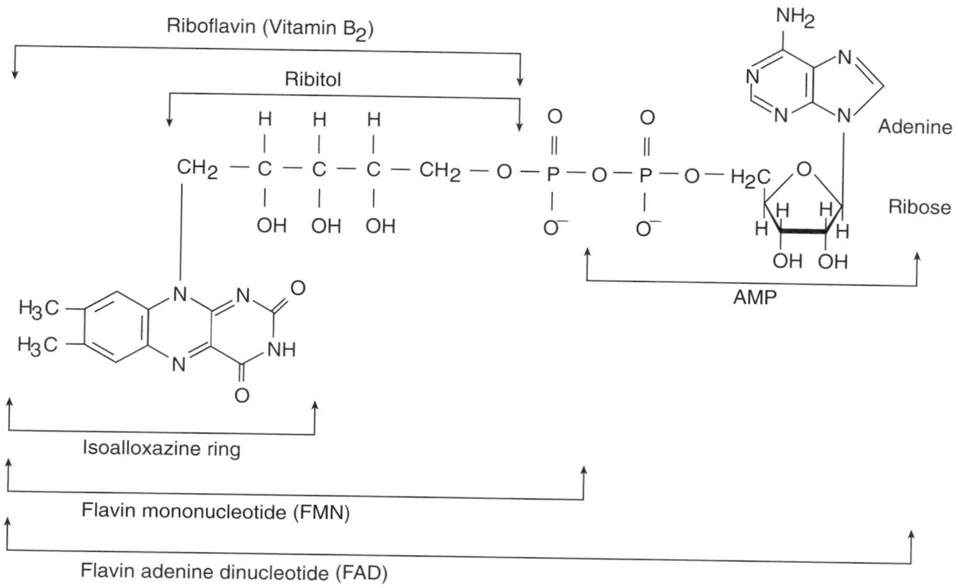

Figure 11.5. Flavin coenzymes.

In some ways, coenzyme A has structural features similar to those of NAD$^+$ and FAD. It, too, has a nucleotide-like component, contains a pyrophosphate group, and includes a vitamin. Of the entire molecule, however, the comparatively small sulfhydryl group constitutes the "business end" of the coenzyme.

We abbreviate coenzyme A as *CoA* or as *CoA-SH* to emphasize the functional importance of the SH group. Likewise, we abbreviate acetyl coenzyme A as *acetyl CoA* or as *CH$_3$CO-S-CoA*. The vitamin component of coenzyme A is *pantothenic acid,* a water-soluble B vitamin. Deficiencies of pantothenic acid in humans occur only rarely, probably because of the widespread distribution of the vitamin in natural foods.

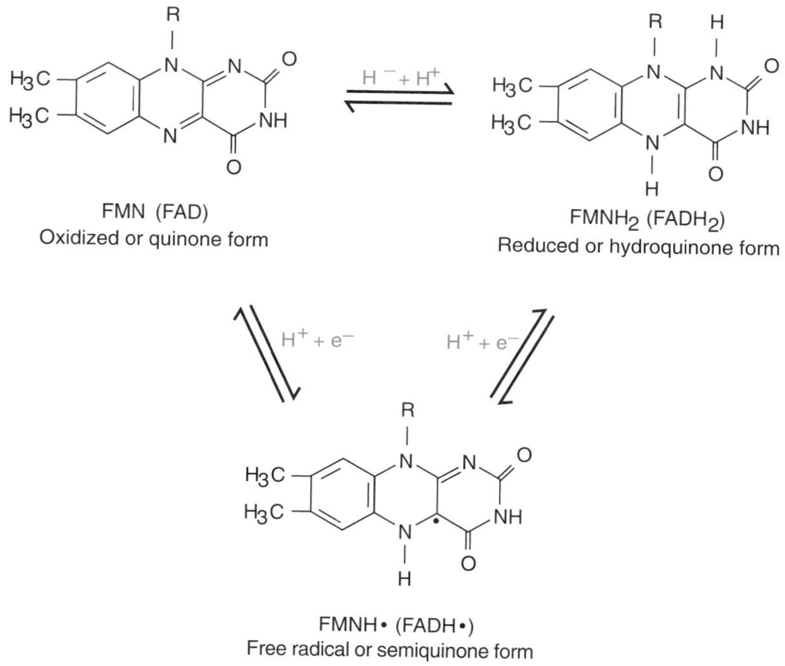

Figure 11.6. Oxidation and reduction of flavin coenzymes.

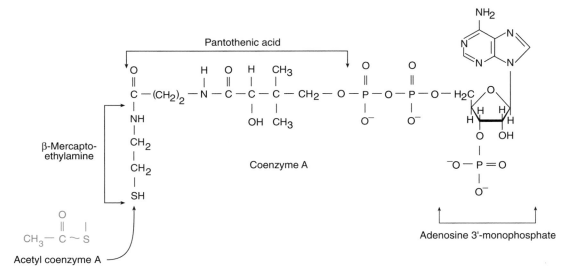

Figure 11.7.   Coenzyme A and acetyl coenzyme A. The thioester bond in acetyl coenzyme A (~) is an energy-rich bond.

Acetyl CoA constitutes not only a key coenzyme but also an *energy-rich compound.* It belongs to the class of special esters, specifically that of *thioesters,* in which sulfur replaces an oxygen atom in the ester bond. Any *acyl CoA* compound, such as *succinyl CoA,* one of the intermediates in the citric acid cycle, constitutes an energy-rich compound. Hydrolysis of the thioester bond (~) has a highly negative free energy change associated with it:

$$CH_3CO \sim S\text{-}CoA^{4-} + H_2O \rightarrow CH_3COO^- + CoA\text{-}SH^{4-}$$
$$+ H^+ \qquad \Delta G^{\ominus\prime} = -31.4 \text{ kJ mol}^{-1}$$

11.1.4. Lipoic Acid

Lipoic acid functions in acyl transfer reactions. The coenzyme has a heterocyclic, sulfur-containing ring that carries an aliphatic side chain (Figure 11.8A). Lipoic acid occurs in tissues in extraordinarily small amounts. It was first isolated in 1949 (Eli Lilly Co.). By using some *10 tons* of beef liver as starting material, investigators were able to obtain approximately *30 mg* of lipoic acid!

Nutritionists generally classify lipoic acid with the B vitamins because it is a growth factor for some microorganisms. The coenzyme form of the vitamin is **lipoamide (lipoyllysine),** a prosthetic group formed by an amide linkage between the carboxyl group of lipoic acid and the ε-amino group of a lysine residue in the enzyme (Figure 11.8B).

11.1.5. Thiamine Pyrophosphate

Thiamine, or *vitamin B₁,* contains two heterocyclic ring systems—a pyrimidine nucleus and a sulfur- and nitrogen-containing ring called a *thiazole* (Figure 11.9). The coenzyme form of thiamine, **thiamine pyrophosphate,** forms by attachment of a pyrophosphate group to the thi-

azole nucleus. We abbreviate thiamine pyrophosphate as *ThPP* or *TPP,* not to be confused with TTP (thymidine triphosphate). Thiamine pyrophosphate functions in aldehyde transfer reactions and as coenzyme for enzymes catalyzing the decarboxylation of α-keto acids. TPP also serves as coenzyme of transketolases. The deficiency disease associated with vitamin B₁ is *beriberi.*

11.2. SYNTHESIS OF ACETYL COENZYME A

The citric acid cycle consists of a series of eight reactions (Figure 11.10). Initiation of each round of the cycle involves the entrance of a molecule of *acetyl CoA,* which condenses with oxaloacetate to form citrate. Acetyl CoA forms via two major routes from either lipids or carbohydrates. Lipid catabolism yields acetyl CoA directly (Sec-

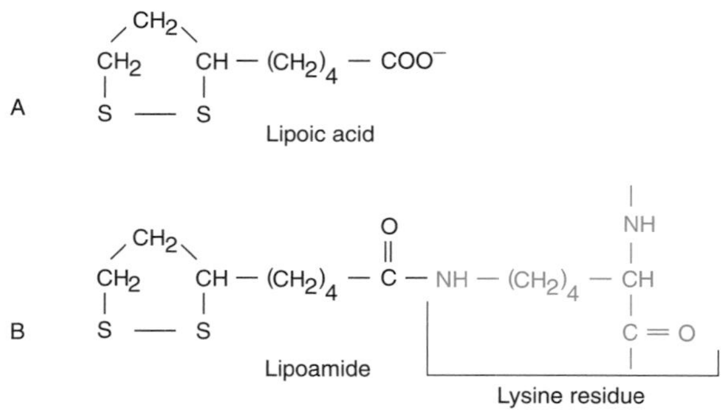

Figure 11.8.   (A) Lipoic acid. (B) Lipoamide (lipoyllysine). In lipoamide, lipoic acid is linked covalently to the ε-NH₂ group of a lysine residue in an enzyme.

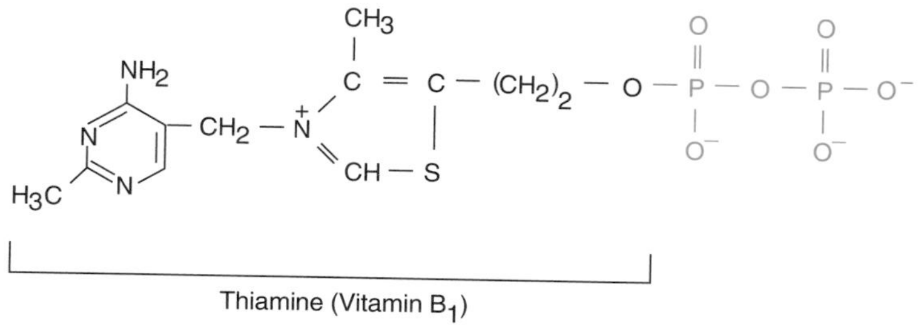

Figure 11.9. Thiamine pyrophosphate. The pyrophosphate group is esterified to a primary alcohol group in thiamine.

tion 13.3). Carbohydrate catabolism produces acetyl CoA by way of pyruvate. Recall that pyruvate constitutes the end product of glycolysis. Following its formation in the cytoplasm, pyruvate moves across the mitochondrial membrane into the matrix via an antiport transport system. In this system, pyruvate transport is coupled to hydroxide-ion transport in the opposite direction. A multienzyme system called the **pyruvate dehydrogenase complex** and located in the mitochondrial matrix then catalyzes the conversion of pyruvate to acetyl CoA.

The pyruvate dehydrogenase complex consists of multiple copies of three enzymes and five coenzymes. The three enzymes are *pyruvate dehydrogenase* (E_1), *dihydrolipoyl transacetylase* (E_2), and *dihydrolipoyl dehydrogenase,* a flavoprotein (E_3). A cubic cluster of E_2 forms the "core" of the complex, with E_1 and E_3 being assembled on the core's surface.

The enzymes form a very large multienzyme system. The *E. coli* pyruvate dehydrogenase complex has a molecular weight of 4.6×10^6 and consists of 24 polypeptide chains of E_1, 24 chains of E_2, and 12 chains of E_3. The mammalian complex has a molecular weight of 8.4×10^6 and contains 120 polypeptide chains of E_1, 60 chains of E_2, and 12 chains of E_3.

Three of the five coenzymes of the pyruvate dehydrogenase complex function as catalysts, and each is associated with a component enzyme: TPP is linked to E_1; lipoic acid is covalently linked to E_2; and FAD forms an integral part of E_3. The remaining two coenzymes, coenzyme A and NAD$^+$, function as substrates and undergo chemical alteration in the course of the reaction. The overall reaction catalyzed by the pyruvate dehydrogenase complex comprises both an oxidation (dehydrogenation) and a decarboxylation (loss of CO_2). Accordingly, we refer to it as an **oxidative decarboxylation.** The reaction has the following stoichiometry:

Note that we place three coenzymes above or below the arrow to indicate their catalytic nature; they are regenerated in their original form at the end of the reaction. The overall reaction is strongly exergonic ($\Delta G^{\circ\prime} = -33.5$ kJ mol^{-1}).

Lipoic acid, attached via an amide bond to a lysine residue of E_2, represents a key feature of the reaction mechanism (Figure 11.11). The resultant **lipoyllysine arm** has a fully extended length of about 14 Å. Researchers believe that this arm functions as a tether, swinging between participating sites on E_1 and E_3. They postulate that at least two arms bind to each E_2 such that the acetyl group can be transferred from one arm to the other.

ATP inhibits the pyruvate dehydrogenase complex, and AMP activates it. In addition, the enzyme system is subject to product inhibition by both NADH and acetyl CoA. These regulatory effects make good metabolic sense. Acetyl CoA initiates the reactions of the citric acid cycle. Oxidative steps of the cycle yield the reduced coenzymes NADH and FADH$_2$; these are oxidized via the electron transport system with the concomitant synthesis of ATP. Because high concentrations of ATP or NADH result in inhibition of the pyruvate dehydrogenase complex, these conditions prevent synthesis of additional and unneeded ATP by the electron transport system. Conversely, low ATP levels and high AMP levels activate the enzyme system, resulting in enhanced ATP synthesis. High concentrations of acetyl CoA inhibit the enzyme system and block synthesis of additional acetyl CoA. In eukaryotes, additional regulation involves covalent modification of E_1; phosphorylated E_1 is inactive whereas the dephosphorylated form is active.

11.3. REACTIONS OF THE CYCLE PROPER

Each of the eight reactions of the citric acid cycle is enzyme-catalyzed. In eukaryotes, all of the reactions take place in the mitochondria.

$$\text{Pyruvate}^- + \text{CoA} - \text{SH}^{4-} + \text{NAD}^+ \xrightarrow[\text{lipoic acid}]{\text{FAD, TPP}}$$

$$\text{acetyl CoA}^{4-} + CO_2 + \text{NADH}$$

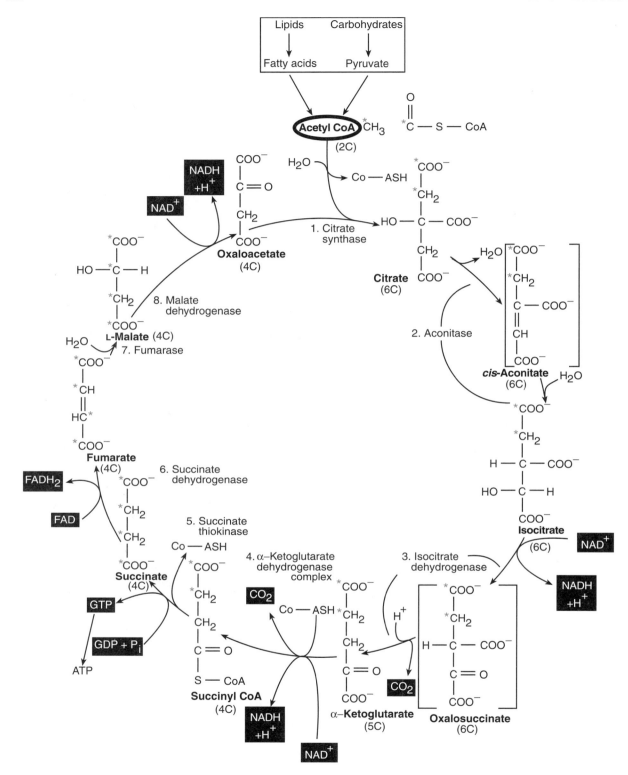

Figure 11.10.   The citric acid cycle. Bracketed compounds designate enzyme-bound intermediates. When carbon atoms entering as acetyl CoA are labeled (asterisk), the label becomes scrambled between the two symmetrical halves of succinate. Fate of the label is shown up to malate. The two molecules of CO_2 that are lost do not come from the acetyl CoA that has just entered the cycle but rather from the oxaloacetate formed during the preceding turn.

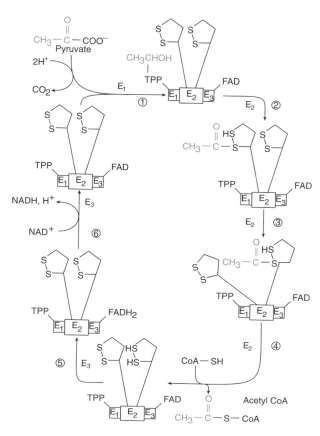

Figure 11.11. Postulated mechanism of the pyruvate dehydrogenase complex. (1) A two-carbon fragment is transferred from pyruvate to E_1; (2) the fragment is transferred to one lipoyllysine arm and simultaneously oxidized to an acetyl group; (3) the lipoyllysine arms swings to transfer the acetyl group to a second arm; (4) the second arm positions the acetyl group for transfer to coenzyme A; (5) E_3 catalyzes the oxidation of the reduced lipoyllysine arm; (6) the reduced flavin is oxidized with NAD^+, forming NADH and H^+.

11.3.1. Citrate Synthase

The first reaction of the cycle proper involves a *condensation* between the acetyl group of acetyl CoA (a two-carbon fragment) and **oxaloacetate** (a four-carbon compound) to form **citrate** (a six-carbon compound). **Citrate synthase** catalyzes this reaction:

$$\begin{array}{c} COO^- \\ | \\ C{=}O \\ | \\ CH_2 \\ | \\ COO^- \end{array} + CH_3{-}\overset{\overset{O}{\|}}{C}{-}S{-}CoA + H_2O \rightarrow$$

Oxaloacetate   Acetyl CoA

$$\begin{array}{c} COO^- \\ | \\ CH_2 \\ | \\ HO{-}C{-}COO^- \\ | \\ CH_2 \\ | \\ COO^- \end{array} + CoA{-}SH + H^+$$

Citrate

The citrate synthase reaction is strongly exergonic (Table 11.1) owing to cleavage of the *energy-rich bond* in acetyl CoA. Citrate synthase is an allosteric enzyme for which ATP, NADH, and succinyl CoA serve as negative effectors. The mammalian enzyme is a dimer, consisting of two identical subunits (MW = 49,000 each).

The citrate synthase reaction represents one of three major regulatory sites of the citric acid cycle. When concentrations of the enzyme's negative allosteric effectors build up, the cycle sequence becomes inhibited. Because

Table 11.1. Reactions of the Citric Acid Cycle

	Enzyme	Mitochondrial location	Type of reaction	$\Delta G^{\circ\prime}$ (kJ mol^{-1})
1.	Citrate synthase	Matrix	Condensation	−32.2
2.	Aconitase	Matrix	Isomerization	+5.0
3.	Isocitrate dehydrogenase	Matrix	Oxidative decarboxylation	−20.9
4.	α-Ketoglutarate dehydrogenase complex	Matrix	Oxidative decarboxylation	−33.5
5.	Succinate thiokinase (succinyl CoA synthase)	Matrix	Hydrolysis/ phosphorylation	−3.3
6.	Succinate dehydrogenase	Inner membrane	Dehydrogenation	~0
7.	Fumarase	Matrix	Hydration	−3.8
8.	Malate dehydrogenase	Matrix	Dehydrogenation	+29.7
			Overall	−59.0

of the cycle's link to the electron transport system, inhibition of citrate synthase results in decreased ATP synthesis. Thus, as in the case of the pyruvate dehydrogenase complex, high concentrations of ATP and NADH prevent synthesis of unneeded ATP.

11.3.2. Aconitase

The **aconitase** reaction has two steps, dehydration of citrate followed by hydration. The intermediate, called *cis-aconitate,* remains enzyme-bound. Both steps are stereospecific and lead to formation of isocitrate. The overall reaction constitutes an isomerization of citrate to **isocitrate:**

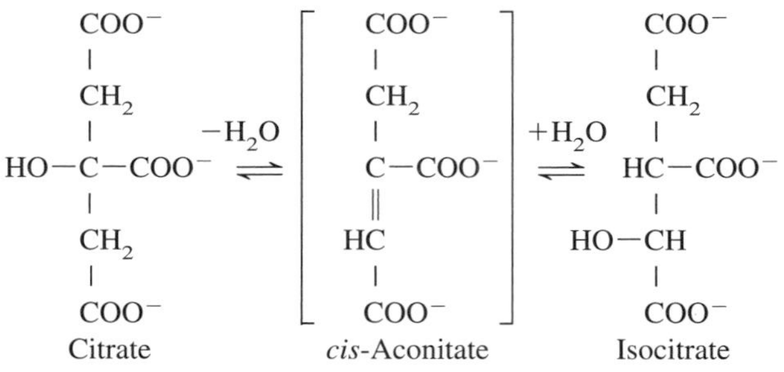

Isomerization results in conversion of a hydroxyl group of a tertiary alcohol at C(3) to a hydroxyl group of a secondary alcohol at C(4), thereby setting the stage for the subsequent decarboxylation step. Aconitase is an iron–sulfur protein and contains a [4Fe-4S] cluster. The mammalian enzyme consists of two identical subunits (MW = 45,000 each).

Investigators have dubbed the postulated mechanism of action of aconitase "ferrous wheel." According to this concept, *cis*-aconitate binds to the enzyme at three sites, including the iron. Stereospecific addition of water, to form either citrate or isocitrate, involves partial rotation of this "ferrous wheel."

The aconitase reaction is of historical interest. When Ogston studied it in 1948, he found that the enzyme catalyzed a reaction with only one of the two identical parts of the symmetric citrate molecule. Consequently, only one of two possible chiral products formed. This finding led him to propose the *polyaffinity theory* for enzyme action (Section 4.2).

Aconitase constitutes the ultimate target for the inhibition produced by *fluoroacetate,* a compound that occurs in some South African plants. Fluoroacetate is one of the most toxic small molecules known. When ingested by animals, it is converted to fluoroacetyl CoA by *acetate thiokinase* (Figure 11.12). Citrate synthase can use fluoroacetyl CoA as an alternate substrate and catalyzes its conversion to *fluorocitrate,* a powerful inhibitor of aconitase. Fluorocitrate is used as a rodenticide. Its inhibition of aconitase is so strong that ingesting it can be lethal.

Fluoroacetate is an example of a **suicide substrate**—a substance that has no toxicity by itself but that is acted upon by an enzyme because of its similarity to a normal substrate. Enzymatic action converts the substance to a toxic form that inhibits a critical enzyme and causes cell death. The cell has died, or "committed suicide," as a result of a transformation of the originally nontoxic substance.

11.3.3. Isocitrate Dehydrogenase

Isocitrate dehydrogenase also catalyzes an *oxidative decarboxylation.* The enzyme is a pyridine-linked dehydrogenase that uses NAD^+ as coenzyme and forms **α-ketoglutarate,** a five-carbon compound:

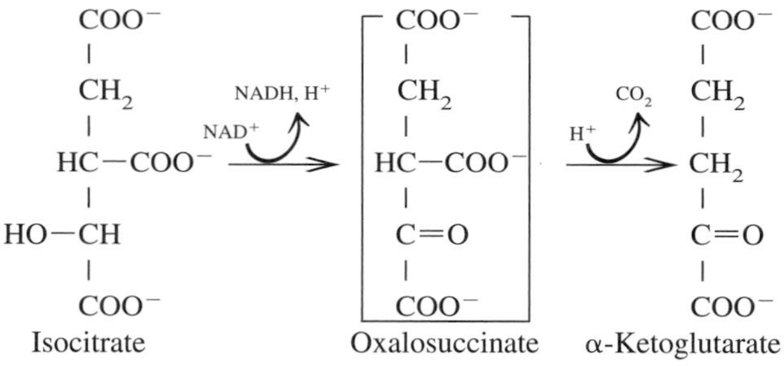

The reaction proceeds through formation of an enzyme-bound intermediate, **oxalosuccinate,** so called be-

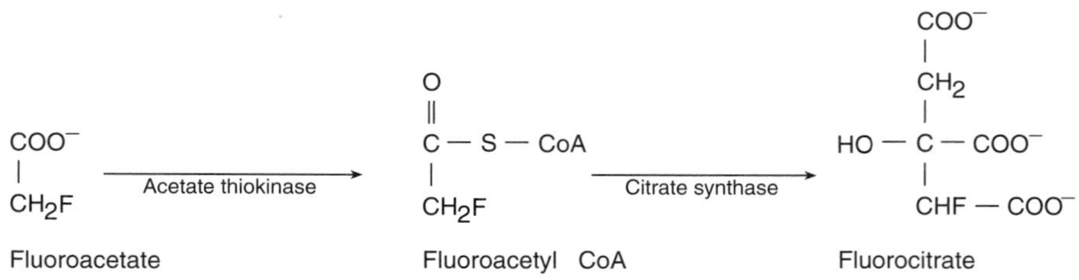

Figure 11.12. Conversion of fluoroacetate to fluorocitrate.

cause it has the structure of a condensation product between succinate and oxalate ($^-$OOC-COO$^-$). Isocitrate dehydrogenase is an allosteric enzyme, and its regulation constitutes the second major control point of the citric acid cycle. ATP and NADH inhibit the enzyme; ADP and NAD$^+$ activate it. As in the case of citrate synthase, these regulatory effects reflect the level of ATP in the cell. The mammalian enzyme is a tetramer ($\alpha_2\beta\gamma$) with a total molecular weight of about 190,000.

Most tissues contain two isocitrate dehydrogenases. The NAD$^+$-linked enzyme occurs only in mitochondria and functions in the citric acid cycle. A second, NADP$^+$-linked dehydrogenase occurs both in mitochondria and in the cytoplasm. Its main function appears to be generating reducing power in the form of NADPH for some cytoplasmic reactions.

11.3.4. α-Ketoglutarate Dehydrogenase Complex

The **α-ketoglutarate dehydrogenase complex** comprises a multienzyme system similar to the pyruvate dehydrogenase complex in both structure and function. Both complexes catalyze oxidative decarboxylations, require the same coenzymes, and involve the same mechanistic steps. The α-ketoglutarate dehydrogenase complex accomplishes the shortening of α-ketoglutarate by one carbon, eliminated as CO_2. Because the product, **succinyl CoA,** has the structure of an acyl CoA, it constitutes an energy-rich compound:

$$
\begin{array}{c}
COO^- \\
| \\
CH_2 \\
| \\
CH_2 \\
| \\
C=O \\
| \\
COO^-
\end{array}
$$

$$\alpha\text{-Ketoglutarate}^{2-} + NAD^+ + CoA{-}SH^{4-} \xrightarrow[\text{lipoic acid}]{\text{FAD, TPP}}$$

$$
\begin{array}{c}
COO^- \\
| \\
CH_2 \\
| \\
CH_2 \\
| \\
O=C{-}S{-}CoA
\end{array}
$$

$$\text{succinyl CoA}^{5-} + NADH + CO_2$$

The α-ketoglutarate dehydrogenase reaction represents the third major control point of the citric acid cycle.

ATP and NADH inhibit the enzyme; ADP and NAD$^+$ activate it. As you saw for the other two regulatory sites of the citric acid cycle, activations and inhibitions reflect the intracellular level of ATP.

11.3.5. Succinate Thiokinase (Succinyl CoA Synthase)

Succinate thiokinase, or *succinyl CoA synthase,* catalyzes the cleavage of succinyl CoA to **succinate** and coenzyme A. Thus, the combined change for coenzyme A in this and the previous reaction equals zero. The exergonic cleavage of the energy-rich bond in succinyl CoA is coupled to the endergonic phosphorylation of guanosine 5′-diphosphate (GDP) to guanosine 5′-triphosphate (GTP):

$$
\begin{array}{cc}
COO^- & COO^- \\
| & | \\
CH_2 & CH_2 \\
| & | \\
CH_2 & CH_2 \\
| & | \\
O=C{-}S{-}CoA & COO^-
\end{array}
$$

$$\text{Succinyl CoA}^{5-} + GDP^{3-} + P_i^{2-} \rightarrow \text{succinate}^{2-} +$$
$$CoA{-}SH^{4-} + GTP^{4-}$$

Scientists believe that the reaction proceeds in three steps. In the first, succinyl phosphate forms from P_i and succinyl CoA. In the second step, the phosphate group is transferred to a histidine residue on the enzyme. Lastly, GTP forms by transfer of the phosphate group from the enzyme to GDP. Because the overall reaction involves breakage of one energy-rich bond and synthesis of another, the free energy change is close to zero. Production of GTP is equivalent to production of ATP, since the phosphate is readily transferred from GTP to ADP in a nucleotide exchange reaction, catalyzed by *nucleoside diphosphate kinase:*

$$GTP + ADP \rightleftarrows GDP + ATP$$

In plants and bacteria, ATP forms directly from succinyl CoA and ADP. Synthesis of GTP and ATP, coupled to cleavage of the energy-rich bond in succinyl CoA, represents another instance of *substrate-level phosphorylation,* similar to that occurring in glycolysis.

Note the difference between the pyruvate dehydrogenase complex and the α-ketoglutarate dehydrogenase complex as regards use of the energy contained in their respective acyl CoA compounds. Since the energy of acetyl CoA merely drives its condensation with oxaloacetate to form citrate, the energy appears to be "squandered." In reality, it makes this reaction highly exergonic and thereby produces the committed step of the cycle. By contrast, the energy of succinyl CoA drives the synthesis of an ATP equivalent in the form of GTP. Hence, the energy of succinyl CoA is "saved" directly by forming an energy-rich compound.

11.3.6. Succinate Dehydrogenase

Succinate dehydrogenase catalyzes a *dehydrogenation* (i.e., oxidation) that converts succinate to **fumarate.** The enzyme, a flavoprotein, uses FAD as coenzyme. In this stereospecific reaction, fumarate—the *trans* isomer—forms with a 100% yield. Maleate, the *cis* isomer, does not form at all.

$$\begin{array}{cc} COO^- & COO^- \\ | & | \\ CH_2 & CH \\ | & \| \\ CH_2 & HC \\ | & | \\ COO^- & COO^- \end{array}$$

Succinate^{2-} + FAD → fumarate^{2-} + FADH$_2$

Succinate dehydrogenase is an *iron–sulfur protein* (4Fe-4S type), located in the inner mitochondrial membrane, where it constitutes part of *respiratory complex II* (see Section 12.3). The mammalian enzyme is a dimer of two unequal subunits having molecular weights of 30,000 and 70,000. Because of its structural similarity to succinate, researchers have used *malonate* ($^-OOC-CH_2-COO^-$) as a competitive inhibitor of succinate dehydrogenase.

11.3.7. Fumarase

Fumarase, like succinate dehydrogenase, catalyzes a stereospecific reaction. In this reaction, only one of the two enantiomers of *malic acid* forms, namely, L-malate. The mammalian enzyme is a tetramer of four identical subunits, having a molecular weight of 49,000 each. Fumarase catalyzes a hydration reaction whereby water adds across the double bond of fumarate:

$$\begin{array}{cc} COO^- & COO^- \\ | & | \\ CH & HO-CH \\ \| & | \\ HC & CH_2 \\ | & | \\ COO^- & COO^- \end{array}$$

Fumarate^{2-} + H$_2$O → L-malate^{2-}

11.3.8. Malate Dehydrogenase

Malate dehydrogenase catalyzes the last reaction of the citric acid cycle. The enzyme, a pyridine-linked dehydrogenase, uses NAD$^+$ as coenzyme. The mammalian enzyme is a dimer of identical subunits having a molecular weight of 35,000 each. Malate dehydrogenase catalyzes a *dehydrogenation* (i.e., oxidation) that converts malate to oxaloacetate:

$$\begin{array}{cc} COO^- & COO^- \\ | & | \\ HO-CH & C=O \\ | & | \\ CH_2 & CH_2 \\ | & | \\ COO^- & COO^- \end{array}$$

L-Malate^{2-} + NAD$^+$ → oxaloacetate^{2-} + NADH + H$^+$

Oxaloacetate can combine with a second molecule of acetyl CoA and start another round of the cycle.

11.3.9. Overall Reaction

By *combining the steps described in Sections 11.3.1 through 11.3.8,* we obtain the overall reaction of the citric acid cycle:

$$\text{Acetyl CoA}^{4-} + 3NAD^+ + FAD^{2-} + 2H_2O + GDP^{3-} + P_i^{2-}$$
$$\downarrow$$
$$\text{CoA}-SH^{4-} + 3NADH + 2H^+ + FADH_2^{2-} + GTP^{4-} + 2CO_2$$

When we *include the conversion of pyruvate to acetyl CoA,* the overall reaction becomes

$$\text{Pyruvate}^- + 4NAD^+ + FAD^{2-} + 2H_2O + GDP^{3-} + P_i^{2-}$$
$$\downarrow$$
$$4NADH + 2H^+ + FADH_2^{2-} + GTP^{4-} + 3CO_2$$

11.4. MAJOR FEATURES OF THE CYCLE

Figure 11.10 illustrates key aspects of the citric acid cycle. The cycle is "fed" by entrance of acetyl CoA that condenses with oxaloacetate to form citrate. After releasing the coenzyme A component in the first step, the equivalent of the acetyl group is oxidized to two molecules of CO$_2$ in subsequent steps. Insofar as carbon atoms are concerned, complete balance exists between the input into the cycle and the output from it. However, the carbons released as CO$_2$ are not actually those of the incoming acetyl group. Instead, they originate from the oxaloacetate of the previous round. You can demonstrate this experimentally by using suitably labeled acetyl CoA (see Figure 11.10).

The citric acid cycle contains four oxidation steps, each involving removal of two hydrogen atoms from a substrate in the form of a hydride ion and a proton (H^- + H^+). Thus, a total of 8H are removed as $4H^-$ and $4H^+$. Three of these steps require pyridine-linked dehydrogenases that use NAD^+ as coenzyme ($3NAD^+ \rightarrow 3NADH$ + $3H^+$). One step requires *succinate dehydrogenase,* a flavoprotein that uses FAD as coenzyme (FAD $\rightarrow$ $FADH_2$). Succinate dehydrogenase is embedded in the inner mitochondrial membrane. All other cycle enzymes occur in the mitochondrial matrix.

Operation of the cycle results in reduction of the coenzymes NAD^+ and FAD to NADH and $FADH_2$. The reduced coenzymes must be oxidized to regenerate NAD^+ and FAD. Cells cannot tolerate an accumulation of NADH and $FADH_2$, since then the citric acid cycle, glycolysis, and other metabolic systems requiring NAD^+ or FAD would cease. Oxidation of NADH and $FADH_2$ by means of the electron transport system, located in the inner mitochondrial membrane, results in ATP synthesis via oxidative phosphorylation. Because NADH is loosely bound to its dehydrogenase, it can diffuse from the matrix to the inner membrane to be oxidized. $FADH_2$, however, is tightly bound to its dehydrogenase and cannot be oxidized in this fashion. This explains why succinate dehydrogenase is the only cycle enzyme located directly in the inner membrane.

The last reaction of the electron transport system, Eq. (12.5), requires molecular oxygen. Because of this reaction, and because of the citric acid cycle's obligatory link to the electron transport system, the cycle constitutes an aerobic pathway. Even though none of the steps of the cycle proper requires oxygen directly, the cycle can operate only under aerobic conditions.

The citric acid cycle has intimate links to respiration, both to the exhalation of carbon dioxide and to the inhalation of oxygen. The CO_2 produced by the two decarboxylation reactions of the cycle is largely exhaled and constitutes the bulk of respiratory CO_2. NADH and $FADH_2$, produced by the cycle, are oxidized by means of molecular oxygen in the electron transport system. The required oxygen constitutes the bulk of respiratory oxygen. Thus, the citric acid cycle produces most of the respiratory carbon dioxide and uses, indirectly, most of the respiratory oxygen.

When NADH and $FADH_2$ become oxidized via the electron transport system, the final reaction involves the reduction of oxygen to water. Hence, when we write the overall reaction for the citric acid cycle linked to the electron transport system, we must modify the equations

in Section 11.3.9 to include water as another product of the overall reaction. The combined two metabolic systems yield two products derived from the acetyl group of acetyl CoA: CO_2 *and* H_2O. Carbon dioxide forms in the citric acid cycle, and water forms in the electron transport system. In other words, the combined action of the citric acid cycle and the electron transport system results in *complete oxidation of the acetyl group to carbon dioxide and water.*

All of the intermediates of the cycle constitute weak acids. The first few are *tricarboxylic acids,* and the remaining ones are *dicarboxylic acids.* Conversion of succinyl CoA to succinate results in formation of GTP. Because GTP can phosphorylate ADP (yielding ATP), formation of GTP represents an effective production of a molecule of ATP.

You may wonder why the oxidation of a relatively simple acetyl group requires such a complex cycle in biological systems. The reason is that straight oxidation of acetic acid requires drastic conditions, incompatible with survival of a living cell. To circumvent the problem, a longer but gentler way evolved in the course of evolution that allows cells to carry out the necessary oxidation within the intracellular environment.

11.5. ENERGETICS AND CONTROL

11.5.1. Coupled Reactions

Biochemical standard free energy changes for individual steps of the citric acid cycle are given in Table 11.1. The overall reaction has a highly negative free energy change; it is strongly exergonic and proceeds spontaneously. If we include the pyruvate dehydrogenase reaction ($\Delta G^{\circ\prime}$ = -33.5 kJ mol^{-1}) in the overall reaction, the free energy change becomes even more favorable. Beginning with pyruvate, the entire sequence has a free energy change of $\Delta G^{\circ\prime}$ = -92.5 kJ mol^{-1}. Starting with pyruvate, the sequence includes four strongly exergonic reactions, four reactions for which $\Delta G^{\circ\prime}$ is close to zero, and one reaction (malate dehydrogenase) that has a pronounced positive free energy change. The latter reaction proceeds because of its coupling to the exergonic citrate synthase reaction. Indeed, all of the cycle's reactions constitute sets of *coupled reactions.* The product of one reaction serves as a reactant for the next; each product functions as a *common intermediate* of two reactions. In the citric acid cycle, any two coupled reactions are catalyzed by *different* enzymes, except those involving aconitase and isocitrate dehydrogenase.

Occurrence of strongly exergonic reactions means that these steps are irreversible, at least under biochemical standard conditions, ensuring progression of the sequence in the order indicated.

11.5.2. Efficiency

Most of the energy from acetyl CoA catabolism is produced by reducing NAD^+ and FAD in the citric acid cycle and subsequently oxidizing the reduced coenzymes via the electron transport system. As you will see (Section 12.3), oxidation of NADH yields 3 ATP, and oxidation of $FADH_2$ yields 2 ATP (Section 12.4). Additionally, one ATP forms via GTP produced in the succinate thiokinase reaction. Accordingly, the total number of molecules (moles) of ATP formed per molecule (mole) of acetyl coenzyme A processed via the citric acid cycle in conjunction with the electron transport system is:

$$3 \text{ (NADH, H}^+\text{) lead to:} \quad \frac{3 \text{ (NADH, H}^+\text{)}}{\text{Acetyl CoA}} \times \frac{3 \text{ ATP}}{\text{NADH, H}^+} = 9 \text{ ATP formed}$$

$$1 \text{ FADH}_2 \text{ leads to:} \quad \frac{1 \text{ FADH}_2}{\text{Acetyl CoA}} \times \frac{2 \text{ ATP}}{\text{FADH}_2} = 2 \text{ ATP formed}$$

$$1 \text{ GTP leads to:} \quad \frac{1 \text{ GTP}}{\text{Acetyl CoA}} \times \frac{1 \text{ ATP}}{\text{GTP}} = 1 \text{ ATP formed}$$

$$\text{Total} \quad 12 \text{ ATP formed}$$

Oxidation of acetyl CoA has a $\Delta G^{\circ\prime}$ value of -870 kJ mol^{-1} (see Problem 11.5). Hydrolysis of ATP has a $\Delta G^{\circ\prime}$ value of -30.5 kJ mol^{-1}. We can, therefore, calculate the efficiency of energy conservation for the *combined* operation of the citric acid cycle and the electron transport system as follows:

$$\frac{(30.5 \text{ kJ mol}^{-1} \times 12) \, 100}{870 \text{ kJ mol}^{-1}} = 42.1\%$$

This represents an impressive efficiency in view of the complexity of the combined systems. Keep in mind, though, that this constitutes only a rough estimate since we have based our calculations on $\Delta G^{\circ\prime}$ rather than on $\Delta G'$ values.

11.5.3. Regulatory Ratios

As you saw, control of the citric acid cycle proper occurs at three steps involving the enzymes citrate synthase, isocitrate dehydrogenase, and α-ketoglutarate dehydrogenase complex (Figure 11.13). An important additional control point is located outside the cycle, at the conversion of pyruvate to acetyl CoA (pyruvate dehydrogenase complex). At all of these points, major control becomes exerted through the ATP/ADP and $NADH/NAD^+$ ratios.

An actively metabolizing cell uses up energy rapidly. Such a cell has a low ATP concentration and a small ATP/ADP ratio. The cell also has a low NADH concentration; NADH must be rapidly oxidized via the electron transport system in order to generate the required energy in the form of ATP. Thus, the cell has a small $NADH/NAD^+$ ratio.

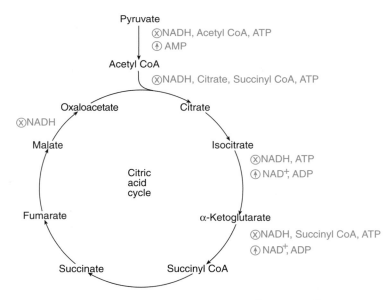

Figure 11.13.   Controls of the citric acid cycle. $\otimes$ Inhibitor, $\uparrow$ activator.

In a resting cell, the situation is reversed. The cell needs and uses relatively small amounts of energy. Such a cell has fairly high concentrations of both ATP and NADH so that the ATP/ADP and NADH/NAD$^+$ ratios have large values.

Both ATP and NADH inhibit the citric acid cycle. High concentrations block the operation of the cycle and, at the same time, slow down the electron transport system by limiting the supply of protons and electrons transferred to it from the cycle. ADP and NAD$^+$, on the other hand, serve as activators of the citric acid cycle and, at the same time, accelerate the electron transport system.

11.5.4. Anaplerotic Reactions

When the citric acid cycle functions in *catabolism,* acetyl CoA enters the cycle, and its acetyl moiety condenses with oxaloacetate to form citrate. Subsequent steps of the cycle result in regeneration of oxaloacetate. This means that the concentration of oxaloacetate, and that of all other cycle intermediates, *remains constant.* Processing of acetyl CoA by intermediates of the citric acid cycle is analogous to processing of metal by machines of a moving belt system for producing car bodies. The metal introduced is altered and shaped to form a car body, but the basic machinery of the moving belt system remains unchanged.

A different situation exists when the citric acid cycle

functions in *anabolism.* Under these conditions, cycle intermediates are siphoned off from the pool and used as reactants for synthesis of biomolecules. Progress of such anabolic reactions lowers the concentrations of cycle intermediates. It becomes necessary to replenish the intermediates and restore their concentrations to normal levels. Special reactions called **anaplerotic reactions** (from the Greek, meaning "to fill up") accomplish this task. Typically, anaplerotic reactions involve a *carboxylation* step in which CO_2 is taken up by a compound. The conversion of pyruvate (CH_3-CO-COO$^-$) to oxaloacetate ($^-$OOC-CH_2-CO-COO$^-$), catalyzed by *pyruvate carboxylase,* illustrates anaplerotic reactions:

$$\text{Pyruvate}^- + CO_2 + ATP^{4-} + H_2O \rightarrow$$
$$\text{oxaloacetate}^{2-} + ADP^{3-} + P_i^{2-} + 2H^+$$

Pyruvate carboxylase requires **biotin,** a member of the vitamin B complex. The coenzyme form of biotin, much like that of lipoic acid, consists of the vitamin attached via an amide bond to the ϵ-amino group of lysine in an enzyme (Figure 11.14). The complex is called **biocytin (biotinyllysine)** and serves as a carrier of CO_2.

Pyruvate carboxylase is an allosteric enzyme for which acetyl CoA serves as positive allosteric effector. Consider what this means for operation of the citric acid cycle. When the cycle functions in anabolism, concentra-

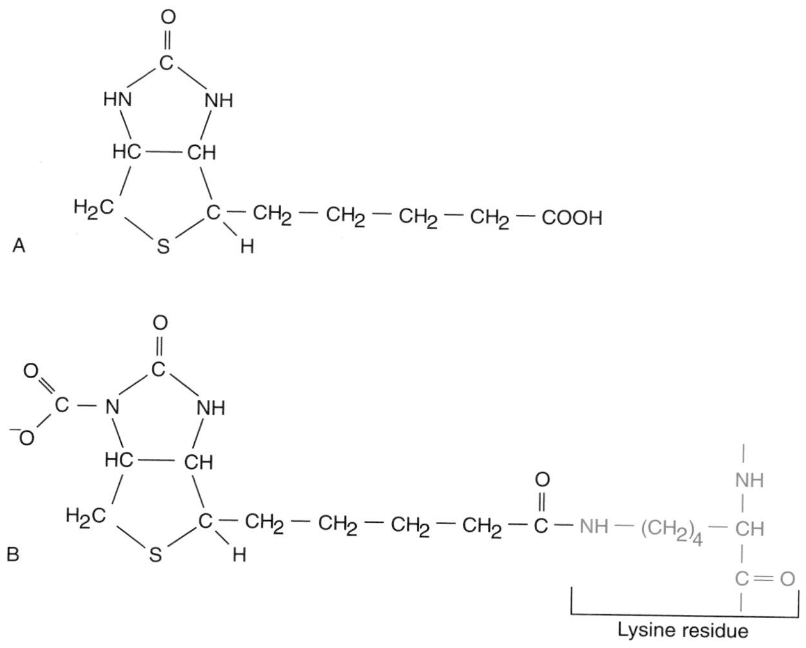

Figure 11.14. (A) Biotin. (B) Carboxybiocytin (carboxybiotinyllysine). In biocytin, biotin is linked covalently to the ϵ-NH$_2$ group of a lysine residue in an enzyme. Biocytin serves as a carrier of CO_2.

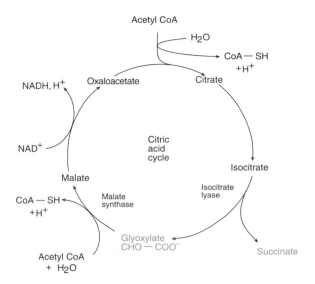

Figure 11.15.   The glyoxylate cycle.

sults in conversion of pyruvate to oxaloacetate, thereby raising the level of all cycle intermediates and allowing the cycle to metabolize the accumulated acetyl CoA.

11.6. GLYOXYLATE CYCLE

In animals, acetyl CoA cannot be used for direct synthesis of carbohydrates. Such synthesis requires conversion of acetyl CoA to pyruvate, followed by conversion of pyruvate to glucose via *gluconeogenesis*. In animals, this pathway is blocked because the reaction catalyzed by the pyruvate dehydrogenase complex has a highly negative free energy change and cannot be reversed.

Some plants and bacteria, however, can use acetyl CoA for direct synthesis of carbohydrates. In these organisms, carbohydrate synthesis proceeds via a modified citric acid cycle called the **glyoxylate cycle** (Figure 11.15). In plants, the glyoxylate cycle occurs in specialized cytoplasmic organelles named **glyoxysomes.** The cycle, proposed by Krebs and Kornberg in 1957, employs some citric acid cycle reactions as well as two unique reactions. One, catalyzed by **isocitrate lyase,** results in cleavage of isocitrate to succinate and *glyoxylate*. The other, catalyzed by **malate synthase,** leads to condensation of acetyl CoA

tions of cycle intermediates decrease. Acetyl CoA, derived from catabolism, cannot be processed as rapidly as it forms and begins to accumulate. As the concentration of acetyl CoA builds up, it begins to exert its allosteric effect on pyruvate carboxylase. Activation of this enzyme re-

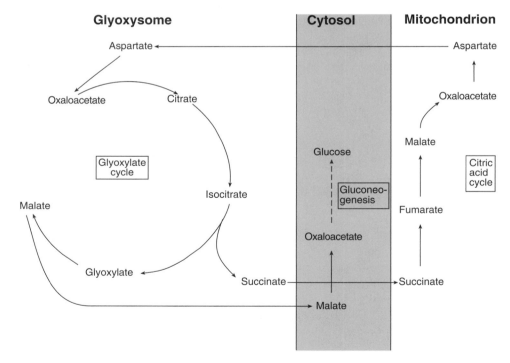

Figure 11.16.   Transfer of metabolites between the glyoxysome, the mitochondrion, and the cytosol. Combined operation of the glyoxylate cycle, the citric acid cycle, and gluconeogenesis requires transfer of metabolites between these three compartments. Malate dehydrogenase catalyzes the conversion of malate to oxaloacetate in both the cytosol and the mitochondrion. Aspartate aminotransferase catalyzes the interconversion of aspartate and oxaloacetate (transamination) in both the glyoxysome and the mitochondrion.

with glyoxylate to form malate. These two steps effectively bypass the two decarboxylation reactions of the citric acid cycle.

Another significant difference between the glyoxylate and citric acid cycles is that the glyoxylate cycle requires input of *two* molecules of acetyl CoA. In fact, the cycle achieves a net conversion of two two-carbon fragments (the acetyl moiety of acetyl CoA) to one four-carbon compound (succinate). You can see this from the balanced equation:

$$2 \text{ Acetyl CoA}^{4-} + 2\text{H}_2\text{O} + \text{NAD}^+ \rightarrow$$
$$\text{succinate}^{2-} + 2 \text{ CoA—SH}^{4-} + \text{NADH} + 3\text{H}^+$$

No carbon atoms of acetyl CoA are released as CO_2 during the operation of the glyoxylate cycle. The succinate formed in the glyoxysomes is transported into the mitochondria, where citric acid cycle reactions convert it to oxaloacetate. Because oxaloacetate serves as precursor for glucose synthesis by gluconeogenesis, succinate formed in the glyoxylate cycle leads to net synthesis of carbohydrates. However, gluconeogenesis takes place in the cytosol, and oxaloacetate, formed in the mitochondria, cannot pass directly across the mitochondrial membrane. To link these different pathways requires transfer of metabolites between three compartments: the glyoxysome, the mitochondrion, and the cytosol. As you can see from Figure 11.16, transfers of succinate, aspartate, and malate provide the necessary links.

Plants and bacteria, therefore, use acetate both as an energy source (citric acid cycle) and as a carbon source for synthesis of carbohydrates (glyoxylate cycle). However, the glyoxysomal pool of acetyl CoA, which drives carbohydrate synthesis via gluconeogenesis, is entirely separate from the mitochondrial pool, which drives generation of energy via the citric acid cycle.

SUMMARY

The citric acid cycle functions in both catabolism and anabolism. It constitutes the major pathway for oxidation of nutrients and for the production of metabolic energy. It also serves as an important source of intermediates for biosynthetic reactions. The cycle consists of eight consecutive, enzyme-catalyzed reactions. Any two successive reactions are energetically coupled. In addition to NAD^+ and FAD, enzymes of the cycle use three other coenzymes—coenzyme A, lipoic acid, and thiamine pyrophosphate.

The citric acid cycle is "fed" by entry of acetyl CoA. Lipid catabolism produces acetyl CoA directly. Carbohydrate catabolism yields pyruvate, which is converted to acetyl CoA by the pyruvate dehydrogenase complex, a multienzyme system.

One turn of the cycle leads to loss of two molecules of CO_2 and eight hydrogen atoms. Six hydrogens are removed in the form of 3 ($NADH$, H^+) and two in the form of $FADH_2$. These reduced coenzymes enter the electron transport system, where they are oxidized and lead to synthesis of ATP via oxidative phosphorylation. In addition, the cycle yields a molecule of GTP (ATP) by substrate-level phosphorylation. The total yield of ATP from operation of the citric acid cycle and the electron transport system is 12 molecules of ATP per molecule of acetyl CoA entering the cycle.

Major controls of the cycle occur at three steps and involve primarily the ratios ATP/ADP and $NADH/NAD^+$. ATP and NADH inhibit the cycle; ADP and NAD^+ activate it. When anabolic reactions lower the levels of cycle intermediates, anaplerotic reactions come into play that restore these concentrations to their normal levels. Plants and bacteria use a modified form of the citric acid cycle, called the glyoxylate cycle, that allows them to use acetyl CoA directly for synthesis of carbohydrates.

SELECTED READINGS

Barron, J. T., Kopp, S. J., Tow, J., and Parrillo J. E., Fatty acid, tricarboxylic acid cycle metabolites, and energy metabolism in vascular smooth muscle, *Am. J. Physiol.* 267:H764–H769 (1994).

Beevers, H., The role of the glyoxylate cycle, in *The Biochemistry of Plants,* Vol. 4 (P. K. Stumpf and E. E. Conn, eds.), Academic Press, New York (1980).

Beylot, M., Soloviev, M. V., David, F., Landau, B. R., and Brunengraber, H., Tracing hepatic gluconeogenesis relative to citric acid cycle activity *in vitro* and *in vivo, J. Biol. Chem.* 270:1509–1514 (1995).

DiDonato, L., Des Rosiers, C., Montgomery, J. A., David, F., Garneau, M., and Brunengraber, H., Rates of gluconeogenesis and citric acid cycle in perfused livers, assessed from the mass spectrometric assay of the carbon-13-labeling pattern of glutamate, *J. Biol. Chem.* 268:4170–4180 (1993).

Drozdov, L. N., *et al.,* The optimal structure of the tricarboxylic acid cycle multienzyme system of *E. coli* for the cases of cell growth on various carbon sources, *Biokhimiya (Moscow)* 59:368–380 (1994).

Durschlag, H., *et al.,* Structural changes of citrate synthase upon ligand binding and upon denaturation, *Prog. Colloid Polym. Sci.* 93:222–230 (1993).

Hiromasa, Y., Aso, Y., Yamashita, S., and Aikawa, Y., Homogeneity of the pyruvate dehydrogenase multienzyme complex from *Bacillus stearothermophilus, J. Biochem (Tokyo)* 117:467–470 (1995).

Kleczkowski, L. A, Kinetics and regulation of the NAD(P)H-dependent glyoxylate-specific reductase from spinach leaves, *Z. Naturforsch. C* 50:21–28 (1995).

Kornberg, H. L., Tricarboxylic acid cycles, *BioEssays* 7:236–238 (1987).

Krebs, H. A., The history of the tricarboxylic acid cycle, *Perspect. Biol. Med.* 14:154–170 (1970).

Mason, G. F., *et al.,* Simultaneous determination of the rates of the TCA cycle, glucose utilization, α-ketoglutarate/glutamate exchanges, and glutamine synthesis in human brain by NMR, *J. Cereb. Blood Flow Metab.* 15:12–25 (1995).

Patel, M. S., and Roche, T. E., Molecular biology and biochemistry of pyruvate dehydrogenase complexes, *FASEB J.* 4:3224–3233 (1990).

REVIEW QUESTIONS

A.   Define each of the following terms:

   Amphibolic pathway
   Anaplerotic reaction
   Biocytin
   Lipoamide
   Oxidative decarboxylation
   Pyruvate dehydrogenase complex
   Succinyl CoA
   Suicide substrate

B.   Differentiate between the two terms in each of the following pairs:

   Succinate thiokinase/
      succinate dehydro-
      genase
   Isocitrate lyase/isocitrate
      dehydrogenase
   Succinate/oxalosuccinate
   Citrate/isocitrate

   Citric acid cycle/
      glyoxylate cycle
   Malate synthase/
      malate dehydro-
      genase
   Coenzyme A/acetyl
      coenzyme A
   Lipoic acid/lipoyllysine

C.   (1) Write out the eight individual reactions that, together, constitute the citric acid cycle. Include the names of the enzymes, abbreviations for the coenzymes, and names and structures of the reactants and products.
(2) What are the major functions of the citric acid cycle?
(3) What is the function of the glyoxylate cycle and how is it achieved?

PROBLEMS

11.1.   Compare the ratios of NADH/NAD$^+$ and ATP/ADP in heart muscle during periods of sleep and jogging.

11.2.   (a) Prepare a plot of 1/v versus 1/[S] for succinate dehydrogenase (v = velocity; [S] = substrate concentration). Draw 2 graphs for the following conditions on the same set of axes: (1) no inhibition; (2) in the presence of malonate. (b) Would it be possible to decrease the inhibition of succinate dehydrogenase by malonate in a liver homogenate by adding oxaloacetate? If so, explain the reason for this effect.

11.3.   Write the overall equations for the following partial sequences in both the citric acid cycle and the glyoxylate cycle: (a) citrate → malate; (b) acetyl CoA → citrate; (c) isocitrate → succinate.

11.4.* Use the data in Table 11.1 to calculate the relative concentrations of citrate and isocitrate at pH 7.0.

11.5.* Calculate $\Delta G^{\circ\prime}$ for the oxidation of acetyl CoA by coupling the following two reactions:

$$CH_3COOH + CoA\text{-}SH \rightleftarrows CH_3CO\text{-}S\text{-}CoA + H_2O$$
$$\Delta G^{\circ\prime} = +31.4 \text{ kJ mol}^{-1}$$
$$CH_3COOH + 2O_2 \rightleftarrows 2CO_2 + 2H_2O$$
$$\Delta G^{\circ\prime} = -837.0 \text{ kJ mol}^{-1}$$

11.6. (a) Why does it make good metabolic sense to have ATP and NADH function as inhibitors of the citric acid cycle? (b) Most of this inhibition is concentrated in the first half of the cycle (Figure 11.13). Is that advantageous? Why?

11.7. Adding the dicarboxylic acids succinate, fumarate, and oxaloacetate has been shown to increase the rate of CO_2 production by liver homogenates. What is the reason for this effect?

11.8. Can you achieve *net synthesis* of oxaloacetate by adding acetyl CoA to a cell-free extract that contains only the enzymes, coenzymes, and intermediates of the citric acid cycle?

11.9. According to Figure 11.13, AMP serves as an activator of the pyruvate dehydrogenase complex. Why is this metabolically desirable?

11.10.* A researcher labels acetyl CoA completely with ^{14}C in each of the two carbons of its acetyl group and uses the acetyl CoA to study the citric acid cycle. What fractional intensity of released $^{14}CO_2$ can the researcher expect after one, two, and three complete turns of the citric acid cycle? Assume that any acetyl CoA entering the cycle in the second and third turns does *not* carry any radioactive label.

11.11.* Suppose that a mutation in *E. coli* resulted in an altered aconitase that *binds citrate symmetrically.* What fraction of the labeled carbon atoms, introduced as the acetyl group of acetyl CoA, would be released as labeled CO_2 in one complete turn of the citric acid cycle? (Hint: Refer to Figure 11.10.)

11.12. What type of inhibition do you expect to be involved when fluorocitrate acts on aconitase? Given that fluorocitrate is a potent inhibitor, would its K_i be large or small?

11.13. How many molecules of ATP can be synthesized by the electron transport system *per molecule of acetyl CoA* (a) in one complete turn of the citric acid cycle and (b) via the glyoxylate cycle? For (b), consider only the mitochondrial reactions shown in Figure 11.16.

11.14.* What is the minimum [citrate]/[isocitrate] ratio required for the aconitase reaction to proceed spontaneously at pH 7.0? (Refer to Table 11.1.)

11.15. Succinyl CoA and citrate both inhibit the enzymes involved in their own synthesis (Figure 11.13). Name this type of inhibition.

11.16. Using the data of Table 11.1 and a value of $\Delta G^{\circ\prime} = -30.5 \text{ kJ mol}^{-1}$ for the hydrolysis of GTP to GDP and P_i, calculate $\Delta G^{\circ\prime}$ for the hydrolysis of succinyl CoA to succinate and CoA-SH.

11.17. Based on your calculation from the previous problem and the data of Table 11.1, calculate the efficiency of substrate-level phosphorylation (GTP synthesis) for the combined α-ketoglutarate dehydrogenase complex and succinate thiokinase reactions.

11.18. How many ATPs could theoretically be formed from coupling the first three reactions of the citric acid cycle, based on $\Delta G^{\circ\prime}$ values?

11.19. Assume that biosynthesis of purines has lowered the α-ketoglutarate concentration. How will this affect the concentrations of isocitrate and fumarate?

11.20.* If labeled oxaloacetate ($^-OOC^*\text{--}CH_2\text{--}CO\text{--}COO^-$) were to be processed through the citric acid cycle, would the CO_2 released during the first turn be labeled or not? If labeled, at what step is it released?

11.21. A student was asked to list some examples of anaplerotic reactions. The student listed the glyoxylate cycle as one answer. Is that correct? Why or why not?

11.22. Malic enzyme catalyzes the following reaction:

$$Malate^{2-} + NADP^+ \rightarrow pyruvate^- + CO_2 + NADPH$$

Could the addition of malic enzyme to a liver homogenate be used to lower the concentration of malate in the citric acid cycle?

11.23.* The hydroxyethyl group linked to TPP in Figure 11.11 is known as *active acetaldehyde.* Active acetaldehyde forms by loss of the hydrogen bonded to carbon in the thiazole ring of TPP (see Figure 11.9), forming a carbanion that attacks the carbonyl carbon of pyruvate. An input of $2H^+$ and a loss of CO_2 yields the final product. Write out the structures for the sequence of events just described. What does this mechanism tell you about the role of vitamin B_1 in TPP?

11.24. None of the reactions of the citric acid cycle requires oxygen as a reactant. Why, then, does the cycle constitute an aerobic metabolic pathway?

Electron Transport and Oxidative Phosphorylation

12

Free energy changes of chemical reactions are related to a number of parameters: entropy and enthalpy changes (Eq. 9.1), equilibrium constants (Eq. 9.2), and oxidation–reduction potentials. Of these, equilibrium constants and oxidation–reduction potentials are of primary importance for computing free energy changes of metabolic reactions. Relationships between free energy changes and equilibrium constants were explored in Chapter 9. Now we will focus on the relationships between free energy changes and oxidation–reduction potentials.

As you know, *oxidation–reduction* (or redox) reactions form the core of stage III of catabolism or cellular respiration (Figure 8.2). Oxidation of metabolites in the citric acid cycle results in removal of hydride ions (H^- or H:) and protons that are channeled into the *electron transport system* and ultimately reduce molecular oxygen to water. Energy derived from this flow of electrons leads to ATP synthesis via *oxidative phosphorylation.*

Electron transport and oxidative phosphorylation constitute the culminating events of cellular respiration. The degradative pathways of carbohydrates, lipids, and proteins converge into this final stage that provides the major metabolic synthesis of ATP in aerobic organisms. Cell respiration has been the subject of intensive research since the days of Lavoisier (1777).

In eukaryotes, electron transport and oxidative phosphorylation occur in the mitochondria, the "power houses" of the cell. Mitochondria were first described in the period 1850–1890. Techniques for isolating these organelles were perfected in the late 1940s, and between 1948 and 1950 Eugene Kennedy and Albert Lehninger discovered that the citric acid cycle, fatty acid oxidation, and oxidative phosphorylation take place in the mitochondria.

12.1. OXIDATION–REDUCTION POTENTIALS

12.1.1. Standard Redox Potentials

The nature of an overall oxidation–reduction reaction depends on the electron affinities of the oxidants in the two component half-reactions. *The half-reaction whose oxidant has the greater affinity for electrons will proceed as a reduction; the other half-reaction will proceed as an oxidation.* We describe the electron affinity of an oxidant by an **oxidation–reduction (redox) potential** for the half-reaction involved. In biochemistry, we use *reduction poten-*

tials, redox potentials based on the tendency for *gaining* electrons. Reduction potentials are measured by reference to the hydrogen half-reaction, or *hydrogen electrode:*

$$H^+ + e^- \rightleftarrows \tfrac{1}{2}H_2$$

A hydrogen electrode consists of platinum metal in contact with hydrogen gas and a solution containing hydrogen ions. We assign this electrode a potential of 0.0000 V (volts) when $[H^+] = 1.0M$, the pressure of hydrogen gas is 1.0 atm, and the temperature is 25°C. This constitutes the **standard hydrogen electrode.**

We define standard conditions for other half-reactions as we did for free energy changes (see Section 9.1): Reactant and product concentrations are 1.0M each, the temperature is 25°C, and the pressure is 1.0 atm. Under these conditions, any half-reaction that has a greater tendency to act as a reducing agent (that is, to generate electrons) than the standard hydrogen electrode is credited with a *negative* standard reduction potential. Half-reactions that have smaller tendencies to act as reducing agents are credited with *positive* standard reduction potentials. We designate a *standard reduction potential* as $E°$.

Because standard conditions include $[H^+] = 1.0M$ (pH 0), they are as inappropriate for evaluating biochemical potential changes as they are for evaluating biochemical free energy changes (see Section 9.1 again). We, therefore, use a different reference potential, called the **biochemical standard reduction potential,** and designate it $E°'$.

> $E°'$ is the reduction potential of a half-reaction that develops when all reactants and products are at an initial concentration of 1.0M each, except for protons, the initial concentration of which, unless otherwise specified, is taken as $[H^+] = 10^{-7}M$ (pH 7.0), the temperature is 25°C, and the pressure is 1.0 atm.

As we did for for free energy changes, we can define redox potentials in two ways, conceptually and mathematically. The above statement constitutes the *conceptual definition* of $E°'$ and describes its physical meaning. The *mathematical definition* of $E°'$ is given by the following equation, which permits a calculation of the value of $E°'$:

$$E°' = \frac{2.303RT}{nF}\log K'_{bio} = \frac{2.303RT}{nF}\log \frac{[\text{reductant}]}{[\text{oxidant}]} \quad (12.1)$$

where R is the gas constant (8.314 J deg^{-1} mol^{-1} = 1.987 cal deg^{-1} mol^{-1}), T is the absolute temperature (K), n is the number of electrons in the half-reaction, F is the Fara-

day constant (96,491 J V^{-1} mol^{-1} = 23,062 cal V^{-1} mol^{-1}), and K'_{bio} is the equilibrium constant at pH 7.0.

Generally, K'_{bio} consists of only two terms, [reductant] and [oxidant]; electrons are omitted from the expression, and protons do not appear in it because we calculate K'_{bio} for a fixed pH of 7.0. The concentrations of all components are those existing *at equilibrium.* At 25°C (T = 298.2 K), the coefficient of the logarithmic term equals $0.06/n$.

One can determine biochemical standard reduction potentials by measuring the potential, or *electromotive force,* developed in an electrochemical cell composed of two half-reactions, a standard hydrogen electrode and the half-reaction of interest (see Appendix D, Figure D.2). Table 12.1 lists $E°'$ values for some half-reactions of biological interest.

12.1.2. Actual Redox Potentials

Reduction potentials developed under nonstandard conditions constitute *actual reduction potentials.* The potential corresponding to $E°$ is designated E, and that corresponding to $E°'$ is designated E'. E' is the **biochemical actual**

Table 12.1. Biochemical Standard Reduction Potentials ($E°'$) of Some Half-Reactions of Biological Relevance

Half-reaction[a]	$E°'$ (volts)
Ferredoxin (Fe^{3+}) + e^- ⇌ ferredoxin (Fe^{2+})	−0.43
H$^+$ + e^- ⇌ $\tfrac{1}{2}$H$_2$	−0.42
α-Ketoglutarate + CO$_2$ + 2H$^+$ + 2e^- ⇌ isocitrate	−0.38
NAD$^+$ + 2H$^+$ + 2e^- ⇌ NADH + H$^+$	−0.32
NADP$^+$ + 2H$^+$ + 2e^- ⇌ NADPH + H$^+$	−0.32
1,3-Bisphosphoglycerate + 2H$^+$ + 2e^- ⇌ G-3-P + P$_i$	−0.29
Acetaldehyde + 2H$^+$ + 2e^- ⇌ ethanol	−0.20
Pyruvate + 2H$^+$ + 2e^- ⇌ lactate	−0.19
FAD + 2H$^+$ + 2e^- ⇌ FADH$_2$ (free coenzyme)	−0.18
FMN + 2H$^+$ + 2e^- ⇌ FMNH$_2$ (free coenzyme)	−0.18
Oxaloacetate + 2H$^+$ + 2e^- ⇌ malate	−0.17
Fumarate + 2H$^+$ + 2e^- ⇌ succinate	+0.03
Myoglobin (Fe^{3+}) + e^- ⇌ Myoglobin (Fe^{2+})	+0.05
DHA + 2H$^+$ + 2e^- ⇌ ascorbic acid	+0.06
Cyt b (Fe^{3+}) + e^- ⇌ Cyt b (Fe^{2+})	+0.07
CoQ + 2H$^+$ + 2e^- ⇌ CoQH$_2$	+0.05
Hemoglobin (Fe^{3+}) + e^- ⇌ hemoglobin (Fe^{2+})	+0.17
Cyt c_1 (Fe^{3+}) + e^- ⇌ Cyt c_1 (Fe^{2+})	+0.22
Cyt c (Fe^{3+}) + e^- ⇌ Cyt c (Fe^{2+})	+0.25
Cyt a (Fe^{3+}) + e^- ⇌ Cyt a (Fe^{2+})	+0.29
O$_2$ + 2H$^+$ + 2e^- ⇌ H$_2$O$_2$	+0.30
NO$_3^-$ + 2H$^+$ + 2e^- ⇌ NO$_2^-$ + H$_2$O	+0.42
Cyt a_3 (Fe^{3+}) + e^- ⇌ Cyt a_3 (Fe^{2+})	+0.39
Fe^{3+} + e^- ⇌ Fe^{2+}	+0.77
$\tfrac{1}{2}$O$_2$ + 2H$^+$ + 2e^- ⇌ H$_2$O	+0.82

[a]Abbreviations: G-3-P, glyceraldehyde 3-phosphate; DHA, dehydroascorbic acid; Cyt, cytochrome.

reduction potential. E is identical to E', much as ΔG is identical to $\Delta G'$ (see Section 9.1). There exists only *one* actual reduction potential for a given set of nonstandard conditions, whether we base it on $E°$ and designate it E or base it on $E°'$ and designate it E'. To be consistent with free energy designations, we will use $E°'$ and E' for reduction potentials.

The relationship of E' to $E°'$ is described by the **Nernst equation:**

$$E' = E°' - \frac{2.303RT}{nF} \log \frac{[\text{reductant}]}{[\text{oxidant}]} \qquad (12.2)$$

where [reductant] and [oxidant] represent the *actual initial* (not equilibrium) *concentrations* of the reduced and oxidized forms. Note that just as $\Delta G'$ is related to $\Delta G°'$ by a concentration term, so E' is related to $E°'$ by a similar term.

For redox reactions, free energy changes are related to potentials by two equations:

$$\Delta G°' = -nF\Delta E°' \qquad (12.3)$$

$$\Delta G' = -nF\Delta E' \qquad (12.4)$$

where $\Delta E°'$ and $\Delta E'$ represent the overall change in reduction potential when two half-reactions are combined, and n is the number of electrons generated by one half-reaction and consumed by the other half-reaction.

These equations show that a change in reduction potential is *equivalent* to a change in free energy. It is this interdependence that accounts for the importance of redox reactions in biochemical energetics.

The overall redox reaction resulting from the combination of two half-reactions represents a special case of *coupled reactions* (Section 9.3) in which electrons constitute the *common intermediate.* Electrons are a product of one half-reaction and serve as a reactant for the second half-reaction. How to couple two half-reactions under standard or actual conditions is described in detail in Appendix D. As pointed out there, *what ultimately determines the direction of an overall redox reaction are the intracellular concentrations of reactants and products.*

12.2. BIOLOGICAL ELECTRON CARRIERS

In Chapter 11 you saw that two types of dehydrogenases—pyridine-linked dehydrogenases and flavoproteins—catalyze the oxidation of metabolites in the citric acid cycle. The redox reactions are mediated by NAD^+ and FAD, the coenzymes of these dehydrogenases. When the metabolite undergoes oxidation, the coenzyme undergoes reduction; NAD^+ is reduced to NADH, and FAD is re-

duced to $FADH_2$. The process requires removal of two hydrogen atoms from the metabolite in the form of a hydride ion (H^- or H:) and a proton (H^+), one or both of which are transferred to the coenzyme. Reduction of NAD^+ requires transfer of H^-, and reduction of FAD requires transfer of both H^- and H^+. From NADH and $FADH_2$, the abstracted hydride ions and protons are transferred to other redox coenzymes and special components that function as consecutive **electron carriers.** Electrons flow through this chain of carriers to oxygen. The entire array of specific enzymes, coenzymes, and other components constitutes the **electron transport system (ETS).**

In eukaryotes, the electron transport system is located in the inner mitochondrial membrane. Prokaryotes have similar electron transport systems but composed of different electron carriers. All prokaryotic components are attached to the plasma membrane so that the reactions occur at the cell periphery.

The mitochondrial ETS consists of five groups of biological electron carriers: *pyridine-linked dehydrogenases, flavoproteins, coenzyme Q (ubiquinone), cytochromes,* and *iron–sulfur proteins*. We discussed pyridine-linked dehydrogenases and flavoproteins in Section 11.1. The remaining electron carriers are described below.

12.2.1. Coenzyme Q (Ubiquinone)

Coenzyme Q or **ubiquinone,** abbreviated **CoQ** or **Q,** is a generic term for a group of electron carriers discovered in the late 1950s and characterized by *quinone*-like structures and *ubiquitous* occurrence (Figure 12.1).

Despite their names, ubiquinones are *not* coenzymes and are not linked to proteins. Rather, they comprise lipid-soluble compounds present in the lipid phase of the mitochondrial membrane, where they function as **mobile electron carriers.** Ubiquinones owe their nonpolar, *lipophilic* character to the *isoprenoid side chain.* Like the flavoproteins, ubiquinones can participate in either a one-electron or a two-electron transfer. Ubiquinone forms a free radical that has a stable semiquinone-type structure (Figure 12.2).

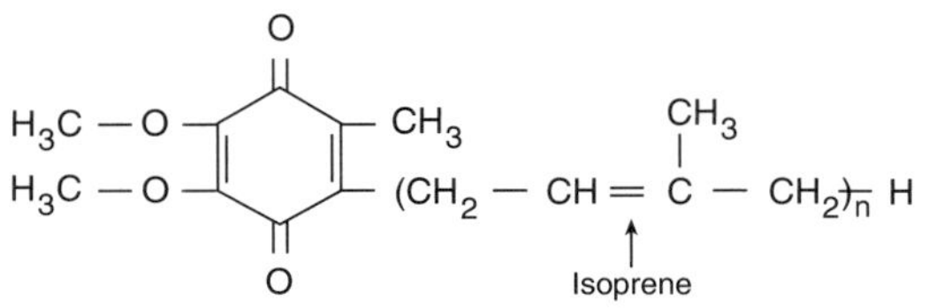

Figure 12.1.   Coenzyme Q (ubiquinone). The length of the isoprenoid chain varies. Typically, n has a value of 10 for eukaryotes and 6 for bacteria.

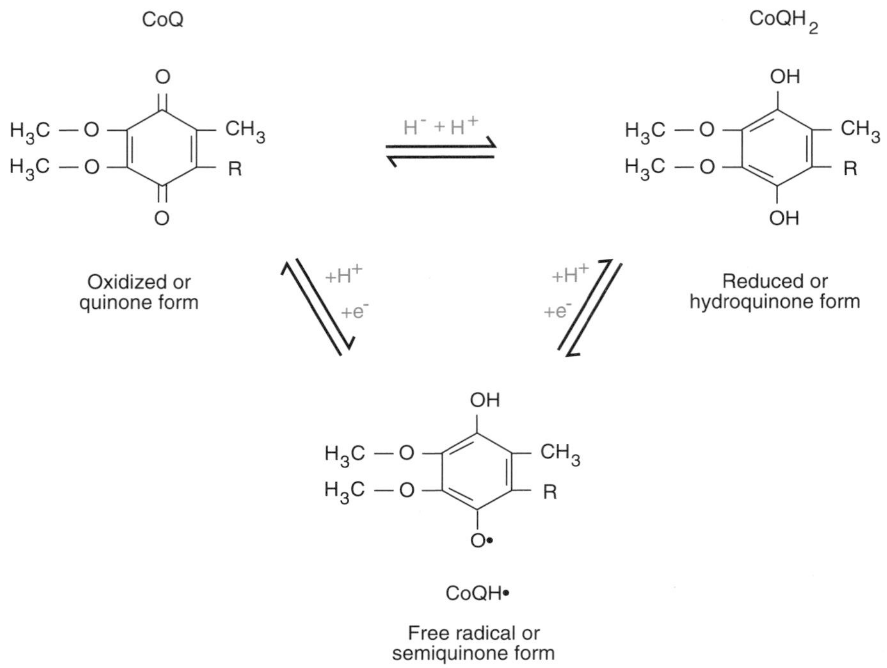

CoQ

CoQH$_2$

Oxidized or
quinone form

Reduced or
hydroquinone form

CoQH•

Free radical or
semiquinone form

Figure 12.2. Oxidation and reduction of coenzyme Q.

12.2.2. Cytochromes

Cytochromes ("cellular pigments") were first identified in the late 19th century. We now know that cytochromes occur in all types of cells. They are located primarily in membranes and, in eukaryotes, specifically in the mitochondrial membrane.

Cytochromes are red-brown, conjugated proteins that contain an *iron–porphyrin complex* or *heme*. Some cytochromes contain a heme identical to that present in hemoglobin and myoglobin. Investigators have identified some 30 different cytochromes. They vary in the substituents of the tetrapyrrole ring system (Figure 12.3), the amino acid sequence of the conjugated protein, and the linkage of heme to protein. Of all the cytochromes, only cytochrome c is readily extracted from the mitochondrial membrane; the other cytochromes are *integral membrane proteins*.

Cytochromes have characteristic spectral properties and absorb strongly in the visible range (Figure 12.4). Cytochromes accept or donate one electron by virtue of reversible oxidation–reduction of the heme iron:

$$Fe^{3+} \underset{-e^-}{\overset{+e^-}{\rightleftharpoons}} Fe^{2+}$$

Although all cytochromes share this fundamental redox reaction, they do not have the same $E^{\circ\prime}$ value (see Table 12.1) because reduction of the iron atom occurs within different environments. The polypeptide chains that surround the hemes of different cytochromes vary with respect to the type, number, and position of charged functional groups. Additionally, cytochromes may differ in the structure of the heme and in its attachment to the polypeptide chain. Hence, the local environment in which the reduction of iron takes place varies from compound to compound. Reduction proceeds more readily in some cytochromes than in others, a property reflected in their nonidentical $E^{\circ\prime}$ values.

12.2.3. Iron–Sulfur Proteins

Iron–sulfur proteins (Fe-S), or **nonheme iron proteins (NHI proteins),** are conjugated proteins that contain complexes of iron and sulfur but no heme. Biochemists recognize several types of iron–sulfur proteins (Figure 12.5), but in all cases each iron atom is coordinated to four sulfur atoms. Some sulfur atoms constitute components of cysteine's sulfhydryl groups; others are "acid-labile" atoms that can be released from the iron–sulfur protein at pH $\approx$ 1 in the form of H_2S. Iron–sulfur proteins undergo oxidation–reduction reactions by virtue of reversible oxidation of their iron. The oxidized and reduced states of all iron–sulfur proteins differ by one charge, regardless of the number of iron atoms they contain. Hence, much like cytochromes, iron–sulfur proteins function in one-electron transfer reactions.

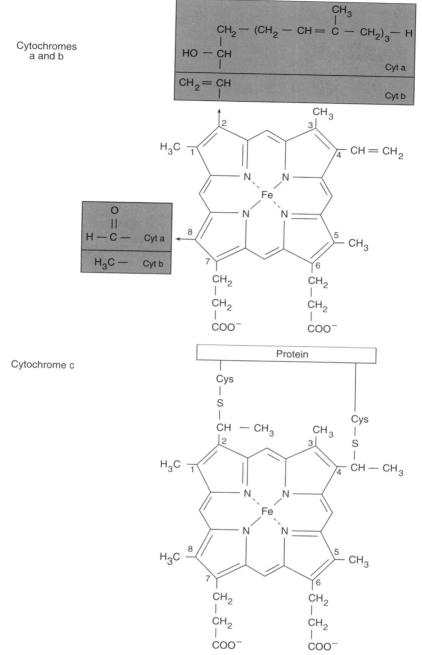

Figure 12.3. Structures of the heme groups of cytochromes *a*, *b*, and *c*. Cytochrome *b* heme is identical to that occurring in hemoglobin and myo-globin (iron-protoporhyrin IX).

12.2.4. Comparing the Electron Carriers

Oxidation–reduction characteristics of cytochromes and iron–sulfur proteins differ from those of pyridine nucleotides, flavins, and ubiquinones in two respects. First, redox reactions of cytochromes and iron–sulfur proteins involve only electrons, not electrons plus varying forms of hydrogen atoms. Second, cytochromes and iron–sulfur proteins are restricted to one-electron transfer reactions. Pyridine nucleotides, on the other hand, require a two-electron transfer (H^-), and flavins and coenzyme Q can function in either one-electron or two-electron transfers.

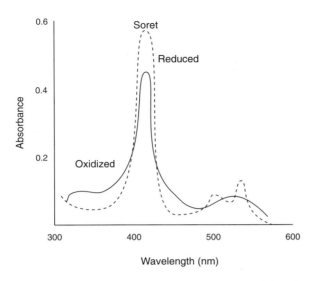

Figure 12.4. Absorption spectra of cytochrome *c* at pH 7.0. The characteristic absorption band at about 400 nm is called the *Soret band*. Other cytochromes have similar absorption spectra.

Oxidation of metabolites in the citric acid cycle generates the electrons that flow through the electron transport system. These redox reactions involve the transfer of two electrons to NAD$^+$ or FAD. As the two electrons pass from NADH or FADH$_2$ to a cytochrome or an iron–sulfur protein, either one electron carrier molecule must be reduced twice consecutively, accepting one electron at a time, or two electron carrier molecules must function in concert and be reduced simultaneously, with each accepting a single electron. Researchers have not yet definitively established which of these two mechanisms applies in each case.

The capacity of flavoproteins and ubiquinones to function as electron carriers in *either* one-electron or two-electron transfer reactions is of crucial importance for operation of the electron transport system. Because of this flexibility, flavoproteins and ubiquinones function as electron conduits between the two-electron donor, NADH, and the one-electron acceptors, cytochromes and iron–sulfur proteins.

12.3. ELECTRON TRANSPORT SYSTEM (ETS)

We are now ready to link together the reactions of the various electron carriers and construct the mitochondrial electron transport system. In this system, electrons flow from metabolite to molecular oxygen, which is reduced to water. Electron flow leads to pumping of protons across the inner mitochondrial membrane. Proton pumping occurs via three complexes located in the membrane and results in establishment of a proton gradient that provides the energy for ATP synthesis.

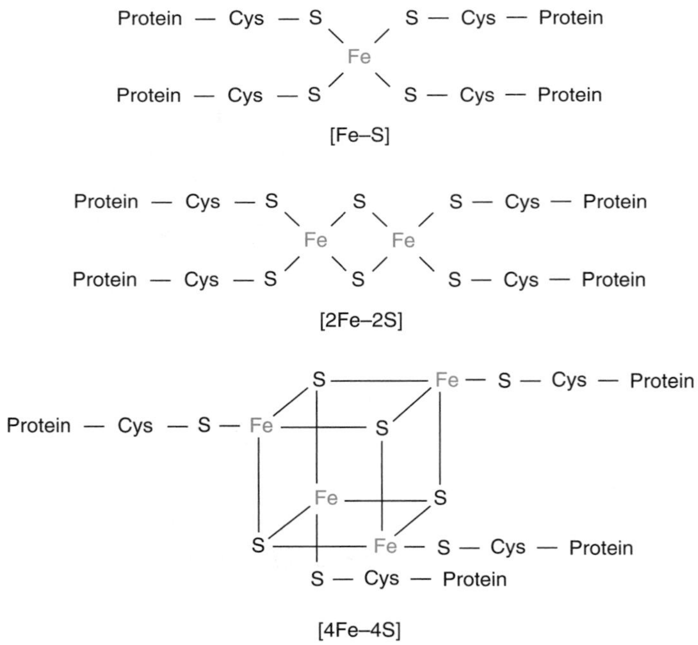

Figure 12.5. Common forms of iron-sulfur clusters in nonheme iron proteins.

12.3.1. Chain of Electron Carriers

12.3.1A. From Metabolite to Cytochrome c.

The initial step in the sequence involves the oxidation of a substrate of a pyridine-linked dehydrogenase. Oxidation of isocitrate, the substrate of isocitrate dehydrogenase and a citric acid cycle intermediate, represents a typical example.

Let us designate the general substrate as some metabolite, MH_2, from which two hydrogens can be removed. Oxidation of MH_2 is *coupled* to the reduction of NAD^+ to NADH:

$$MH_2 + NAD^+ \rightleftharpoons M + NADH + H^+$$

As noted earlier, the cell cannot tolerate extensive depletion of NAD^+. If NAD^+ is not regenerated from NADH, operation of both glycolysis and the citric acid cycle will come to a halt. NAD^+ regeneration is accomplished by coupling the oxidation of NADH to the reduction of a flavoprotein; specifically, flavin mononucleotide (FMN) is reduced to $FMNH_2$ as NADH is oxidized to NAD^+:

$$NADH + H^+ + FMN \rightleftharpoons NAD^+ + FMNH_2$$

Much as the cell cannot tolerate depletion of NAD^+, it can also not tolerate depletion of FMN. To regenerate FMN, electrons flow from $FMNH_2$ to an iron–sulfur protein; the oxidation of $FMNH_2$ is coupled to reduction of an iron–sulfur protein. Because full oxidation of $FMNH_2$ requires removal of two electrons, one iron–sulfur protein must be reduced twice consecutively or two iron–sulfur proteins must be reduced simultaneously:

$$FMNH_2 + 2Fe^{3+}-S \rightleftharpoons FMN + 2Fe^{2+}-S + 2H^+$$

The reduced iron–sulfur protein is reoxidized by transferring its electrons to coenzyme Q. Full reduction of coenzyme Q requires the transfer of two electrons:

$$2Fe^{2+}-S + CoQ + 2H^+ \rightleftharpoons 2Fe^{3+}-S + CoQH_2$$

Following the reduction of coenzyme Q, the remaining electron carriers—another iron–sulfur protein and several cytochromes—can participate only in one-electron transfers. The first electron carrier is a cytochrome, designated cytochrome *b*, that oxidizes $CoQH_2$ back to CoQ. In the process, the ferric iron (Fe^{3+}) of cytochrome *b* is reduced to the ferrous state (Fe^{2+}):

$$CoQH_2 + 2 \text{ Cyt } b (Fe^{3+}) \rightleftharpoons CoQ + 2 \text{ Cyt } b (Fe^{2+}) + 2H^+$$

Reduced cytochrome *b* is oxidized by means of an iron–sulfur protein:

$$2 \text{ Cyt } b (Fe^{2+}) + 2Fe^{3+}-S \rightleftharpoons 2 \text{ Cyt } b (Fe^{3+}) + 2Fe^{2+}-S$$

From the iron–sulfur protein, the electrons flow through a series of cytochromes. The first cytochrome re-oxidizes the iron–sulfur protein and is itself reduced. The reduced cytochrome is then reoxidized by means of another cytochrome that follows it in the series and is reduced in the process. In principle, this reduction and reoxidation of cytochromes could go on until all the potential free energy of the transfer process is expended. Instead, the chain of cytochromes and the entire ETS is brought to a useful termination point by a significant change in the mechanism.

12.3.1B. Cytochrome Oxidase.

The change comes in the form of **cytochrome oxidase**, the last cytochrome in the series. Like all oxidases, cytochrome oxidase catalyzes the *direct combination of a substrate with molecular oxygen*. The cytochrome oxidase step represents one of the few, but critical, biochemical redox reactions in which oxidation occurs by addition of oxygen rather than by removal of hydrogen. Cytochrome oxidase is a multienzyme complex. Mammalian cytochrome oxidase consists of 10 subunits and contains two different cytochromes, *a* and a_3. The enzyme complex also has two essential copper ions (designated Cu_A and Cu_B) that can alternate between +1 and +2 states as they participate in the redox reactions. One copper ion is associated with each heme. Heme *a* is close to Cu_A in subunit II, and heme a_3 is close to Cu_B in subunit I.

The oxidized form of cytochrome oxidase is regenerated from the reduced enzyme by reaction with molecular oxygen. In the process, molecular oxygen is reduced to water. We can formulate the final step of the ETS as follows:

$$2 \text{ Cyt oxidase } (Fe^{2+}) + \tfrac{1}{2}O_2 + 2H^+ \rightarrow$$
$$2 \text{ Cyt oxidase } (Fe^{3+}) + H_2O \qquad (12.5)$$

By being able to use molecular oxygen, an *external source of oxidizing power*, the chain of electron carriers comes to an end. As in all previous steps, the oxidizing electron carrier is reduced. However, in this step, as distinct from all previous steps, depletion of the oxidized form of the final electron carrier (O_2) and accumulation of its reduced form (H_2O) present no problem. There is no need to regenerate oxygen by oxidation of water since the supply of molecular oxygen is virtually unlimited. Likewise, accumulation of water can be tolerated since the water formed is readily removed by excretion.

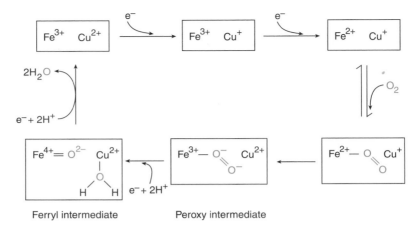

Ferryl intermediate   Peroxy intermediate

Figure 12.6.   Proposed cyclic mechanism for the four-electron reduction of O_2 as catalyzed by cytochrome oxidase at the (heme a_3–Cu_B) center of the enzyme. [Adapted, with permission, from M. Wikstrom and J. E. Morgan, *J. Biol. Chem.* 267:10266–10273 (1992).]

The cytochrome oxidase reaction, Eq. (12.5), constitutes the key reaction of respiration. In addition to providing a useful end to the series of redox steps, the reaction also requires $2H^+$. These protons are available from the oxidation of $CoQH_2$ to CoQ. Inclusion of these protons provides for complete material balance of the sequence. However, note that, although the reaction is stoichiometrically correct, the actual cytochrome oxidase reaction must involve one (not one-half) molecule of oxygen and is, therefore a *four-electron* process:

$$O_2 + 4H^+ + 4e^- \rightarrow 2H_2O$$
$$4 \text{ Cyt oxidase (Fe}^{2+}) + O_2 + 4H^+ \rightarrow$$
$$4 \text{ Cyt oxidase (Fe}^{3+}) + 2H_2O \qquad (12.5a)$$

Reduction of one molecule of oxygen by four electrons must proceed without formation of intermediates in which oxygen is incompletely reduced. Such oxygen forms (see Section 12.5) are extremely toxic to living systems. Figure 12.6 shows the proposed mechanism by which cytochrome oxidase accomplishes its task. Reduced cytochrome c donates its electron to the heme a–Cu_A center of cytochrome oxidase. An electron is then transferred to the heme a_3–Cu_B center of the enzyme, where O_2 is reduced in a cyclic series of steps to two molecules of water.

The cycle begins with the heme a_3–Cu_B center in its fully oxidized state (Fe^{3+}–Cu^{2+}). The first electron relayed from reduced cytochrome c reduces Cu^{2+} to Cu^+. The second electron relayed reduces Fe^{3+} to Fe^{2+}, producing a fully reduced iron–copper center (Fe^{2+}–Cu^+). An oxygen molecule now binds to the iron of the center and abstracts an electron from both center ions to form a *peroxy* intermediate. The peroxide is cleaved when a third

electron and two protons are taken up. At this point, one oxygen atom is bound as water to Cu^{2+} while the second oxygen atom is bound to iron in the *ferryl* state (Fe^{4+}). Input of a fourth electron and two more protons leads to release of oxygen in the form of two water molecules and to regeneration of the fully oxidized iron–copper center (Fe^{3+}–Cu^{2+}). The last two stages of the cycle involve proton pumping across the mitochondrial membrane. Four protons are pumped into the intermembrane space as two electrons flow through the oxidase.

The cytochrome oxidase reaction makes breathing an essential aspect of life for plants and animals. Inhaled oxygen is required to regenerate the oxidized form of cytochrome oxidase and thereby permit operation of the electron transport system. Because of this link to respiration, we also refer to the mitochondrial electron transport

Table 12.2. Reactions of the Electron Transport System[a]

$MH_2 + NAD^+ \rightleftharpoons M + NADH + H^+$
$NADH + H^+ + FMN \rightleftharpoons NAD^+ + FMNH_2$
$FMNH_2 + 2Fe^{3+}-S \rightleftharpoons FMN + 2Fe^{2+}-S + 2H^+$
$2Fe^{2+}-S + CoQ + 2H^+ \rightleftharpoons CoQH_2 + 2Fe^{3+}-S$
$CoQH_2 + 2 \text{ Cyt } b \text{ (Fe}^{3+}) \rightleftharpoons CoQ + 2 \text{ Cyt } b \text{ (Fe}^{2+}) + 2H^+$
$2 \text{ Cyt } b \text{ (Fe}^{2+}) + 2Fe^{3+}-S \rightleftharpoons 2 \text{ Cyt } b \text{ (Fe}^{3+}) + 2Fe^{2+}-S$
$2Fe^{2+}-S + 2 \text{ Cyt } c_1 \text{ (Fe}^{3+}) \rightleftharpoons 2Fe^{3+}-S + 2 \text{ Cyt } c_1 \text{ (Fe}^{2+})$
$2 \text{ Cyt } c_1 \text{ (Fe}^{2+}) + 2 \text{ Cyt } c \text{ (Fe}^{3+}) \rightleftharpoons 2 \text{ Cyt } c_1 \text{(Fe}^{3+}) + 2 \text{ Cyt } c \text{ (Fe}^{2+})$
$2 \text{ Cyt } c \text{ (Fe}^{2+}) + 2 \text{ Cyt } a \text{ (Fe}^{3+}) \rightleftharpoons 2 \text{ Cyt } c \text{ (Fe}^{3+}) + 2 \text{ Cyt } a \text{ (Fe}^{2+})$
$2 \text{ Cyt } a \text{ (Fe}^{2+}) + 2Cu^{2+} \rightleftharpoons 2 \text{ Cyt } a \text{ (Fe}^{3+}) + 2Cu^+$
$2Cu^+ + 2 \text{ Cyt } a_3 \text{ (Fe}^{3+}) \rightleftharpoons 2Cu^{2+} + 2 \text{ Cyt } a_3 \text{ (Fe}^{2+})$
$2 \text{ Cyt } a_3 \text{ (Fe}^{2+}) + \frac{1}{2}O_2 + 2H^+ \rightarrow 2 \text{ Cyt } a_3 \text{ (Fe}^{3+}) + H_2O$

Overall reaction: $MH_2 + \frac{1}{2}O_2 \rightarrow M + H_2O$
Actual reaction: $2MH_2 + O_2 \rightarrow 2M + 2H_2O$

[a]Fe–S, Iron–sulfur protein; Cyt, cytochrome.

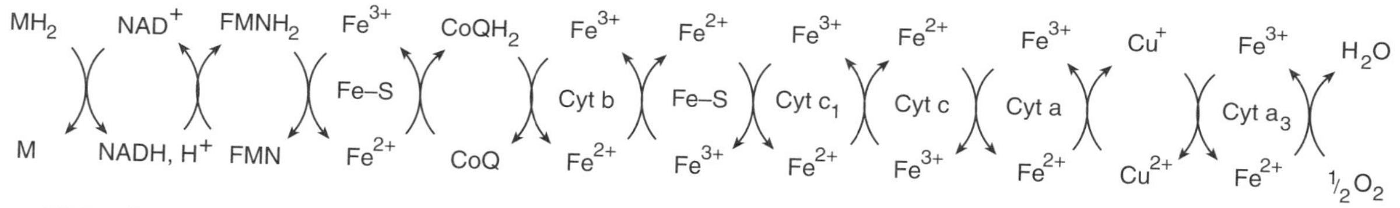

MH2
NAD+
FMNH2
Fe3+
CoQH2
Fe3+
Fe2+
Fe3+
Fe2+
Fe3+
Cu+
Fe3+
H2O
Fe–S
Cyt b
Fe–S
Cyt c1
Cyt c
Cyt a
Cyt a3
M
NADH, H+
FMN
Fe2+
CoQ
Fe2+
Fe3+
Fe2+
Fe3+
Fe2+
Cu2+
Fe2+
½O2

Figure 12.7. Sequence of the electron carriers in the mitochondrial electron transport system (ETS). Fe–S represents iron–sulfur proteins. See Figure 12.6 for the detailed mechanism of cytochrome oxidase.

system as the **respiratory chain.** Cyanide combines avidly with ferric iron (Fe^{3+}) of cytochrome oxidase, thus blocking the electron transport system and the associated synthesis of ATP via oxidative phosphorylation. This is why cyanide is such a strong poison. Carbon monoxide also inhibits cytochrome oxidase; it binds to the ferrous iron (Fe^{2+}) of the enzyme.

12.3.1C. Pathways of Electron Transport. Table 12.2 provides a summary of the entire sequence of reactions. The detailed mechanism corresponding to the last three reactions in the table was described above. Figure 12.7 schematically represents the coupled reactions by means of curved arrows.

Many metabolites are oxidized via the pathway shown in Figure 12.7. Succinate, however, is processed by means of a second pathway in which a flavoprotein (FAD) carries out the initial metabolite oxidation. Reduced $FADH_2$ is then oxidized, to regenerate FAD, by reaction with an iron–sulfur protein. The next electron carrier is a cytochrome *b,* different from that in Figure 12.7, which subsequently reduces coenzyme Q and thereby enters the major pathway. Figure 12.8 outlines the two pathways of electron transport. In the overall scheme, NAD^+ and CoQ serve as central electron-collecting compounds; they channel the electrons removed from metabolites through the remaining portion of the ETS. The four complexes shown in Figure 12.8 are discussed below.

12.3.2. Sequence of ETS Components

Researchers have elucidated the sequence of electron carriers in the ETS by means of three main experimental approaches. These involve (1) characterizing respiratory complexes, (2) using artificial electron acceptors, and (3) using inhibitors.

12.3.2A. Respiratory Complexes. Careful fragmentation of the inner mitochondrial membrane, site of the ETS, has yielded a number of macromolecular aggregates that have the capacity to carry out specific portions of the entire sequence. In particular, investigators have characterized four such functional assemblies, called **respiratory complexes** (Table 12.3). Each complex represents a multienzyme system that catalyzes a number of consecutive reactions. Electron flow through three of these transmembrane complexes (I, III, and IV) leads to proton pumping across the mitochondrial membrane (Figure 12.9). The resultant proton gradient drives ATP synthesis, but actual ATP synthesis occurs at the active site of the enzyme ATP synthase.

Analysis of each complex for its content of electron carriers and for the reactions catalyzed by them has helped to establish the sequence of ETS reactions. These studies assume the likelihood that electron carriers located close to each other participate in reactions that are closely linked mechanistically.

Complex I
MH2 → NAD+ → FMN → Fe–S → CoQ →
ADP +Pi
ATP
Complex III
Cyt b → Fe–S → Cyt c1 → Cyt c →
ADP +Pi
ATP
Complex IV
Cyt a → Cyt a3 → O2
ADP +Pi
ATP
Complex II
FAD → Fe–S → Cyt b
Succinate

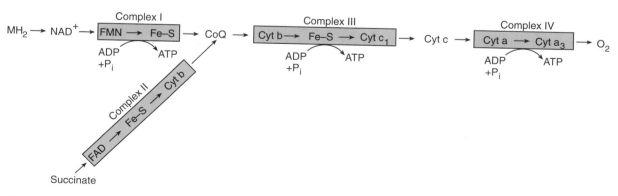

Figure 12.8. The two pathways of electron transport in mitochondria. Cytochrome *b* of Complex II is not identical to that of Complex III.

Table 12.3. Respiratory Complexes of the Electron Transport System

	Complex I	Complex II	Complex III	Complex IV
Name	NADH-CoQ reductase	Succinate-CoQ reductase[a]	Cytochrome reductase	Cytochrome oxidase
Reaction sequence	$NAD^+ \rightarrow CoQ$	Succinate $\rightarrow$ CoQ	$CoQ \rightarrow$ cytochrome c	Cytochrome c $\rightarrow O_2$
Molecular weight	850,000	127,000	280,000	200,000
No. of subunits	26	5	10	13
Fe–S protein	Yes	Yes	Yes	—
$\Delta E^{\circ\prime}$ (volts)	+0.37	+0.02	+0.20	+0.57
ATP synthesis	Yes	—	Yes	Yes

[a]Also known as *succinate dehydrogenase complex*.

12.3.2B. Artificial Electron Acceptors.

A second approach for determining the sequence of electron carriers is based on using **artificial electron acceptors.** An artificial electron acceptor is a compound that, like naturally occurring electron carriers, undergoes oxidation–reduction reactions as a function of its specific reduction potential. When such a compound is added to the ETS, it can be reduced by an electron carrier with a lower reduction potential. In the presence of an artificial electron acceptor, then, electrons passing through the ETS are *diverted* and used to reduce the added compound.

We can pinpoint the site at which this takes place. Siphoning off electrons from the ETS by the added compound results in fewer electrons being available for the remaining portion of the ETS. With fewer electrons, less reduction of electron carriers takes place in the remaining ETS segment. Hence, diversion of electrons leads to a *decrease in the concentration of reduced forms of electron carriers subsequent to the point of electron diversion* (Figure 12.10).

Put somewhat differently, the ratio of the oxidized to the reduced form increases for carriers subsequent to the point of electron diversion. Changes in these ratios can be detected spectrophotometrically, since reduced and oxidized forms of many electron carriers have different absorption properties (see Figure 12.4). By using a variety of artificial electron acceptors, we can elucidate the sequence of electron carriers much as we can elucidate a metabolic pathway by using mutants (Section 8.3). Figure 12.11 indicates the points of action of a number of artificial electron acceptors.

12.3.2C. Inhibitors.

The third approach for determining the sequence of carriers in the ETS involves using inhibitors that *block* specific steps in the sequence (Figure 12.11). The effect of adding an inhibitor is identi-

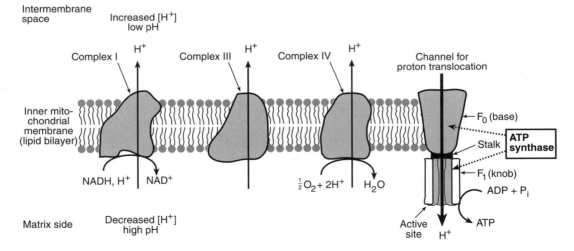

Figure 12.9. Localization of the ETS in the inner mitochondrial membrane. Three respiratory complexes pump protons from the membrane's matrix side to the intermembrane space. The resultant proton gradient drives ATP synthesis, catalyzed by ATP synthase. The enzyme, a transmembrane protein, has an active site on the matrix side and contains a channel through which protons move. Proton translocation is coupled to ATP synthesis. [Adapted from D. R. Ort and N. E. Good, *Trends Biochem. Sci.* 13:467–469 (1988) with kind permission of Elsevier Science-NL, Sara Burgerhartstraat 25, 1055 KV Amsterdam, The Netherlands.]

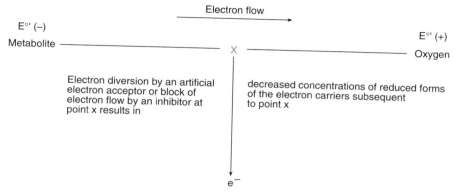

Figure 12.10.   Principle of action of artificial electron acceptors and inhibitors when used with the ETS.

cal to that of adding an artificial electron acceptor. An inhibitor also leads to a *decrease in the concentration of reduced forms of electron carriers subsequent to the point of inhibition.* Therefore, points of action of inhibitors can likewise be pinpointed by spectrophotometric measurements of the electron carriers.

12.3.2D. Reduction Potentials of Electron Carriers.

We generally discuss the sequence of electron carriers in terms of the carriers' biochemical standard reduction potentials ($E^{\circ\prime}$). Table 12.4 lists these, beginning with isocitrate as the metabolite undergoing oxidation to α-ketoglutarate:

$$\text{Isocitrate}^{3-} + NAD^+ \rightarrow \alpha\text{-ketoglutarate}^{2-} + NADH + CO_2$$

In the ETS, electrons flow from a metabolite, the initial electron donor, to oxygen, the terminal electron acceptor. In this flow, each carrier must be a stronger reducing agent than the one following it in the chain; electron flow must be from a smaller to a larger *biochemical actual reduction potential (E')*. As you can see from Table 12.4, the sequence of electron carriers is *essentially* in agreement with that predicted on the basis of their *biochemical standard reduction potentials (E^{\circ\prime})*.

The reason for the agreement between carrier sequence based on E' and that based on $E^{\circ\prime}$ derives from the intracellular concentrations of electron carriers. For most of the electron carriers, the concentrations of their oxidized and reduced forms are similar so that $E^{\circ\prime} \approx E'$ (see Eq. 12.2). Remember, of course, that *what really determine the sequential arrangement of the electron carriers in the ETS are their biochemical actual reduction potentials (E')*, the potentials existing under intracellular conditions of reactant and product concentrations. These must indeed progress consistently from the smaller to the larger potential.

Note what the above tells you about the efficiency of the ETS. An electron carrier has *maximum capacity* for accommodating *either* loss of electrons (oxidation) or addition of electrons (reduction) when the concentration of its reduced form equals that of its oxidized form. Therefore, the fact that $E^{\circ\prime} \approx E'$ means that the ETS operates rather efficiently.

We encountered an analogous situation when we considered the efficiency of a buffer (Section 1.3). Recall that a buffer has *maximum capacity* for addition of *either* acid or base when the concentration of its dissociated form equals that of the undissociated form: $[A^-] = [HA]$.

12.3.3. Energetics of the ETS

Electrons flow through the chain of ETS carriers much as electrons flow through a wire; the flow is a function of the potential drop. In the example of Table 12.4, the total po-

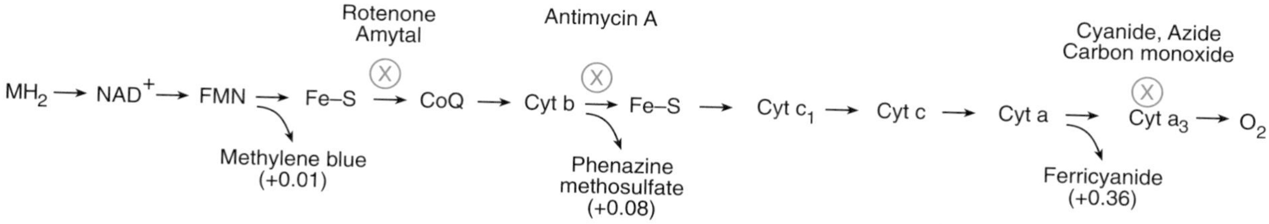

Figure 12.11.   Points of action in the ETS of some specific artificial electron acceptors ($E^{\circ\prime}$ values in parentheses) and inhibitors ($\times$).

Table 12.4. Reduction Potentials of Electron Carriers

Redox pair[a]	$E^{\circ\prime}$ (volts)
α-Ketoglutarate/isocitrate	−0.38
NAD^+/NADH	−0.32
FMN/$FMNH_2$ (enzyme bound)	−0.30
Fe^{3+}–S/Fe^{2+}–S (average)	−0.24
CoQ/$CoQH_2$	+0.05
Cyt b (Fe^{3+})/Cyt b (Fe^{2+})	+0.07
Fe^{3+}–S/Fe^{2+}–S	+0.28
Cyt c_1 (Fe^{3+})/Cyt c_1 (Fe^{2+})	+0.22
Cyt c (Fe^{3+})/Cyt c (Fe^{2+})	+0.25
Cyt a (Fe^{3+})/Cyt a (Fe^{2+})	+0.29
Cu^{2+}/Cu^+ (average)	+0.29
Cyt a_3 (Fe^{3+})/Cyt a_3 (Fe^{2+})	+0.39
$\frac{1}{2}O_2$/H_2O	+0.82

Electron flow ↓

[a]Fe–S, Iron–sulfur protein; Cyt, cytochrome.

tential drop from isocitrate to oxygen equals $+0.82 - (-0.38) = +1.20$ V. This difference in potential is equivalent to a change in free energy (Eq. 12.3). Since $\Delta E^{\circ\prime}$ is positive, $\Delta G^{\circ\prime}$ is negative. The entire ETS, therefore, represents an exergonic set of reactions accompanied by the release of free energy. Remember, though, that intracellularly $\Delta E^\prime$ must be positive in order for $\Delta G^\prime$ to be negative.

12.3.3A. ATP Synthesis.
Free energy released during the operation of the electron transport system drives the synthesis of ATP. We call the combined reactions *oxidative phosphorylation.* Linking the exergonic ETS with the endergonic synthesis of ATP constitutes a special case of *energetically coupled reactions.* ATP synthesis from ADP and P_i requires an input of $+30.5$ kJ mol^{-1} (Table 9.2). In coupling ATP synthesis to the electron transport system, the ETS must generate 30.5 kJ mol^{-1} for every mole of ATP synthesized. You can com-

pute the necessary potential drop, equivalent to 30.5 kJ mol^{-1}, from Eq. (12.3):

$$\Delta G^{\circ\prime} = -n\mathrm{F}\Delta E^{\circ\prime}$$

so that

$$-30,500 = -(2)(96,491)\Delta E^{\circ\prime}$$

and

$$\Delta E^{\circ\prime} = +0.16 \text{ V}$$

Thus, a minimum potential drop of 0.16 V is required for synthesis of one mole (molecule) of ATP per mole (molecule) of metabolite oxidized. This requirement is met at the three respiratory complexes where proton pumping and ATP synthesis occur during oxidative phosphorylation (Figure 12.12). At each complex, the potential drop exceeds 0.16 V.

Because it takes 0.16 V to synthesize one mole of ATP, and because the total potential drop in the above sequence is 1.20 V, there could have occurred, theoretically, a synthesis of $1.20/0.16 \approx 8$ moles of ATP per mole of metabolite oxidized. This calculation assumes a 100% efficiency of energy conservation, an unlikely occurrence, especially in such a complex biological system. The experimentally determined maximum yield is 3 moles of ATP per mole of metabolite (that is, per mole NADH oxidized). Thus, of the 1.2 V available, only $3 \times 0.16 = 0.48$ V serve for ATP synthesis. Hence, we can calculate the efficiency of ATP synthesis as

$$\frac{3 \times 0.16 \text{ Volts}}{1.20 \text{ Volts}} \times 100 = 40\%$$

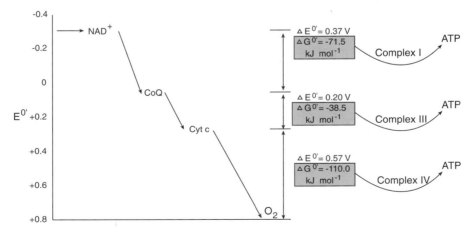

Figure 12.12.   Schematic diagram of the energetics of the ETS. Potential drops at Complexes I, III, and IV generate sufficient energy for ATP synthesis. Actual ATP synthesis occurs at the active site of ATP synthase and is driven by the proton gradient produced by the respiratory complexes.

12.3.3B. Energy Conservation.

The ETS, with its coupled ATP synthesis, illustrates an important principle of energy conservation. In order for ATP synthesis to be efficient, an ideal exergonic pathway should consist of a large number of steps, with each step releasing just enough energy to synthesize one mole of ATP. In the idealized case of isocitrate oxidation, the ETS should consist of eight steps, with each step having a potential drop of 0.16 V.

Two reasons account for constructing an idealized system in this fashion. First, larger potential drops, but not large enough to yield another mole of ATP, result in wasted energy. Assume that the potential drop of a step were 0.20 V, enough energy to synthesize one mole, or one molecule, of ATP (0.16 V) but not enough to synthesize two molecules. Because a fraction of a molecule cannot be synthesized, the excess energy (0.04 V) would be dissipated.

Second, multiples of the required potential drop, in principle available for the synthesis of several moles of ATP, likewise result in wasted energy. Consider a step that has a potential drop of 0.48 V, enough energy to synthesize three moles, or three molecules, of ATP (0.48/0.16 = 3). In order for this to occur, it would be necessary to have the more or less simultaneous interaction of a minimum of 11 entities (three molecules of ADP, three molecules of P_i, three protons, and two molecules of electron carriers). Such multiple molecular collisions are extremely unlikely. In practice, a step with a large potential drop results in synthesis of only one molecule of ATP while the excess energy is dissipated.

The ETS scheme approaches the ideal case in that it consists of a relatively large number of steps, with each step having only a small change in potential and free energy. For identical reasons, other metabolic pathways associated with the synthesis or utilization of ATP typically consist of many steps, with each step involving a relatively small change in free energy.

Keep in mind, however, that we made all of the above calculations using *biochemical standard reduction potentials (E°′)* of the carriers. Our conclusions are, therefore, only tentative. Correct evaluations of the energetics must involve the use of *biochemical actual reduction potentials (E′)*.

12.4. OXIDATIVE PHOSPHORYLATION

12.4.1. P/O Ratio

Oxidative phosphorylation, *synthesis of ATP coupled to operation of the electron transport system,* constitutes the major mechanism for ATP synthesis in all nonphotosynthetic organisms. Coupling of these two processes was discovered early on in the development of biochemistry when researchers found that uptake of oxygen could be correlated with ATP synthesis. Manometric measurements of tissue homogenates and tissue slices show that, as oxygen is taken up, the amount of ATP increases concomitantly. You can assess ATP synthesis by measuring ATP's incorporation of radioactively labeled inorganic phosphate. Under optimal conditions, three *molecules* of phosphate are taken up—that is, three molecules of ATP are synthesized—for every one *atom* of oxygen consumed. We can characterize these results by a *P/O ratio* (or a *P/2e⁻* ratio) of 3. We now know that ATP synthesis is coupled to proton pumping of three respiratory complexes (Figure 12.9).

P/O ratios may have values less than 3 depending upon the point at which metabolite oxidation links up with the ETS. The succinate pathway, for example, connects with the main ETS sequence at the level of CoQ. This means that only Complexes III and IV are available for proton pumping and ATP synthesis. These complexes span a smaller potential drop than the complete ETS, which includes complex I. Under these conditions, only two molecules of ATP are synthesized per atom of oxygen consumed (that is, per $FADH_2$ oxidized), and the P/O ratio equals 2.

A metabolite whose oxidation is linked to the ETS after respiratory complex III yields a P/O ratio of 1. Ascorbate can donate electrons to tetramethyl-*p*-phenylenediamine (TMPD), which can reduce cytochrome *c* directly. Thus, addition of ascorbic acid and TMPD to an electron transport system results in a P/O ratio of 1.

12.4.2. Chemiosmotic Coupling

Coupling of ATP synthesis to the operation of the electron transport system has been the subject of intensive research. The earliest proposal, called the *chemical coupling hypothesis,* suggested a classical mechanism of energetically coupled reactions. According to this hypothesis, operation of the ETS results in formation of covalent *energy-rich compounds* that function as *common intermediates* to drive the synthesis of ATP.

A subsequent proposal, termed the *conformational coupling hypothesis,* postulated that operation of the ETS results in an energized conformation of one or more proteins of the inner mitochondrial membrane. When the membrane proteins return to their low-energy conformation, the energy released energizes the enzyme catalyzing ATP synthesis.

Investigators attempted for many years to isolate the energy-rich intermediates postulated by these hypotheses but were unsuccessful. A completely different mechanism was proposed by Peter Mitchell in 1961 (Nobel Prize,

1978). Mitchell's proposal, termed the **chemiosmotic coupling hypothesis,** has received extensive experimental support and constitutes the currently accepted mechanism of oxidative phosphorylation.

According to this hypothesis, the driving force for ATP synthesis consists of an *electrochemical gradient* that comprises both a pH and an electrical potential component. *Pumping* of protons across the inner mitochondrial membrane by the respiratory complexes sets up a pH gradient across the membrane; the matrix side becomes more basic, and the intermembrane space becomes more acidic. Complexes I, III, and IV contribute to proton gradient formation (Figure 12.13). In Complex III, a molecule of coenzyme Q is involved twice in proton translocation. The resultant mechanism, called the **Q-cycle,** involves the semiquinone form of coenzyme Q. The Q-cycle results in the transport of $4H^+$ across the mitochondrial membrane and the transfer of a pair of electrons from $CoQH_2$ to cytochrome *c*.

Complex II does not contribute to the proton gradient but serves to channel electrons into the ETS at the level of coenzyme Q (Figure 12.14).

Movement of the positive charges across the membrane also generates an electrical potential. The outer (in-termembrane space) side becomes positive relative to the inner (matrix) side, developing a potential of about 0.15 V. The concentration gradient (ΔpH) produces an electrical potential across the membrane. The combined effects of the difference in pH and the difference in potential constitute an electrochemical gradient. Note that the mitochondrial membrane is impermeable to ions such as H^+, OH^-, K^+, and Cl^- whose free diffusion would discharge the potential of such a gradient. We can describe the electrochemical gradient by the equation:

$$\Delta\mu_H = \Delta\psi - 2.303RT\Delta pH/F$$

where $\Delta\mu_H$ is the electrochemical proton potential, $\Delta\psi$ is the membrane potential (potential gradient), the ΔpH term is the proton potential (pH gradient), and F is the Faraday constant.

According to the chemiosmotic hypothesis, dissipation of the electrochemical gradient drives ATP synthesis. Recall that ATP synthesis from ADP requires an input of protons (Figure 9.4). The pH gradient furnishes these protons. As the protons move back across the inner mitochondrial membrane, both the pH gradient and the poten-

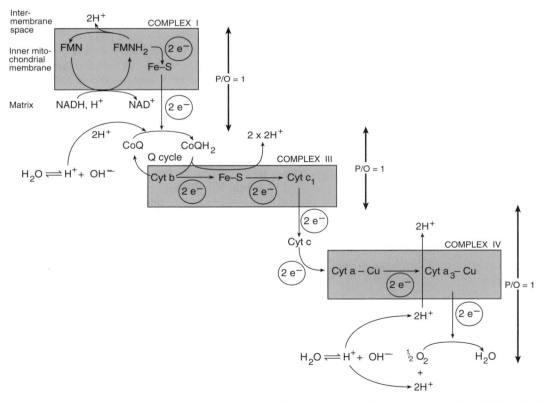

Figure 12.13.   Proposed electron flow through Complexes I, III, and IV. Electrons pass from NADH to coenzyme Q via FMN and Fe-S, then via the Q-cycle to cytochrome *b* and Fe-S, and finally through a series of other cytochromes to oxygen. ②e denotes two one-electron transfers.

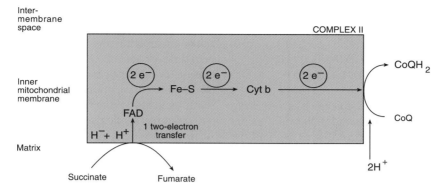

Figure 12.14.   Proposed electron flow through Complex II. (2e) denotes two one-electron transfers.

tial difference across the membrane are being lost. The entire electrochemical gradient is being dissipated at the expense of ATP synthesis. Thus, the gradient serves as the *common intermediate* for the coupled processes of electron transport and ATP synthesis.

12.4.3. ATP Synthase

Actual ATP synthesis occurs at the active site of **ATP synthase,** a complex protein oligomer, located in the inner mitochondrial membrane. Because it consists of two major parts (F_0 and F_1) and serves to couple the ETS and ATP synthesis, the synthase is known as $\mathbf{F_0F_1}$**-coupling factor.** Lastly, because the synthase, like all enzymes, catalyzes both the forward and the reverse reaction, we also refer to the enzyme complex as *ATPase, F_0F_1-ATPase,* or *proton-pumping ATPase.*

The enzyme complex has the appearance of a knob attached to a base by means of a stalk (see Figure 12.9). The base (F_0) is embedded in and spans the membrane; the knob (F_1) protrudes into the matrix. F_0 is a water-insoluble transmembrane protein that consists of four different polypeptide chains and contains a channel for proton translocation. F_1 is a water-soluble peripheral membrane protein. It consists of five different polypeptide chains in the ratio of $\alpha_3\beta_3\gamma\delta\epsilon$ and contains the active site. The stalk consists of two proteins.

Total molecular weight of the F_0F_1-ATPase is 450,000. *Oligomycin,* an antibiotic produced by *Streptomyces,* is an inhibitor of the enzyme. The antibiotic binds to a subunit of F_0, thereby interfering with proton transport through F_0. Investigators believe that proton movement through the channel in F_0 drives ATP synthesis by ATP synthase. Exactly how the enzyme accomplishes the synthesis is not yet clear. It appears that the mechanism does not consist of merely shifting the equilibrium toward ATP synthesis (Figure 9.4). Instead, researchers have proposed that the flow of protons triggers a conformational

change in the enzyme, possibly by means of cooperative interactions among multiple identical active sites.

12.5. CONTROL MECHANISMS

12.5.1. Respiratory Control

Under most physiological conditions, synthesis of ATP is *tightly coupled* to respiration and to operation of the ETS. ATP synthesis is absolutely dependent upon the flow of electrons, and electrons do not normally flow through the ETS unless ADP is being phosphorylated to ATP (see Section 12.5.3 for an exception). We call this regulatory phenomenon **respiratory control,** and its essence lies in the intracellular level of ADP. Active, energy-consuming cells use up ATP and accumulate ADP. The resultant high level of ADP stimulates respiration and enhances the activity of the ETS. Conversely, in resting and well-nourished cells, ATP accumulates at the expense of ADP. Depletion of ADP limits respiration and decreases the activity of the ETS.

Respiratory control constitutes one of the major overall control mechanisms in metabolism. It represents the link between cellular requirements for ATP and the rate at which foodstuffs are oxidized via the ETS.

12.5.2. Energy Charge

Both hydrolysis and synthesis of ATP involve, in addition to ATP, either ADP or AMP. These three adenine nucleotides are interconvertible by means of **adenylate kinase,** which catalyzes a *disproportionation reaction* in which a single reactant yields two different products:

$$2ADP \rightleftarrows ATP + AMP$$

A cell's capacity to carry out ATP-driven reactions depends on the relative concentrations of AMP, ADP, and ATP. A mathematical measure of this capacity was proposed by Daniel Atkinson and is called the **energy charge:**

$$\text{Energy charge} = \frac{[\text{ATP}] + \frac{1}{2}[\text{ADP}]}{[\text{ATP}] + [\text{ADP}] + [\text{AMP}]} \quad (12.6)$$

The numerator of this expression consists of the two energy-rich forms of adenine nucleotides. We multiply the ADP concentration by 0.5 because ADP has one-half the number of energy-rich bonds present in ATP. The denominator represents the total concentration of all adenine nucleotides. Values of the energy charge vary from 0 (all AMP) to 1.0 (all ATP). Most normal cells operate in an energy charge range of 0.8–0.9.

The energy charge exerts its control in metabolism through allosteric regulation of specific enzymes by AMP, ADP, and ATP. You saw previously that high concentrations of ATP inhibit energy-generating pathways such as glycolysis and the citric acid cycle. High concentrations of ADP or AMP, on the other hand, tend to stimulate such pathways. Thus, a large energy charge reflects high intracellular levels of ATP and inhibits ATP-generating pathways. A small energy charge indicates high intracellular levels of ADP and AMP and stimulates ATP-generating pathways.

12.5.3. Uncouplers of Oxidative Phosphorylation

A number of compounds are known that can disconnect electron transport from ATP synthesis. Such **uncouplers** permit electron transport to proceed but prevent ATP synthesis. Figure 12.15 shows the structure of two uncouplers, *2,4-dinitrophenol* and carbonylcyanide-*p*-trifluoromethoxyphenylhydrazone. You can understand their mechanism of action on the basis of chemiosmotic coupling. Both uncouplers are nonpolar and therefore readily pass through the inner mitochondrial membrane. They are

2,4-Dinitrophenol

Carbonylcyanide-*p*-trifluoromethoxyphenylhydrazone

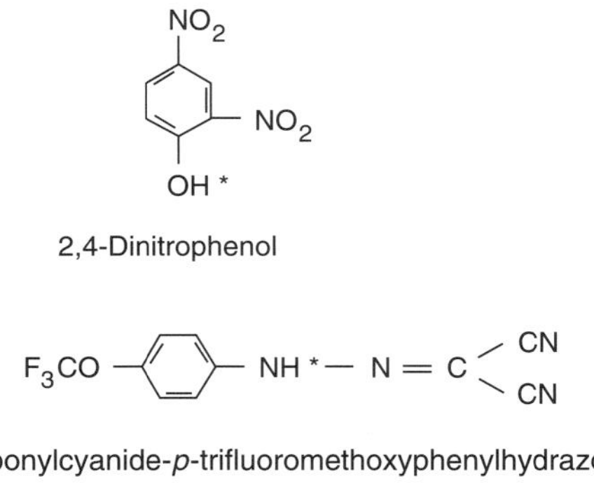

Figure 12.15.   Some uncouplers of oxidative phosphorylation. The asterisk designates an ionizable proton.

also weak acids and exist as anions at physiological pH. The anionic form of the uncoupler binds protons on the acidic (intermembrane space) side of the membrane. It then diffuses across the membrane and releases the protons on the alkaline (matrix) side. In so doing, the uncoupler leads to dissipation of the electrochemical gradient, resulting in loss of the capacity to synthesize ATP.

Some antibiotics, such as valinomycin and gramicidin A (see Figures 6.34 and 6.35), also function as uncouplers. You can explain their action by chemiosmotic coupling as well. Recall that these antibiotics act as *ionophores* and mediate transport of cations across the mitochondrial membrane. Movement of cations across the membrane diminishes the transmembrane potential. Movement of protons decreases the pH gradient as well. One or both of these factors lead to dissipation of the electrochemical gradient and a resultant loss of the capacity to synthesize ATP.

Some uncoupling of oxidative phosphorylation occurs naturally in newborn mammals that lack hair (for example, humans), hibernating animals, and those adapted to cold weather. In these instances, uncoupling of ATP synthesis is desirable because dissipation of the electrochemical gradient generated by the ETS leads to heat production. The process occurs under hormonal control and in special tissues, called *brown fat* or *brown adipose tissue,* located in the neck and upper back. The name derives from the fact that the tissues are rich in cytochromes. Mitochondria of brown fat have become specialized to generate heat from the oxidation of fatty acids.

12.5.4. Incomplete Reduction of Oxygen

As you saw, the complete reduction of oxygen by cytochrome oxidase requires four electrons (Eq. 12.5a). However, other reactions of oxidative metabolism frequently produce partially reduced forms of oxygen. Any source of electrons, such as the thiol group of cysteine or the reduced form of vitamin C, can readily reduce oxygen to form oxygen radicals. One-, two-, and three-electron reductions of O_2 yield the *superoxide anion* $(O_2 \cdot^-)$, *hydrogen peroxide* (H_2O_2), and the *hydroxyl radical* $(OH\cdot)$, respectively:

$$O_2 + e^- \rightarrow O_2 \cdot^-$$
$$O_2 + 2e^- + 2H^+ \rightarrow H_2O_2$$
$$O_2 + 3e^- + 3H^+ \rightarrow H_2O + OH\cdot$$

Incompletely reduced forms of oxygen are extremely reactive; they constitute powerful oxidizing agents that possess high toxicity for living systems. The hydroxyl radical, in particular, represents the most potent oxidizing agent known and the most active mutagen produced by ionizing radiation.

These highly active forms of oxygen must be converted to less reactive ones if the organism is to survive.

Several major self-defense mechanisms protect aerobic cells from the ravages of incompletely reduced oxygen. These involve the enzymes *superoxide dismutase, catalase,* and *peroxidase.*

The primary mode for detoxifying the superoxide anion involves its conversion to hydrogen peroxide by action of superoxide dismutase. This enzyme catalyzes a *dismutation reaction*—a reaction in which two identical substrates have different fates. In the superoxide dismutase reaction, one superoxide anion undergoes oxidation while the other undergoes reduction:

$$O_2 \cdot^- + O_2 \cdot^- + 2H^+ \rightarrow H_2O_2 + O_2$$

The superoxide anion can also give rise to hydrogen peroxide via a second pathway. Protonation of $O_2 \cdot^-$ yields the *hydroperoxyl radical,* $HO_2 \cdot$, the conjugate acid of the superoxide anion:

$$O_2 \cdot^- + H^+ \rightarrow HO_2 \cdot$$

Two hydroperoxyl radicals combine spontaneously to form hydrogen peroxide:

$$HO_2 \cdot + HO_2 \cdot \rightarrow H_2O_2 + O_2$$

Hydrogen peroxide formed from the superoxide anion, or produced by other metabolic reactions, is detoxified enzymatically. Catalase catalyzes the decomposition of hydrogen peroxide without requiring a second substrate

$$2H_2O_2 \rightarrow 2H_2O + O_2$$

while peroxidase destroys hydrogen peroxide with the aid of an added electron donor (AH_2)

$$AH_2 + H_2O_2 \rightarrow 2H_2O + A$$

Glutathione peroxidase represents an important peroxidase and one of a small number of enzymes that contain selenium. The enzyme occurs in erythrocytes, where it catalyzes the decomposition of H_2O_2 coupled to the oxidation of glutathione (GSH):

$$2GSH + H_2O_2 \rightarrow GSSG + 2H_2O$$

Glutathione peroxidase performs an essential function in protecting erythrocytes against the accumulation of peroxides. The enzyme glutathione reductase subsequently catalyzes the regeneration of GSH from GSSG.

The major mechanism for formation of the hydroxyl radical requires both a superoxide anion and hydrogen peroxide:

$$O_2 \cdot^- + H_2O_2 \rightarrow OH \cdot + OH^- + O_2$$

Accordingly, effective scavenging of $O_2 \cdot^-$ and H_2O_2 not only removes these two harmful oxidizing agents, but also prevents their giving rise to the even more dangerous hydroxyl radical. However, in the absence of a 100% scavenging efficiency, and as a result of other reactions, some hydroxyl radicals may form in living systems. In that event, they are likely to participate in three main reactions:

—oxidation of metal ions (M) to higher oxidation states:

$$OH \cdot + M^{n+} \xrightarrow{\quad} (M\ OH)^{n+} \xrightarrow{H^+} M^{n+1} + H_2O$$

—abstraction of a hydrogen from a C-H bond to produce water and an organic radical

—addition to a double bond to form secondary radicals

Because of the hazards that incompletely reduced forms of oxygen pose for humans, they have been implicated by some in the development of cancer and other ailments. Consequently, the claim has been made that dietary supplements of antioxidant vitamins (A, C, and E) are beneficial in the battle against these diseases.

12.6. BALANCE SHEET OF CARBOHYDRATE CATABOLISM

Having covered the electron transport system, the final stage of aerobic metabolism, we can now evaluate the overall energetics of carbohydrate catabolism. To do so, we need to consider the combined operation of glycolysis, the citric acid cycle, and the electron transport system.

The maximum energy yield from glycolysis, under anaerobic conditions, is a net of 2 ATP formed per molecule of glucose catabolized, for the sequence glucose to pyruvate or lactate (Section 10.3). The maximum energy yield from the citric acid cycle/ETS, functioning aerobically, is 12 ATP formed per molecule of acetyl CoA catabolized (Section 11.5). A net energy yield of *2 ATP* from the catabolism of a *six-carbon compound* in glycolysis is much smaller than a yield of *12 ATP* from the catabolism of a *two-carbon compound* in the citric acid cycle. As pointed out earlier, the reason for the difference in energy yield lies in the extent to which the metabolite undergoes oxidation. Conversion of glucose to pyruvate or lactate represents a small degree of oxidation, but conversion of the acetyl group of acetyl CoA to CO_2 and H_2O represents complete oxidation.

Now consider the total energy yield of carbohydrate

catabolism under optimal circumstances in which glycolysis proceeds under aerobic conditions. Pyruvate, produced by glycolysis, forms acetyl CoA. Catabolism of acetyl CoA via the citric acid cycle and its link to the ETS results in ATP synthesis. Additionally, NADH produced in glycolysis is oxidized via the electron transport system and also leads to production of ATP. Combined operation of glycolysis, the citric acid cycle, and the ETS represents full catabolism of carbohydrates and results in *complete oxidation of glucose to CO_2 and H_2O.*

12.6.1. Theoretical Energy Yields

Pyruvate, produced in the cytosol, represents the end product of glycolysis under aerobic conditions. Catabolism of glucose to pyruvate yields a net of 2 ATP. Pyruvate is transported across the mitochondrial membrane and converted *inside the mitochondria* to acetyl CoA by the pyruvate dehydrogenase complex. Acetyl CoA enters the citric acid cycle, located *inside the mitochondria,* and leads to production of 12 ATP/acetyl CoA. Conversion of pyruvate to acetyl CoA by the pyruvate dehydrogenase complex (Section 11.2) results in formation of NADH, which is oxidized via the ETS, located *inside the mitochondria,* yielding 3 ATP/NADH. Since glucose → 2 pyruvate, a total of 2 acetyl CoA (24 ATP) and 2 NADH (6 ATP) are produced per molecule of glucose.

Additionally, 2 NADH are produced per molecule of glucose in the glyceraldehyde 3-phosphate dehydrogenase reaction, the first step of stage II in glycolysis. Assume for a moment that the glycolytic NADH is readily oxidized via the ETS, yielding 3 ATP/NADH, or a total of 6 ATP/glucose (see below for a correction). In that case, the maximum energy yield of carbohydrate metabolism would be as follows:

Glucose → pyruvate (glycolysis)	2 ATP
2 (NADH, H⁺) from glycolysis (via ETS)	6 ATP
2 (NADH, H⁺) from pyrvate → acetyl CoA (via ETS)	6 ATP
2 Acetyl CoA (citric acid cycle)	24 ATP
Total	38 ATP

Thus, 38 molecules (moles) of ATP are formed per molecule (mole) of glucose catabolized. Because $\Delta G^{\circ\prime} = -2870$ kJ mol^{-1} for glucose oxidation and $\Delta G^{\circ\prime} = -30.5$ kJ mol^{-1} for ATP hydrolysis, carbohydrate catabolism would have an efficiency of

$$\frac{(38)\,(30.5\text{ kJ mol}^{-1})}{2870\text{ kJ mol}^{-1}} \times 100 = 40\%$$

similar to that of the ETS alone (Section 12.3). By comparison, the maximum efficiency of a steam engine, operating between the boiling point of water and room temperature, is only about 22%. Keep in mind, though, that these calculations are based on $\Delta G^{\circ\prime}$ values.

12.6.2. Shuttle Systems

A yield of 38 ATP/glucose represents a *maximum* value. The actual number of ATP produced may be smaller because of a *compartmentation* problem that we ignored in our calculations. Glycolysis occurs in the cytosol, but the ETS is located inside the mitochondria, in the inner membrane. As stressed above, conversion of pyruvate to acetyl CoA, catabolism of acetyl CoA via the citric acid cycle, and oxidation of NADH produced by the pyruvate dehydrogenase complex all occur inside the mitochondria and result in ATP production as outlined. By contrast, the NADH generated by glycolysis cannot be oxidized directly by the ETS as we assumed. For NADH to be oxidized via the ETS, it must first be transported from the cytosol into the mitochondria. Yet transport is impossible, because the membranes of animal mitochondria are *impermeable* to NADH and NAD⁺. How then can glycolytic NADH be used?

To circumvent this compartmentation problem in animal tissues (yeast mitochondria are permeable to NADH and NAD⁺), a number of **shuttle mechanisms** have evolved that accomplish oxidation of NADH in an indirect way. Figure 12.16 outlines one such mechanism, called the **glycerol phosphate shuttle.** The shuttle occurs in certain muscle and nerve cells, and its operation involves two compounds, *dihydroxyacetone phosphate,* a glycolytic intermediate, and *glycerol 3-phosphate.*

In this mechanism, NADH generated in glycolysis is oxidized in the *cytosol* by action of an *NAD⁺-linked dehydrogenase.* Oxidation of NADH is coupled to reduction of dihydroxyacetone phosphate to glycerol 3-phosphate. The latter diffuses to the *inner mitochondrial membrane,* where it undergoes oxidation to dihydroxyacetone phosphate by action of an *FAD-linked dehydrogenase.* The active site of the mitochondrial dehydrogenase is on the outer face of the inner mitochondrial membrane. As glycerol 3-phosphate undergoes oxidation, FAD is reduced to FADH$_2$.

The net result of these reactions amounts to a *reduction of FAD to FADH$_2$ inside the mitochondria* at the expense of an *oxidation of NADH to NAD⁺ in the cytosol.* In effect, electrons are transported across the mitochondrial membrane while carrier concentrations remain constant. FADH$_2$, formed inside the mitochondria, enters the ETS directly but yields only 2 ATP (P/O = 2) compared to the 3 ATP (P/O = 3) produced by NADH. Thus, the cell pays a price for operation of this shuttle; instead of deriving 3 ATP/glycolytic NADH, it only derives 2 ATP.

A different shuttle, the **malate–aspartate shuttle,** functions in liver and heart tissue (Figure 12.17). It may appear from the figure that this shuttle operates without

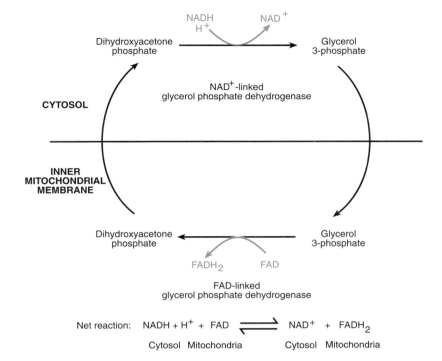

Net reaction: NADH + H$^+$ + FAD $\rightleftharpoons$ NAD$^+$ + FADH$_2$

Cytosol Mitochondria   Cytosol Mitochondria

Figure 12.16.   The glycerol phosphate shuttle.

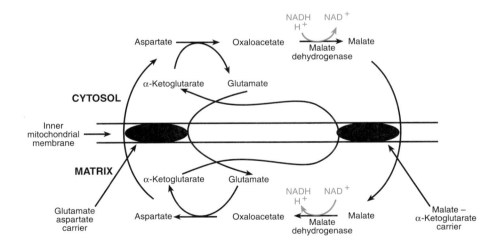

Net reaction: NADH $\rightleftharpoons$ NADH

Cytosol   Matrix

Figure 12.17.   The malate–aspartate shuttle.

loss of efficiency, producing mitochondrial NADH at the expense of cytosolic NADH. This is incorrect. Operation of the shuttle involves a proton motive force estimated to be equivalent to about 0.5 ATP. Thus, the true ATP production of the shuttle amounts to about 2.5 ATP per NADH (P/O = 2.5).

12.6.3. Actual Energy Yields

In view of the need for some type of shuttle mechanism, the true energy yield of carbohydrate catabolism is less than the theoretical yield calculated above and varies depending on the shuttle involved. Assuming operation of the *glycerol phosphate shuttle,* we calculate the maximum energy yield of carbohydrate catabolism under aerobic conditions as follows:

Glucose $\rightarrow$ pyruvate (glycolysis)	2 ATP
2 (NADH, H$^+$) from glycolysis (via glycerol phosphate shuttle and ETS)	4 ATP
2 (NADH, H$^+$) from pyruvate $\rightarrow$ acetyl CoA(via ETS)	6 ATP
2 Acetyl CoA (citric acid cycle)	24 ATP
Total	36 ATP

Hence, the efficiency of energy conservation is:

$$\frac{(36)\,(30.5\ \text{kJ mol}^{-1})}{2870\ \text{kJ mol}^{-1}} \times 100 = 38\%$$

Likewise, we calculate the maximum energy yield, assuming operation of the malate–aspartate shuttle (P/O = 2.5), as follows:

Glucose $\rightarrow$ pyruvate (glycolysis)	2 ATP
2 (NADH, H$^+$) from glycolysis (via malate-asparate shuttle and ETS)	5 ATP
2 (NADH, H$^+$) from pyruvate $\rightarrow$ acetyl CoA(via ETS)	6 ATP
2 Acetyl CoA (citric acid cycle)	24 ATP
Total	37 ATP

Hence, the efficiency of energy conservation is:

$$\frac{(37)\,(30.5\ \text{kJ mol}^{-1})}{2870\ \text{kJ mol}^{-1}} \times 100 = 39\%$$

SUMMARY

We define the biochemical standard reduction potential ($E^{\circ\prime}$) as the potential of a reaction (at pH 7) in which initial reactant and product concentrations are 1.0M each. We define the biochemical actual reduction potential (E') as the potential of a reaction (at pH 7) in which the initial concentrations of reactants and products are not 1.0M each. $E^{\circ\prime}$ is related to the equilibrium constant of the reaction, and we compute E' by means of the Nernst equation. Changes in reduction potentials are equivalent to changes in free energy.

A redox reaction comprises two half-reactions. The half-reaction having the smaller reduction potential represents the stronger reducing agent, provides electrons to the other half-reaction, and proceeds as an oxidation. We deduce the direction of an overall redox reaction from $E^{\circ\prime}$ or E' values of the component two half-reactions. E' values alone determine the directions of overall reactions under intracellular conditions.

In cellular respiration, metabolites are oxidized by removal of hydride ions and protons. The electrons are then passed along a series of electron carriers. The mitochondrial electron transport system (ETS) contains five types of electron carriers: pyridine-linked dehydrogenases, flavoproteins, ubiquinones, cytochromes, and iron–sulfur proteins. Some carriers have the capacity for both one- and two-electron transfers; other carriers only transfer either one or two electrons. Two pathways of electron transport occur; in both, a metabolite serves as the initial electron donor, and molecular oxygen functions as the ultimate electron acceptor.

The sequence of electron carriers in the ETS was elucidated by isolating macromolecular complexes catalyzing portions of the entire sequence and by using artificial electron acceptors and inhibitors. The change in free energy resulting from the change in reduction potential along the ETS drives ATP synthesis. Oxidative phosphorylation refers to this coupling of ATP synthesis and electron transport. According to the chemiosmotic coupling hypothesis, an electrochemical gradient functions as a common intermediate for coupling the two processes.

Maximally, three molecules of ATP are synthesized per atom of oxygen consumed, yielding a P/O ratio of 3. P/O ratios less than 3 result when metabolites are linked to the ETS at other than the initial point. ATP synthase, a complex membrane-bound enzyme, accomplishes the actual synthesis of ATP. The electrochemical gradient produced by proton pumping of three respiratory complexes drives ATP synthesis. Uncouplers of oxidative phosphorylation permit electron transport to occur but prevent ATP synthesis. Incompletely reduced species of oxygen form during oxidative metabolism and are eliminated by specific enzymatic mechanisms. ATP yield from complete oxidation of carbohydrates depends on the shuttle involved in the oxidation of glycolytic NADH. When the glycerol phosphate shuttle is operative, a maximum number of 36 ATP is produced per molecule of glucose oxidized to CO_2 and H_2O.

SELECTED READINGS

Abrahams, J. P., Leslie, A. G. W., Lutter, R., and Walker, J. E., Structure at 2.8Å resolution of F_1-ATPase from bovine heart mitochondria, *Nature (London)* 370:621–628 (1994).

Bensasson, R. V., Land, E. J., and Truscott, T. G., *Excited States and Free Radicals in Biology and Medicine,* Oxford University Press, Oxford (1993).

Fetter, J. R., *et al.,* Possible proton relay pathways in cytochrome *c* oxidase, *Proc. Natl. Acad Sci. USA* 92:1604–1608 (1995).

Fridovich, I., Superoxide radical and superoxide dismutases, *Annu. Rev. Biochem.* 64:97–112 (1995).

Gray, H. B., and Winkler, J. R., Electron transfer in proteins, *Annu. Rev. Biochem.* 65:537–561 (1996).

Mitchell, P., Keilin's respiratory chain concept and its chemiosmotic consequences, *Science* 206:1148–1159 (1979).

Musser, S. M., and Stowell, M. H. B., Cytochrome *c* oxidase: Chemistry of a molecular machine, *Adv. Enzymol. Relat. Areas Mol. Biol.* 71:79–208 (1995).

Pedersen, P. L., and Amzel, L. M., ATP synthases: Structure, reaction center, mechanism, and regulation of one of nature's most unique machines, *J. Biol. Chem.* 268:9937–9940 (1993).

Racker, E., From Pasteur to Mitchell: A hundred years of bioenergetics, *Fed. Proc.* 39:210–215 (1980).

Rees, D. C., and Farrelly, D., Biological electron transfer, in *The Enzymes,* 3rd ed. (D. S. Sigman and P. D. Boyer, eds.), Vol. 19, pp. 38–97, Academic Press, New York (1990).

Stenesh, J., *Core Topics in Biochemistry,* Cogno Press, Kalamazoo, MI (1993).

Trumpower, B. L., and Gennis, R. B., Energy transduction by cytochrome complexes in mitochondrial and bacterial respiration: The enzymology of coupling electron transfer reactions to transmembrane proton translocation, *Annu. Rev. Biochem.* 63:675–716 (1994).

REVIEW QUESTIONS

A. Define each of the following terms:

Oxidative phosphorylation	Uncoupler
Respiratory complex	Shuttle mechanism
Q-cycle	Respiratory control
Iron–sulfur protein	Coenzyme Q

B. Differentiate between the two terms in each of the following pairs:

$E°'/E'$	Cytochrome/cytochrome oxidase
Glycerol phosphate shuttle/ malate–aspartate shuttle	Respiratory complex/ respiratory chain

C. (1) Outline the chemiosmotic coupling hypothesis.
(2) What is the effect of adding an inhibitor or an artificial electron acceptor to the electron transport system? Does the P/O ratio change in the presence of such a compound? How can you determine the point of action of the compound?
(3) Explain how efficiency of energy conservation, in terms of ATP synthesis in oxidative phosphorylation, can be calculated by using either free energy changes or changes in redox potentials. Why should an ideal catabolic pathway consist of a large number of steps?
(4) How do the reduction potentials (E') of the electron carriers change along the ETS? Why must this be so?
(5) What steps comprise the four-electron reduction of O_2 catalyzed by cytochrome oxidase? What is the significance of the cytochrome oxidase reaction?

PROBLEMS

12.1. What effects does a relatively low concentration of 2,4-dinitrophenol have on the rate of electron transport (oxygen consumption) and on the P/O ratio?

12.2. A researcher maintains a solution of cytochrome c at a potential of 0.50 V, a temperature of 25°C, and a pH of 7.0. What percent of cytochrome c will be in the reduced form?

12.3. Write the overall reaction that will occur when the following two half-reactions are coupled under biochemical standard conditions:

Fumarate^{2-} + 2H$^+$ + 2e$^-$ ⇌ succinate^{2-}
$$E^{\circ\prime} = +0.03 \text{ Volts}$$

Dehydroascorbic acid + 2H$^+$ + 2e$^-$ ⇌ ascorbic acid
$$E^{\circ\prime} = +0.06 \text{ Volts}$$

Calculate the value of the overall $\Delta E^{\circ\prime}$, $\Delta G^{\circ\prime}$, and K'_{bio}.

12.4. What is the total biochemical standard free energy change ($\Delta G^{\circ\prime}$) when 1.50 g of succinate (MW = 117) undergoes reaction according to the overall equation of the previous problem?

12.5. Assume that you couple the two half-reactions of Problem 12.3 at pH 7.0, using the following initial concentrations: [fumarate] = 5.0 × 10^{-1}M; [succinate] = 2.0 × 10^{-2}M; [dehydroascorbic acid] = 1.0 × 10^{-4}M; [ascorbic acid] = 3.0 × 10^{-1}M. (a) What overall reaction will take place under these conditions? (b) Calculate the change in reduction potential ($\Delta E'$) and the change in free energy ($\Delta G'$) for the overall reaction.

12.6.* What [succinate]/[fumarate] ratio is required in order to drive the reaction in the opposite direction to that deduced for Problem 12.5 when the initial concentration of dehydroascorbic acid is 5.0 × 10^{-4}M and that of ascorbic acid is 2.0 × 10^{-1}M? (Hint: Set $E'_{succinate} = E'_{ascorbate}$.)

12.7. Calculate the [lactate]/[pyruvate] ratio when the half-reaction describing this system has an E' value of zero (see Table 12.1).

12.8.* Cyanide is a powerful poison because it binds avidly to the Fe^{3+} of cytochrome oxidase. Hemoglobin (Fe^{2+}) has a relatively low affinity for cyanide, but methemoglobin (Fe^{3+}) binds cyanide strongly. Sublethal cyanide poisoning may be reversed by immediate administration of nitrites, strong oxidizing agents capable of oxidizing the Fe^{2+} in hemoglobin to the Fe^{3+} state. Why is this treatment effective?

12.9.* What [succinate]/[fumarate] ratio would be required to enable succinate dehydrogenase to use NAD$^+$ as a coenzyme instead of its normal FAD under biochemical standard conditions?

12.10. Why is it that some soft drinks, with ice, can help keep you warm even on a very cold day? Why is this not the case with all soft drinks (diet drinks)?

12.11. What is the energy charge for a system in which adenine nucleotides are present entirely as ADP?

12.12. Which cell is likely to have a higher energy charge, one at rest or one actively metabolizing? Why?

12.13. Calculate the efficiency of energy trapping, in the form of ATP synthesis from ADP, when succinate is oxidized by FADH$_2$ via the second pathway of electron transport. Base your calculation on the $E^{\circ\prime}$ values of the pathway. The $\Delta G^{\circ\prime}$ of ATP hydrolysis is -30.5 kJ mol^{-1}.

12.14. Refer to Figure 12.8 and Table 12.1 and calculate the efficiency of energy trapping, in the form of ATP synthesis from ADP, for respiratory complexes I, III, and IV. In other words, consider the following sequences under biochemical standard conditions:

(a). Isocitrate ⟶ CoQ
(b). CoQ ⟶ cytochrome c
(c). Cytochrome c ⟶ O$_2$

12.15.* Succinate dehydrogenase and cytochrome oxidase are incubated, in the presence of oxygen, with cytochrome c, succinate, and coenzyme Q. What overall oxidation–reduction reaction would you expect to take place under these conditions?

12.16.* Calculate the net yield of ATP molecules per molecule of maltose for the complete oxidation of maltose to CO$_2$ and H$_2$O under aerobic conditions, assuming operation of: (a) the glycerol phosphate shuttle; (b) the malate–aspartate shuttle. Hint:

$$\text{Maltose} + H_2O \xrightarrow{\text{maltase}} \text{D-glucose} + \text{D-glucose}$$

12.17. Electron carriers of a prokaryotic ETS have $E^{\circ\prime}$ values as shown. What is the maximum number of ATP molecules that could be synthesized, per molecule of metabolite, when ATP synthesis is coupled to the operation of this ETS under biochemical standard conditions? What is the likely number of ATP molecules actually obtained? The $\Delta G^{\circ\prime}$ for the hydrolysis of ATP is -30.5 kJ mol^{-1}.

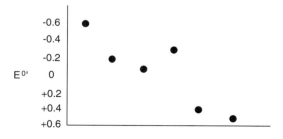

12.18. If the system described in the previous problem is known to operate at an efficiency of energy conservation of 30%, how many ATP molecules would be synthesized per molecule of metabolite?

12.19. Which of the following oxidation–reduction reactions will proceed as written under biochemical standard reaction conditions? (See Table 12.1.) Hb = hemoglobin.

(a) Ferredoxin (Fe^{2+}) + Hb (Fe^{3+}) →
ferredoxin (Fe^{3+}) + Hb (Fe^{2+})

(b) Myoglobin (Fe^{2+}) + NAD^+ →
myoglobin (Fe^{3+}) + NADH + H^+

(c) NAD^+ + pyruvate$^-$ → NADH + H^+ + lactate$^-$

12.20. The $E^{\circ\prime}$ of covalently bound FAD is +0.05 V. The $E^{\circ\prime}$ values of NAD^+ and succinate are listed in Table 12.1. On this basis, explain why the coenzyme of succinate dehydrogenase is FAD rather than NAD^+.

12.21.* The compound 2,4-dinitrophenol was at one time prescribed as a weight-reducing drug. People soon realized that this constitutes an *extremely dangerous* method of weight control, and the compound was no longer used after some deaths occurred. (a) Why was this compound chosen as a weight-reducing drug? (b) Why did some deaths occur in using it? (c) Ingestion of 2,4-dinitrophenol also led to an increase in body temperature and to profuse sweating. What is the explanation for these side effects?

12.22.* Adding dicyclohexylcarbodiimide (DCCD) to actively respiring mitochondria leads to a decrease in both the rate of the ETS (oxygen consumption) and the rate of ATP production (ATP synthesis). Adding 2,4-dinitrophenol restores the oxygen consumption to its normal level, but ATP synthesis remains inhibited. How can the effect of DCCD be explained? Why does addition of 2,4-dinitrophenol lead to the observed results?

12.23.* Oxidative phosphorylation of a mitochondrial sample led to a $1.2 \times 10^{-4}M$ increase in the concentration of ATP. It also led to a decrease in the concentration of NADH; this amounted to a decrease of 0.25 in the absorbance at 340 nm (light path = 1.0 cm). Given that the extinction coefficient (ϵ) of NADH is $6.2 \times 10^3\ M^{-1}\ cm^{-1}$, calculate the number of ATP molecules synthesized per molecule of NADH oxidized. Note that $A = \epsilon c l$, where A is absorbance, ϵ is the extinction coefficient, c is the concentration, and l is the length of the light path (see Appendix C).

12.24. The compounds 2,6-dichlorophenol indophenol and tetramethyl-*p*-phenylenediamine (TMPD) are two artificial electron acceptors used in studies of the ETS. Their $E^{\circ\prime}$ values are +0.22 and +0.26, respectively. Predict the points in the ETS where they are likely to act under biochemical standard conditions.

12.25. What is the expected P/O ratio when the ETS is: (a) inhibited at cytochrome *c*; (b) inhibited at coenzyme Q; (c) provided with ascorbic acid and tetramethyl-*p*-phenylenediamine (TMPD)?

12.26.* A male adult requires a minimum of about 7500 kJ per day (Table 8.6). Assuming that all of this energy derives from ATP hydrolysis ($\Delta G^{\circ\prime} = -30.5$ kJ mol^{-1}), calculate the number of grams of ATP (MW = 507) that must be hydrolyzed per day. Given that the actual amount of ATP in the individual is about 50 g, compute the approximate number of times that all of the body's ATP must be hydrolyzed and resynthesized.

12.27.* Calculate the *net* yield of ATP molecules per molecule of metabolite (underlined) for the following cases (refer to Chapters 10 and 11). The reactions occur under aerobic conditions and involve participation of the glycerol phosphate shuttle.

(a) Oxidation of <u>glyceraldehyde 3-phosphate</u> to acetyl CoA

(b) Complete oxidation of <u>fructose 6-phosphate</u> to CO_2 and H_2O

(c) Complete oxidation of <u>3-phosphoglycerate</u> to CO_2 and H_2O.

12.28.* What would be the theoretical energy yield, in terms of ATP molecules per molecule of glucose, if glucose is phosphorylated by hexokinase, the glucose 6-phosphate is catabolized via the pentose phosphate pathway, and the NADPH of the pentose phosphate pathway is oxidized via the ETS, yielding the same number of molecules of ATP as are produced by NADH? (a) Assume that NADPH is formed inside the mitochondria; (b) assume that NADPH is formed in the cytosol and the glycerol phosphate shuttle applies.

Lipid Metabolism

Lipid metabolism, like that of carbohydrates, can be divided into five broad areas—*digestion, transport, storage, degradation,* and *biosynthesis.* Since **fats** (acylglycerols) constitute most of an organism's lipids and are the major dietary form of lipids, this chapter will focus primarily on the metabolism of fats.

Fat digestion occurs in the aqueous environment of the intestine and requires the action of water-soluble **lipases,** enzymes that catalyze the hydrolysis of fats. Digestion also requires solubilization of the non-polar fats. **Bile salts,** present at high concentrations in the bile (which empties into the intestine), solubilize fats by emulsifying them. Following their digestion (Section 8.4), most fats enter the lymphatic circulation as **chylomicrons** that subsequently enter the bloodstream. Short-chain fatty acids are absorbed directly from the intestine into the circulatory system.

Transport of fats by the blood and intracellular fluids also requires their solubilization. Formation of *lipoproteins* provides water-soluble macromolecular aggregates owing to the presence of proteins and amphipathic lipids. Formation of fatty acid/serum albumin complexes solubilizes free fatty acids. **Serum albumin** is a monomeric protein (MW = 66,000) and the most abundant of all plasma proteins. It occurs at a concentration of about 4 g/100 ml of plasma and constitutes about 50% of total plasma protein. Serum albumin can bind up to 10 fatty acid molecules per molecule of protein.

In addition to being incorporated into lipoproteins and albumin complexes, lipids are transported as **ketone bodies,** products of fatty acid catabolism that serve as an energy source under some conditions. Fatty acid metabolism comprises two major pathways: their stepwise degradation by means of *β-oxidation* and their stepwise synthesis by means of *fatty acid biosynthesis.*

13.1. STORAGE OF FATS

13.1.1. Depot Fat

Lipids, particularly fats, constitute the main storage form of energy in animals. Because lipids have the highest caloric value of all nutrients (37.7 kJ/g or 9.0 kcal/g), using them for storage provides a decided advantage. For animals to store an energetically equivalent amount of either carbohydrate or protein would necessitate deposition of considerably larger quantities of substance. Also, carbohydrate and protein stores would be even larger because the polar character of these substances would result in inclusion of water in the deposits.

We refer to stored fat as **depot fat** or **adipose tissue,** and dietary lipids must be converted to it prior to storage. The composition of stored fat is always characteristic of the organism, regardless of the source of dietary lipids. We can distinguish human, sheep, pig, and other animal fats by the relative proportions and compositions of their *mono-, di-,* and *triacylglycerols.* To produce an organism's unique mixture of acylglycerols from dietary fats requires a variety of chemical reactions, including changes in fatty acid chain length and saturation. Transformations of dietary fats, termed "working over," occur predominantly in the liver.

Storage and transport of fats are interrelated processes (Figure 13.1). When fat has to be catabolized to produce usable energy in the form of ATP, fatty acids are removed from adipose tissue and transported to the liver for degradation. We refer to the release of fatty acids from depot fat as **mobilization.** The process requires the action of enzymes. *Lipases* and *phospholipases* catalyze hydrolysis of ester bonds in acylglycerols and glycerophospholipids, respectively (Figure 13.2). Released fatty acids undergo catabolism in the mitochondria, and glycerol is subject to degradation in the cytosol.

A hormonally controlled enzyme cascade (Figure 13.3), similar to the cascades in regulation of carbohydrate metabolism (see Figures 10.26 and 10.30), regulates fatty acid mobilization. Once free fatty acids have been released inside *adipocytes* (cells of adipose tissue), they dif-

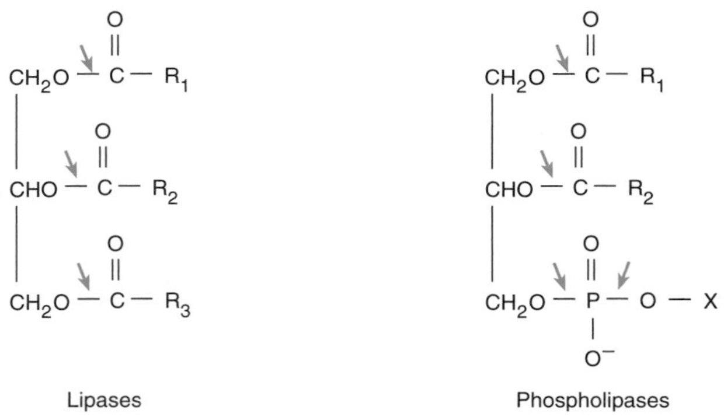

Figure 13.2.   Points of attack of lipases and phospholipases.

fuse across cell membranes, become bound to serum albumin, and are transported to the tissues.

13.1.2. Fatty Liver

Excessive mobilization of fatty acids may lead to development of **fatty liver,** a liver that has been infiltrated by fat cells so that portions have become nonfunctional fatty tissue. Fatty livers may occur in diabetics suffering from insulin deficiency. Diabetics cannot metabolize glucose properly and must use other nutrients as sources of energy. Typically, this results in an overreliance on fat metabolism, including excessive fatty acid mobilization and excessive fatty acid metabolism in the liver.

Fatty liver may also result from exposure to chemicals such as carbon tetrachloride and pyridine. These compounds destroy liver cells and lead to their replacement by fatty tissue. Dietary deficiencies of *choline* and *methionine,* called **lipotropic agents,** may likewise result in formation of fatty liver because of an effect on lipid transport.

Recall that choline occurs as a component of *phosphatidyl choline,* one of the *phospholipids.* Its synthesis involves the carbon skeleton of serine and requires a transfer of three methyl groups from *S*-adenosylmethionine (Figure 13.4). **S-Adenosylmethionine (SAM),** which forms by reaction of methionine and ATP, serves as a donor of methyl groups for many biological methylations. We can describe phosphatidyl choline biosynthesis by a schematic sequence, which indicates the metabolic fate of the methyl group:

Methionine → *S*-adenosylmethionine → choline → phosphatidyl choline

Inadequate supplies of methionine and choline in the diet lead to insufficient synthesis of phosphatidyl choline in particular and of phospholipids in general. De-

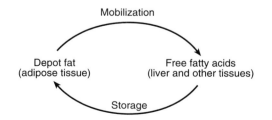

Figure 13.1.   The link between mobilization and storage of fatty acids.

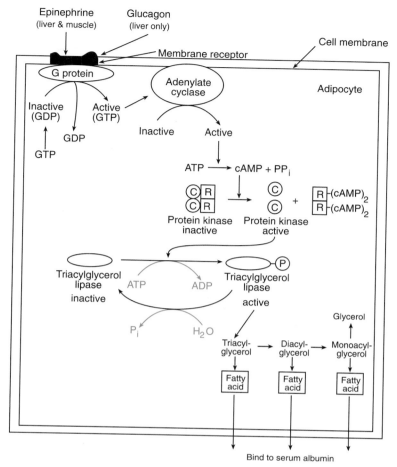

Figure 13.3. The hormonally controlled enzyme cascade regulating fatty acid mobilization. Triacylglycerol lipase catalyzes the hydrolysis of fatty acids linked to C(1) and C(3) of glycerol, yielding mono- and diacylglycerols that are then acted upon by mono- and diacylglycerol lipases.

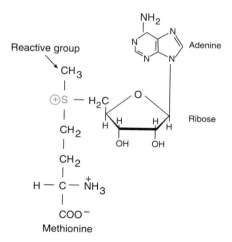

Figure 13.4. *S*-Adenosylmethionine (SAM).

creased phospholipid synthesis results in a deficiency of lipoproteins. Because lipoproteins constitute a major form in which lipids are transported from the liver, decreased lipoprotein concentrations may lead to excessive lipid accumulation in the liver, ultimately producing fatty liver.

13.2. FAT CATABOLISM

13.2.1. Degradation of Glycerol

Fat catabolism begins with the hydrolysis of fats to glycerol and fatty acids. These two structural components are metabolized separately. Glycerol catabolism occurs in liver cytosol and involves conversion to a glycolytic intermediate. First, *glycerokinase* catalyzes a phosphorylation of glycerol to glycerol 3-phosphate:

$$\begin{array}{ccc} CH_2OH & & CH_2OH \\ | & & | \\ H-C-OH & & H-C-OH \\ | & & | \\ CH_2OH & & CH_2O\circledP \end{array}$$

Glycerol + ATP^{4-} ⟶ Glycerol 3-phosphate^{2-}
+ ADP^{3-} + H$^+$

Glycerol 3-phosphate dehydrogenase then catalyzes the oxidation of glycerol 3-phosphate to dihydroxyacetone phosphate. This reaction constitutes part of the *glycerol phosphate shuttle.*

$$\begin{array}{ccc} CH_2OH & & CH_2OH \\ | & & | \\ H-C-OH & & C=O \\ | & & | \\ CH_2O\circledP & & CH_2O\circledP \end{array}$$

Glycerol 3-phosphate^{2-} ⟶ Dihydroxyacetone phosphate^{2-}
+ NAD$^+$ + NADH + H$^+$

Dihydroxyacetone phosphate enters glycolysis, where it is converted to glyceraldehyde 3-phosphate, an intermediate of both the glycolytic and gluconeogenic pathways. Glycolysis leads to pyruvate and amino acids, and gluconeogenesis leads to glucose. Thus, glycerol, a component of *fats,* can be converted to both *proteins* and *carbohydrates.* Such interconversions are characteristic of the interrelationships in metabolism and illustrate the principle that one type of nutrient may readily be converted to several others.

Glycerol catabolism can provide energy. NADH, produced in the glyeraldehyde 3-phosphate dehydrogenase reaction of glycolysis, can be oxidized to NAD$^+$ by the electron transport system via specific shuttles that link the compartments in which these pathways occur.

13.2.2. Knoop's Hypothesis

A clue to the mechanism of fatty acid catabolism came from some ingenious experiments carried out by F. Knoop at the turn of the century (1904). In those days, radioactive isotopes were not yet in use, radioactivity having only recently been discovered, and labeled compounds were not available. Yet Knoop designed the first "tracer" experiment in biochemistry. He did this by tagging the methyl carbon of both even- and odd-numbered fatty acids with a phenyl group. He then fed these "labeled" fatty acids to dogs and analyzed the urine excreted. Under these conditions, the phenyl group is not metabolized and is excreted in the form of specific organic compounds.

Knoop found that dogs fed even-numbered fatty acid derivatives excreted the phenyl group in the form of *phenylaceturic acid;* those fed odd-numbered fatty acid derivatives excreted the phenyl group in the form of *hippuric acid* (Figure 13.5). Phenylaceturic acid and hippuric acid were known to be formed, respectively, from *phenylacetic acid* and *benzoic acid* by coupling each with a molecule of *glycine.* Based on these experiments, Knoop proposed that fatty acids are degraded in a stepwise fashion, by successive cleavage of two-carbon fragments (in the form of acetate), beginning with the carboxyl end of the molecule.

Coupling of phenylacetic acid and benzoic acid to glycine and excretion of the products illustrates an organ-

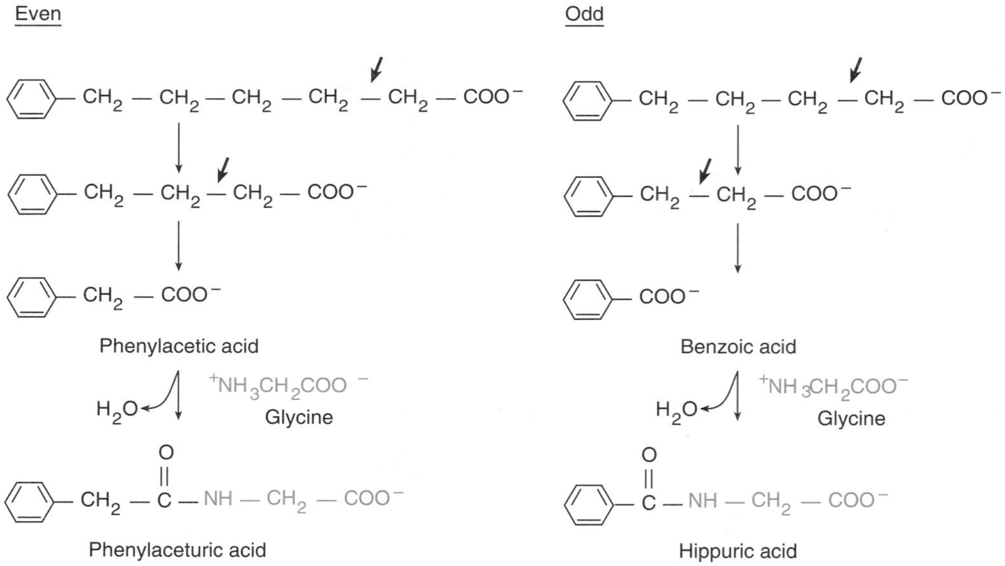

Figure 13.5.   Illustration of Knoop's experiments on the oxidation of phenyl-labeled fatty acids. Arrows indicate the presumed points of cleavage.

ism's capacity to divest itself of foreign toxic compounds by *detoxification*. In this process, enzymatic reactions convert such compounds to less harmful substances or couple them to naturally occurring metabolites. The products of these reactions are then excreted. For example, use of benzoic acid as a food preservative poses little danger because benzoic acid is readily detoxified to hippuric acid, followed by excretion.

13.3. BETA OXIDATION OF FATTY ACIDS

Subsequent work corroborated Knoop's brilliant deductions. Fatty acid degradation, called **β-oxidation,** does occur by successive cleavage of two-carbon fragments from the carboxyl end. However, the modern view differs from Knoop's hypothesis in three ways:

1. The two-carbon fragment is removed as *acetyl CoA,* not as acetate.
2. All intermediates in the reaction sequence are bound to coenzyme A.
3. Initiation of the degradation requires ATP hydrolysis.

β-Oxidation occurs principally in the mitochondria and consists of five enzymatic steps:

1. Activation: Thiokinase (acyl CoA synthase)
2. Dehydrogenation: Acyl CoA dehydrogenase
3. Hydration: Enoyl CoA hydratase
4. Dehydrogenation: L-3-Hydroxyacyl CoA dehydrogenase
5. Cleavage: Thiolase (β-ketothiolase)

Fatty acid activation takes place in the endoplasmic reticulum or the outer mitochondrial membrane. The remaining four reactions occur in the mitochondrial matrix and are linked to fatty acid activation by means of a specific carrier system. Figure 13.6 shows the entire reaction sequence.

13.3.1. Individual Reactions

13.3.1A. Thiokinase. The first step in the β-oxidation of fatty acids involves a conversion of the fatty acid to a chemically more reactive form. The reaction, called **fatty acid activation,** is catalyzed by **thiokinase** or **acyl CoA synthase.** Three types of thiokinases occur, specific for short-, medium-, and long-chain fatty acids, re-

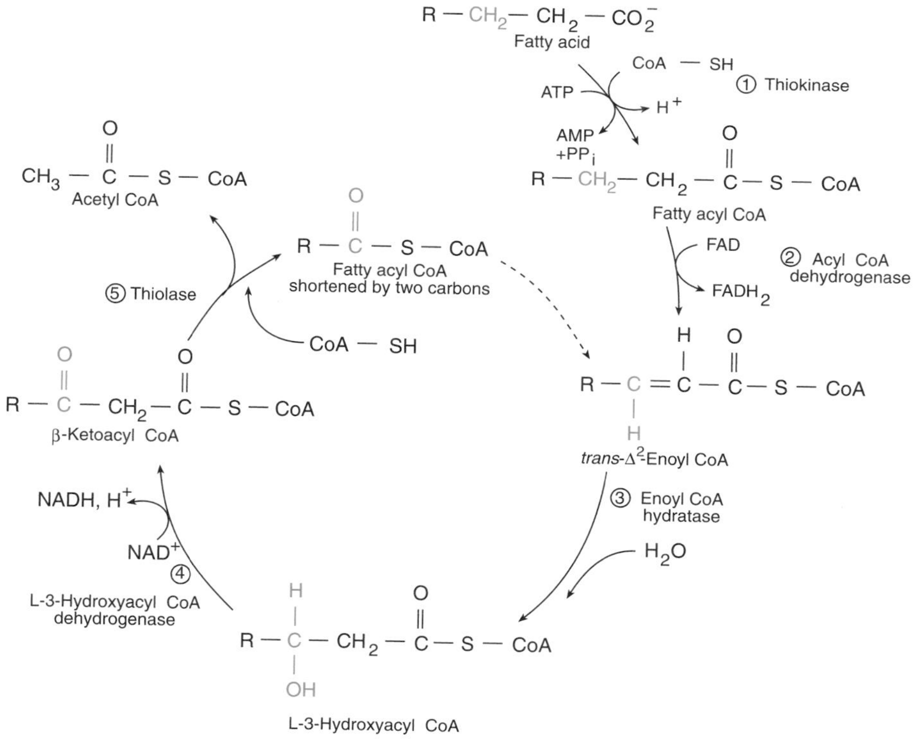

Figure 13.6. The β-oxidation cycle of saturated fatty acids. One turn of the cycle produces a fatty acyl CoA, shortened by two carbons from the one that entered the cycle. The two carbons are split off as acetyl CoA.

spectively. Thiokinases act on both saturated and unsaturated fatty acids. Fatty acid activation requires cleavage of an energy-rich bond in ATP. We can write the coupled reactions as follows:

$$\overset{(2)}{\text{Fatty acid}^-} + \text{ATP}^{4-} + \text{H}^+ \rightleftharpoons \overset{(1)}{\text{fatty acyl-AMP}^-} + \overset{(1)}{\text{PP}_i^{3-}}$$

$$\overset{(1)}{\text{Fatty acyl-AMP}^-} + \text{HS-CoA}^{4-} \rightleftharpoons \overset{(1)}{\text{fatty acyl CoA}^{4-}} + \text{AMP}^{2-} + \text{H}^+$$

Overall reaction:

$$\overset{(2)}{\text{Fatty acid}^-} + \text{ATP}^{4-} \ \text{HS-CoA}^{4-} \rightleftharpoons$$

$$\overset{(1)}{\text{Fatty acyl CoA}^{4-}} + \overset{(1)}{\text{AMP}^{2-}} + \text{PP}_i^{3-} \qquad \Delta G^{\circ\prime} = -0.8 \text{ kJ mol}^{-1}$$

The common intermediate, **fatty acyl adenylate (fatty acyl-AMP)**, consists of a fatty acid molecule attached to AMP via a mixed acid anhydride linkage (Figure 13.7). Because of the anhydride structure, the fatty acid–AMP link constitutes an energy-rich bond. Accordingly, fatty acyl adenylates are *energy-rich compounds*. Hydrolysis of the fatty acid–AMP bond results in a reaction with a highly negative biochemical standard free energy change ($\Delta G^{\circ\prime}$).

In the above three equations, we have indicated the number of energy-rich bonds per compound in parentheses. Note that each reaction has an equal number of energy-rich bonds for the reactants and the products. Hence all three reactions are readily reversible, with biochemical standard free energy changes close to zero. The overall reaction is driven toward formation of fatty acyl CoA because of a fourth, related reaction involving *pyrophosphatase*. This enzyme catalyzes the strongly exergonic hydrolysis of pyrophosphate (PP_i) to inorganic phosphate (P_i):

$$\text{PP}_i^{3-} + \text{H}_2\text{O} \rightarrow 2\text{P}_i^{2-} + \text{H}^+ \qquad \Delta G^{\circ\prime} = -33.1 \text{ kJ mol}^{-1}$$

By coupling PP_i hydrolysis to fatty acid activation, the overall reaction proceeds spontaneously with a free energy change of $\Delta G^{\circ\prime} = -33.9 \text{ kJ mol}^{-1}$. Using stearic acid (18 carbons, saturated) to illustrate β-oxidation, we have for the fatty acid activation step:

$$\text{CH}_3-(\text{CH}_2)_{14}-\overset{\beta}{\text{CH}_2}-\overset{\alpha}{\text{CH}_2}-\text{COO}^- + \text{ATP}^{4-} + \text{HS}-\text{CoA}^{4-} \rightleftharpoons$$
Fatty acid

$$\text{CH}_3-(\text{CH}_2)_{14}-\overset{\beta}{\text{CH}_2}-\overset{\alpha}{\text{CH}_2}-\overset{\text{O}}{\overset{\|}{\text{C}}}-\text{S}-\text{CoA}^{4-} + \text{AMP}^{2-} + \text{PP}_i^{3-}$$
Fatty acyl CoA

13.3.1B. Carnitine Carrier System.
Because fatty acid activation occurs in the endoplasmic retic-

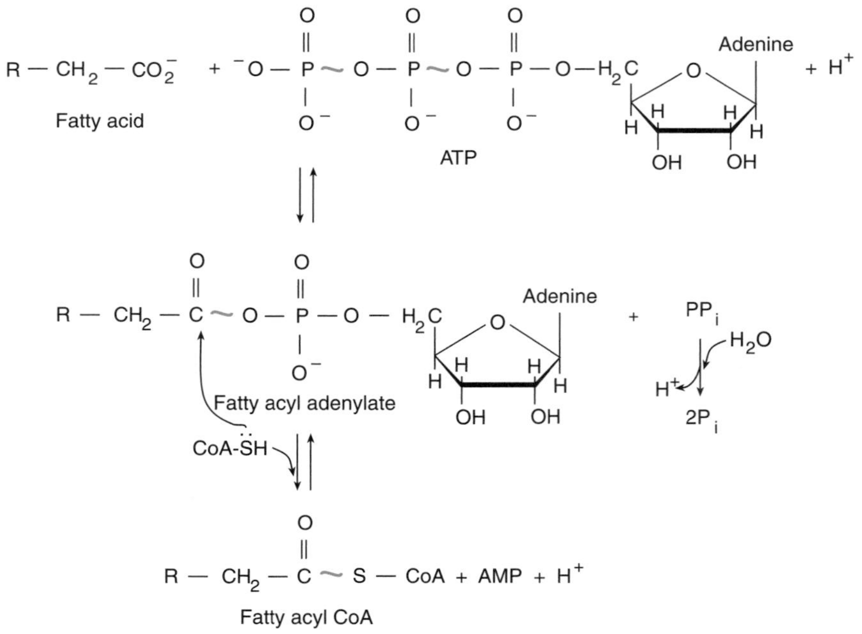

Figure 13.7.   Fatty acid activation. Pyrophosphatase-catalyzed hydrolysis of PP_i drives the reaction to completion.

ulum or the outer mitochondrial membrane, the reaction product must be transported from there *into* the mitochondrial matrix, where subsequent steps of β-oxidation take place. Because fatty acyl CoA cannot cross the inner mitochondrial membrane directly, its movement requires a transport system. The specific carrier of fatty acyl CoA is **carnitine,** a low-molecular-weight compound derived from lysine (Figure 13.8).

The transport mechanism, diagrammed in Figure 13.9, consists of four steps:

1 Cytoplasmic activation of a fatty acid to a fatty acyl CoA, as described above.
2 Passage of fatty acyl CoA through the outer mitochondrial membrane and into the intermembrane space. There the fatty acyl group is transferred to carnitine, forming *acyl carnitine,* and coenzyme A is released back to its extramitochondrial pool.
3 Movement of acyl carnitine from the intermembrane space, across the inner membrane, into the matrix. In the matrix, the fatty acid is transferred to a molecule of coenzyme A from the mitochondrial pool.
4 Free carnitine exits the matrix through the inner mitochondrial membrane.

Transfer of the fatty acid from acyl carnitine to coenzyme A proceeds without additional ATP expenditure because the acyl carnitine bond is sufficiently energy-rich.

Operation of this transport mechanism means that cells maintain two separate pools of coenzyme A, in the cytoplasm and inside the mitochondria, that participate in different metabolic processes. Mitochondrial coenzyme A serves in catabolism of fatty acids (β-oxidation), while cytoplasmic coenzyme A functions in fatty acid biosynthesis. Cells maintain similar separate cytoplasmic and mitochondrial pools of ATP and NAD$^+$.

The **carnitine carrier system** provides a control point of β-oxidation. Malonyl CoA is an allosteric inhibitor of carnitine acyltransferase I and prevents transfer of fatty acyl CoA into mitochondria. Because of malonyl CoA's role as an intermediate in fatty acid biosynthesis, it exerts a dual effect: it stimulates fatty acid synthesis and inhibits fatty acid oxidation.

Other than the effect of malonyl CoA, no specific enzyme control point has yet been identified in β-oxidation. The only other known control of fatty acid catabolism consists of the hormonal effects on fatty acid mobilization (Figure 13.3).

13.3.1C. Acyl CoA Dehydrogenase. The second step of β-oxidation involves an oxidation (*dehydrogenation*) of fatty acyl CoA, catalyzed by **acyl CoA dehydrogenase.** Three types of acyl CoA dehydrogenases occur, specific for short-, medium-, and long-chain fatty acids, respectively. All three enzyme types are flavoproteins and carry tightly bound FAD. The product formed is an α,β-unsaturated fatty acyl CoA. Acyl CoA dehydrogenase produces only the *trans* isomer, designated *trans-Δ²-enoyl CoA:*

$$CH_3-(CH_2)_{14}-\overset{\beta}{CH_2}-\overset{\alpha}{CH_2}-\overset{O}{\overset{\|}{C}}-S-CoA^{4-} + FAD \rightleftharpoons$$
Fatty acyl CoA

$$CH_3-(CH_2)_{14}-\overset{H}{\overset{|}{\underset{|}{\overset{\beta}{C}}}}=\overset{\alpha}{\underset{H}{C}}-\overset{O}{\overset{\|}{C}}-S-CoA^{4-} + FADH_2$$
trans-Δ²-Enoyl CoA

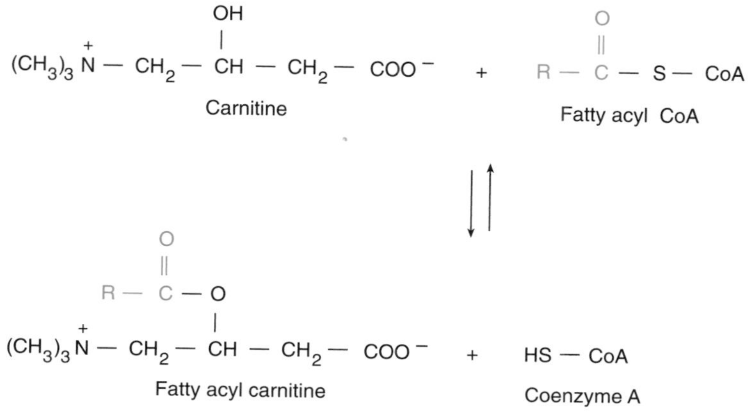

Figure 13.8.   Carnitine and its reversible reaction with fatty acyl CoA.

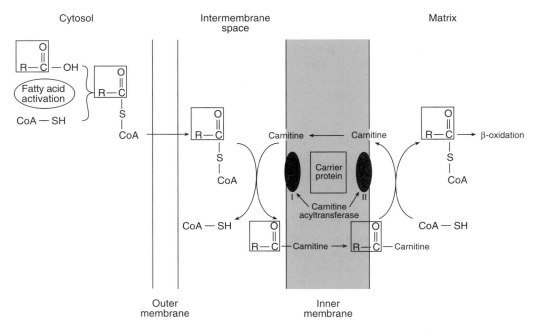

Figure 13.9.   The carnitine carrier system. Translocation is mediated by a carrier protein. Carnitine acyltransferases I and II are located, respectively, on the external and internal surfaces of the inner mitochondrial membrane.

13.3.1D. Enoyl CoA Hydratase. The third reaction of β-oxidation consists of a *hydration,* catalyzed by **enoyl CoA hydratase.** The hydratase requires the *trans* isomer produced in the previous step and catalyzes addition of water across the α,β-double bond. Hydratase action results in formation of a β-alcohol fatty acyl CoA. Enoyl CoA hydratase is stereospecific so that only the L-isomer, *L-3-hydroxyacyl CoA,* is produced:

$$CH_3-(CH_2)_{14}-\overset{\underset{|}{H}}{\underset{H}{C}}\overset{\beta}{=}\overset{\alpha}{C}-\overset{\overset{O}{\|}}{C}-S-CoA^{4-} + H_2O \rightleftharpoons$$

trans-Δ^2-Enoyl CoA

$$CH_3-(CH_2)_{14}-\overset{\overset{OH}{\underset{|}{|}}}{\underset{H}{\overset{\beta}{C}}}-\overset{\alpha}{CH_2}-\overset{\overset{O}{\|}}{C}-S-CoA^{4-}$$

L-3-Hydroxyacyl CoA

13.3.1E. L-3-Hydroxyacyl CoA Dehydrogenase. A second oxidation, or *dehydrogenation,* follows the hydration step. This reaction is catalyzed by a pyridine-linked dehydrogenase, **L-3-hydroxyacyl CoA dehydrogenase,** that uses NAD$^+$ as coenzyme. The enzyme has an absolute specificity for L-3-hydroxyacyl CoA and forms *β-ketoacyl CoA:*

$$CH_3-(CH_2)_{14}-\overset{\overset{OH}{\underset{|}{|}}}{\underset{H}{\overset{\beta}{C}}}-\overset{\alpha}{CH_2}-\overset{\overset{O}{\|}}{C}-S-CoA^{4-} + NAD^+ \rightleftharpoons$$

L-3-Hydroxyacyl CoA

$$CH_3-(CH_2)_{14}-\overset{\overset{O}{\|}}{\overset{\beta}{C}}-\overset{\alpha}{CH_2}-\overset{\overset{O}{\|}}{C}-S-CoA^{4-} + NADH + H^+$$

β-Ketoacyl CoA

13.3.1F. Thiolase. The last reaction of β-oxidation consists of a *cleavage* of the fatty acid molecule at its β-carbon. We call the reaction **thiolysis** by analogy with other cleavage reactions involving water (hydrolysis) or phosphoric acid (phosphorolysis). Thiolytic cleavage is catalyzed by **thiolase** (or **β-ketothiolase**) and consists of a nucleophilic attack by the thiol sulfur of coenzyme A on the electron-deficient carbon of the keto group to form a shortened fatty acyl CoA and a molecule of acetyl CoA. The fatty acid residue in the fatty acyl CoA represents the original fatty acid *shortened by two carbons;* the two carbons removed form the acetyl group of acetyl CoA:

$$CH_3-(CH_2)_{14}-\overset{O}{\overset{\|}{C}}\overset{\beta}{-}\overset{\alpha}{CH_2}-\overset{O}{\overset{\|}{C}}-S-CoA^{4-} + H\,S-CoA^{4-} \rightleftharpoons$$

β-Ketoacyl CoA

$$CH_3-(CH_2)_{14}-\overset{O}{\overset{\|}{C}}\overset{\beta}{-}S-CoA^{4-} + \overset{\alpha}{CH_3}-\overset{O}{\overset{\|}{C}}-S-CoA^{4-}$$

Fatty acyl CoA   acetyl CoA

Following thiolysis, the shortened fatty acyl CoA recycles through the last four steps of β-oxidation—from acyl CoA dehydrogenase through thiolase. The second thiolytic cleavage yields another fatty acyl CoA, in which the original fatty acid has now been shortened by four carbons, as well as another molecule of acetyl CoA. These steps are repeated, each time shortening the original fatty acid by two carbons, *removed as acetyl CoA.* The ATP-requiring activation step is needed only once to form the coenzyme A derivative of the original fatty acid. Because each turn of the cycle yields a fatty acyl CoA molecule, each turn produces an already activated shortened fatty acid that can immediately reenter the cyclic reaction sequence.

We refer to the degradative pathway as β-oxidation (also called *β-oxidation cycle* or *spiral*) because of the progressive oxidation that occurs at the β-carbon of the fatty acid. The β-carbon changes from the reduced carbon of a methylene group (CH_2) to the oxidized carbon of a keto group (CO):

$$-\underset{\beta}{C}H_2-\underset{\alpha}{C}H_2- \rightarrow -\underset{\beta}{C}H=\underset{\alpha}{C}H- \rightarrow -\overset{OH}{\underset{\beta}{\overset{|}{C}}H}-\underset{\alpha}{C}H_2- \rightarrow$$

$$-\overset{O}{\overset{\|}{\underset{\beta}{C}}}-\underset{\alpha}{C}H_2- \rightarrow -\overset{O}{\overset{\|}{\underset{\beta}{C}}}-SCoA$$

The overall reaction for the complete degradation of stearic acid (18 carbons) is

$$\text{Stearic acid}^- + 8FAD + 8NAD^+ + 8H_2O$$
$$+ 9CoA-SH^{4-} + ATP^{4-}$$
$$\downarrow$$
$$9\,\text{Acetyl CoA}^{4-} + 8FADH_2 + 8NADH + 8H^+ + AMP^{2-} + PP_i^{3-}$$

Once acetyl CoA forms by β-oxidation, it can enter the citric acid cycle. Catabolism of acetyl CoA via the citric acid cycle/electron transport system results in complete oxidation of the acetyl group to CO_2 and H_2O. Hence, a combination of β-oxidation and the citric acid cycle/ETS leads to *complete oxidation of fatty acids to carbon dioxide and water.*

13.3.2. Energetics

β-Oxidation of fatty acids leads to the production of large amounts of energy in the form of ATP. The energy derives from $FADH_2$ and NADH, produced during β-oxidation and subsequently oxidized via the ETS. Recall that β-oxidation takes place in the mitochondrial matrix and that the ETS is located in the inner mitochondrial membrane. Hence, $FADH_2$ and NADH formed during β-oxidation can be reoxidized directly via the ETS; no shuttle is required. Oxidation of NADH yields 3 ATP (P/O = 3). $FADH_2$ is oxidized through the intermediacy of an *electron-transfer flavoprotein (ETF),* an inner membrane flavoprotein. From ETF, electrons flow through an iron–sulfur protein (Fe-S) to coenzyme Q, resulting in the following electron carrier sequence:

$$FADH_2 \rightarrow ETF \rightarrow Fe\text{-}S \rightarrow CoQ$$

Because $FADH_2$ enters the ETS at the level of CoQ, oxidation of $FADH_2$ yields only 2 ATP (P/O = 2), rather than the 3 ATP obtained with NADH (P/O = 3).

Consider the β-oxidation of palmitic acid, a 16-carbon saturated fatty acid. Palmitic acid is broken down completely to acetyl CoA. In that process, it passes through the β-oxidation cycle seven times, yielding eight molecules of acetyl CoA. Since the catabolism of one acetyl CoA via the citric acid cycle/ETS yields 12 ATP, we have the following:

8 Acetyl CoA (citric acid cycle)	8 ×12	= 96 ATP
7 $FADH_2$ (ETS)	7 × 2	= 14 ATP
7 (NADH, H^+) (ETS)	7 × 3	= 21 ATP
		131 ATP

From this value we must subtract the 2 ATP equivalents expended in β-oxidation. One ATP equivalent is used directly for formation of fatty acyl CoA (ATP → AMP + PP_i). The other ATP equivalent is used indirectly, since driving fatty acid activation to completion requires hydrolysis of the energy-rich bond in PP_i ($PP_i \rightarrow 2P_i$). Thus, in terms of *net yield of energy-rich bonds,* β-oxidation of palmitic acid produces 129 such bonds.

You can see that β-oxidation of one fatty acid molecule produces a very large number of ATP molecules. This explains why fats represent such a good source of energy. Complete oxidation of fatty acids to CO_2 and H_2O yields greater amounts of energy than the corresponding oxidation of carbohydrates because fatty acids are more highly reduced than carbohydrates. In fatty acids, almost all of the carbons are methylene carbons (CH_2) whereas most of the carbons in carbohydrates carry an OH group (CHOH). Thus, stearic acid

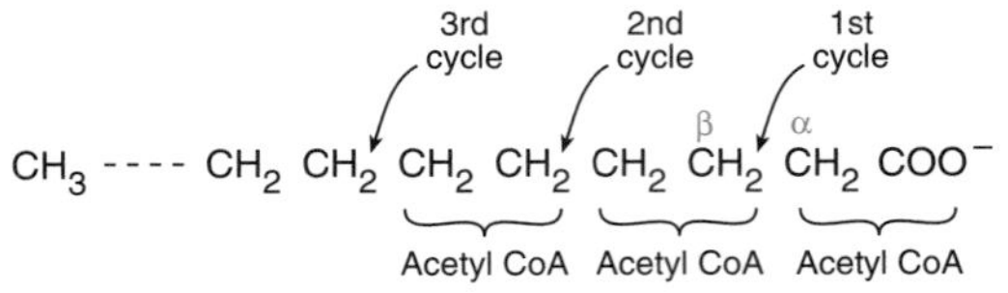

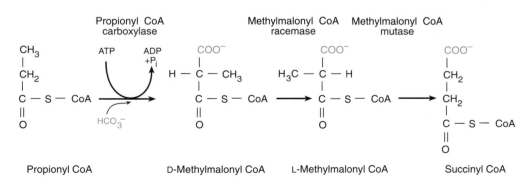

Figure 13.10.   Stepwise shortening of fatty acids during β-oxidation.

($C_{18}H_{36}O_2$) has a C:H:O ratio of 1:2:0.11 while the corresponding ratio for glucose ($C_6H_{12}O_6$) is 1:2:1. Glucose has nine times more oxygen per carbon than stearic acid. The more reduced the state of a compound, the greater is the number of hydride ions and protons abstracted from it in the process of complete oxidation. The larger number of hydride ions and protons, when channeled into the ETS, generate a larger amount of energy in the form of ATP.

Because even a small amount of fat yields considerable amounts of energy, fats constitute ideal substances for energy storage. By the same token, elimination of even a small amount of fat (weight loss) requires a great deal of exercise (energy expenditure). A change of dietary habits will, of course, also help in this regard.

Using the above yield of ATP, we can calculate the efficiency of β-oxidation. Biochemical standard free energy changes ($\Delta G^{\circ\prime}$) for complete oxidation of palmitic acid to CO_2 and H_2O and for hydrolysis of ATP are -9791 kJ mol^{-1} and -30.5 kJ mol^{-1}, respectively. On this basis, the efficiency equals

$$\frac{(129 \times 30.5 \text{ kJ mol}^{-1})}{9791 \text{ kJ mol}^{-1}} \times 100 = 40\%$$

a value similar to the efficiencies we calculated for other metabolic pathways. Remember, however, that a calculation of the actual intracellular efficiency must be based on $\Delta G^{\prime}$ values.

13.3.3. Even- and Odd-Numbered Fatty Acids

Degradation by means of β-oxidation applies to both even- and odd-numbered fatty acids (Figure 13.10). Even-numbered fatty acids are degraded *completely to acetyl CoA*. An even-numbered fatty acid having n carbons yields $n/2$ molecules of acetyl CoA but passes through the cycle only $(n/2) - 1$ times. For example, stearic acid (18 carbons) yields nine acetyl CoA but passes through the cycle only eight times.

Odd-numbered fatty acids also yield large quantities of acetyl CoA when subjected to β-oxidation. However, the final thiolytic cleavage produces acetyl CoA and *propionyl CoA* (CH_3-CH_2-CO-S-CoA). An odd-numbered fatty acid having n carbons yields $(n - 3)/2$ acetyl CoA plus propionyl CoA, and passes through the cycle ($n - 3)/2$ times. Propionyl CoA undergoes a three-step conversion to succinyl CoA (Figure 13.11), an intermediate of the citric acid cycle. The first enzyme in this reaction sequence requires biotin as coenzyme, and the third enzyme requires 5′-deoxyadenosylcobalamin, a derivative of vitamin B_{12} (cyanocobalamin), as coenzyme.

13.3.4. Unsaturated Fatty Acids

β-Oxidation of unsaturated fatty acids requires the action of as many as three auxiliary enzymes, depending on the fatty acid and the organism (Figure 13.12). Two problems must be addressed.

First, most naturally occurring unsaturated fatty acids contain only *cis* double bonds. In order to serve as substrate for enoyl CoA hydratase, such *cis* double bonds must be converted to the *trans* configuration, a reaction catalyzed by **enoyl CoA isomerase.**

Second, a double bond may have to be eliminated in order to convert the molecule to a better substrate for enoyl CoA hydratase. The enzyme **dienoyl CoA reductase** catalyzes reduction of a double to a single bond and

Figure 13.11.   Conversion of propionyl CoA to succinyl CoA.

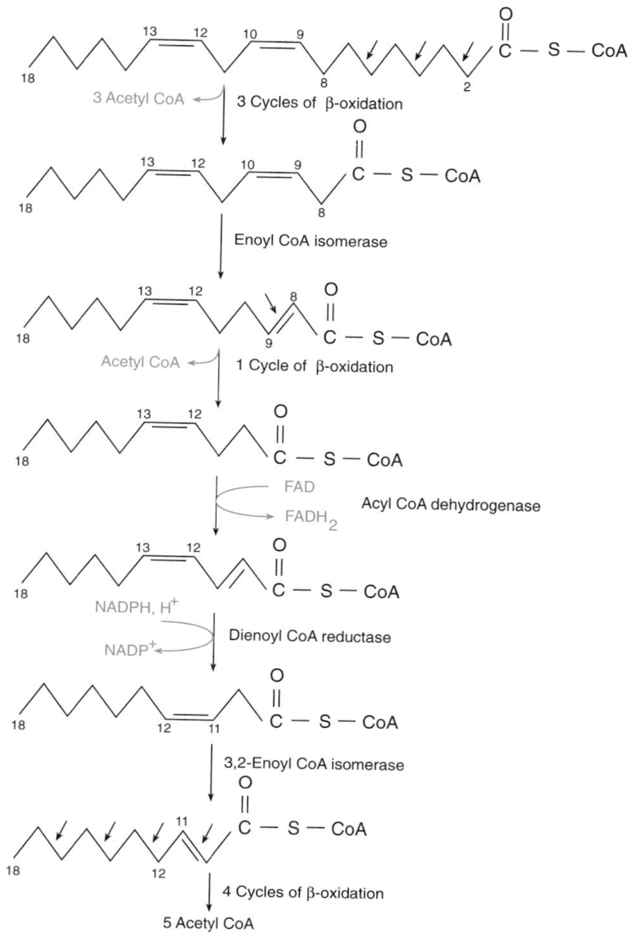

Figure 13.12. β-Oxidation of an unsaturated fatty acid (linoleic acid) in mammals. The original numbering of carbon atoms has been maintained throughout for clarity.

uses NADPH as coenzyme. A third auxiliary enzyme, **3,2-enoyl CoA isomerase,** is required in mammals but not in *Escherichia coli.*

13.4. KETONE BODIES

13.4.1. Metabolic Fates of Acetyl CoA

Next to ATP, acetyl CoA probably represents the most important low-molecular-weight biomolecule of metabolism. Its four major metabolic fates exemplify its pivotal role (Figure 13.13):

1. It represents the form in which carbohydrates, lipids, and some amino acids enter the citric acid cycle. Catabolism of these nutrients via the combined action of the citric acid cycle and the electron transport system results in their complete oxidation to CO_2 and H_2O.

2. It constitutes the source of all the carbons for the biosynthesis of cholesterol. Because cholesterol is the parent compound of steroids, acetyl CoA serves as the precursor of all steroids.

3. It functions as the precursor for the biosynthesis of fatty acids.

4. It serves as the precursor for the biosynthesis of *acetone, acetoacetate,* and *β-hydroxybutyrate,* three compounds known as *ketone bodies.*

13.4.2. Properties of Ketone Bodies

The term *ketone bodies* is somewhat misleading. Acetone, acetoacetate, and β-hydroxybutyrate constitute chemical compounds, not "bodies." Moreover, only two of the compounds are actually ketones; the third—β-hydroxybutyrate—is not a ketone. All three ketone bodies derive from acetyl CoA and form primarily in the liver.

Ketone bodies represent low-molecular-weight, water-soluble forms of lipid-based energy since they can serve as substrates for generating energy in place of glucose in muscle and brain tissue. When functioning in this capacity, ketone bodies undergo conversion to acetyl CoA, which enters the citric acid cycle. Some tissues, notably the heart, commonly derive large amounts of their energy requirements from ketone bodies. Other tissues, especially the brain, can also use ketone bodies as a source of energy under certain conditions.

Brain cells form tight junctions in capillaries that prevent passive entry into the brain of water-soluble substances. This selective filtration system separates the brain from the general circulation and is known as the **blood–brain barrier.** Water-soluble compounds pass this barrier only if they can be moved across by specific transport systems, and fatty acids cannot pass the blood–brain barrier. Glucose is the normal fuel of the brain. However, upon starvation, the brain can adapt to use ketone bodies as a source of energy. With prolonged starvation, acetoacetate can provide 75% of the energy needs of the brain.

You could call synthesis of ketone bodies an "overflow pathway." Such a pathway normally has minor significance but becomes important when concentrations of specific substances increase and metabolites "spill over" into the secondary pathway. In the case of ketone bodies, the pathway becomes accentuated *when the concentration of acetyl CoA builds up,* which occurs when its formation exceeds its consumption. As the overflow pathway begins to be used, concentrations of ketone bodies in the blood

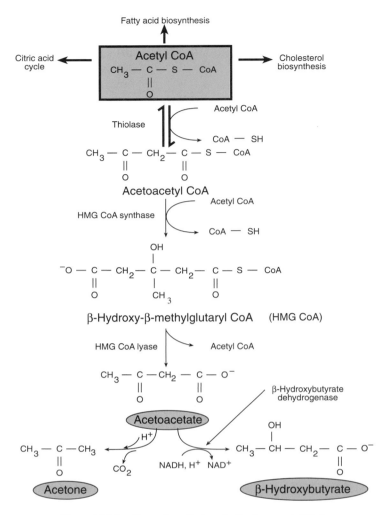

Figure 13.13.   Formation of ketone bodies from acetyl CoA.

begin to increase (*ketonemia*). If the condition persists, large quantities of ketone bodies appear in the urine (*ketonuria*). The generalized condition, called **ketosis,** is of clinical concern; it must be treated to avoid serious pathological complications.

13.4.3. Ketosis and Its Implications

Several conditions can lead to accumulation of acetyl CoA and development of ketosis. These conditions include starvation, diabetes (insulin deficiency), and a carbohydrate-deficient diet.

During starvation, the body has to rely on energy stores of glycogen and adipose tissue. Glycogen stores are small and rapidly exhausted. At that point, extensive mobilization of fatty acids from adipose tissue takes place. Subsequent oxidation of the fatty acids generates large amounts of acetyl CoA. Recall that citric acid cycle intermediates, such as oxaloacetate, have fixed levels of concentration. Therefore, only a limited amount of oxaloacetate is available for combination with acetyl CoA to form citrate in the first reaction of the citric acid cycle. Because there is insufficient oxaloacetate to combine with all of the acetyl CoA formed, excess acetyl CoA begins to accumulate. Thus, extensive reliance on lipid catabolism during starvation leads to an overabundance of acetyl CoA.

Both diabetes caused by insulin deficiency and a carbohydrate-deficient diet have the same effect as starvation. In both instances, the body compensates for a deficiency of carbohydrate metabolism by increased lipid metabolism, and acetyl CoA may accumulate faster than it is consumed. In all three cases, *faulty carbohydrate metabolism* becomes expressed as an *abnormality of lipid metabolism.*

Why is excessive formation of ketone bodies of serious clinical concern? The answer lies in an acid–base reaction that involves ketone bodies. Sodium bicarbonate ($NaHCO_3$), termed the **alkaline reserve** of the body, constitutes one of a small number of compounds in metabolism that have strong basic character and the capacity to raise pH. Basicity of $NaHCO_3$ results from both the Na^+ content and the production of OH^- by hydrolysis:

$$HCO_3^- + H_2O \rightarrow H_2CO_3 + OH^-$$
$$\downarrow$$
$$CO_2 + H_2O$$

In contrast, metabolism abounds in acidic compounds that can lower pH. Examples include the many weak acids you encountered as intermediates in the citric acid cycle and glycolysis. Consequently, sodium bicarbonate has a crucial role in maintaining physiological pH values.

When ketone bodies accumulate, acetoacetic acid and β-hydroxybutyric acid react with sodium bicarbonate, forming salts:

$$CH_3-CO-CH_2-COOH + NaHCO_3 \rightarrow$$
Acetoacetic acid
$$CH_3-CO-CH_2-COO^-Na^+ + CO_2 + H_2O$$
Sodium acetoacetate

The salts formed by these reactions (e.g., sodium acetoacetate) are excreted. Not only has sodium bicarbonate been destroyed, but its sodium ions are also lost from the body. Both effects combine to impair seriously the body's capacity to counteract pH-lowering effects of acidic metabolites. As a result, the overall pH tends to drop, causing *acidosis*. Additionally, the salts formed must be dissolved so that they can be excreted in the urine. Hence, development of acidosis is accompanied by loss of water or *dehydration.*

Symptoms of ketosis become more pronounced as time goes on. In advanced stages, acetone may actually be detected in the breath of afflicted individuals. If not treated, ketosis may lead to a progression of symptoms from nausea to depression of the central nervous system, severe dehydration, deep coma, and ultimately death. Clearly, ketosis cannot be allowed to persist and must be treated promptly.

13.5. FATTY ACID BIOSYNTHESIS

Fatty acid synthesis occurs when the organism needs to store excess energy derived from the diet. The process takes place in the cytosol and involves progressive lengthening of a fatty acid chain by two carbons derived from acetyl CoA. The fatty acids subsequently combine with glycerol and become stored as acylglycerols until needed.

Acetyl CoA is generated inside the mitochondria: the pyruvate dehydrogenase complex catalyzes the conversion of pyruvate to acetyl CoA, and β-oxidation produces acetyl CoA in the thiolase reaction. Before this mitochondria-generated acetyl CoA can participate in fatty acid synthesis, it needs to be transported from the mitochondria to the cytosol. However, the inner mitochondrial membrane is essentially impermeable to acetyl CoA. Accordingly, a shuttle mechanism functions in lieu of direct transport. The pertinent mechanism, called the **tricarboxylate transport system** (Figure 13.14), uses citrate as carrier of the acetyl group. Inside the mitochondria, citrate forms from acetyl CoA and oxaloacetate in the first reaction of the citric acid cycle. In the cytosol, cleavage of citrate, catalyzed by *citrate lyase,* regenerates both acetyl CoA and oxaloacetate:

$$Citrate^{3-} + ATP^{4-} + CoA-SH^{4-} \rightarrow$$
$$Acetyl\ CoA^{4-} + ADP^{3-} + P_i^{2-} + oxaloacetate^{2-}$$

Oxaloacetate leads to malate or pyruvate, either one of which can be transported back into the mitochondria. Under some conditions, the shuttle results in production of NADPH that can be used for reductive reactions of fatty acid synthesis. If malate returns to the mitochondria, no NADPH forms, but if malate is converted to pyruvate by *malic enzyme,* NADPH forms in the cytosol. In the latter case, one molecule of NADPH is produced per molecule of acetyl CoA transported, a substantial fraction of the total required for fatty acid synthesis. For example, synthesis of palmitate requires 8 molecules of acetyl CoA and 14 molecules of NADPH. Transport of 8 molecules of acetyl CoA from the mitochondria to the cytosol generates 8 molecules of NADPH, leaving 6 molecules of NADPH to be obtained from other sources. The remaining 6 molecules of NADPH can be furnished by the pentose phosphate pathway in the liver or by malic enzyme in adipose tissue.

13.5.1. Acetyl CoA Carboxylase

Synthesis of fatty acids begins with formation of **malonyl CoA** from acetyl CoA. The reaction, catalyzed by **acetyl CoA carboxylase,** constitutes the *committed step* of fatty acid synthesis.

The acetyl CoA carboxylase of prokaryotes, such as *E. coli,* comprises a complex of three separate proteins. One protein, called **biotin carboxyl carrier protein**

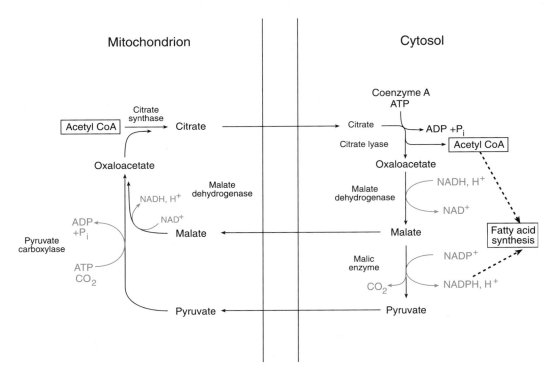

Figure 13.14.   The tricarboxylate transport system.

(**BCCP**), serves as a carrier of *biotin*. Biotin is linked covalently to an ϵ-amino group of a lysine residue in the protein, forming *biocytin* (see Figure 11.14). The remaining two proteins of acetyl CoA carboxylase are enzymes, **biotin carboxylase** and **transcarboxylase**. In *E. coli,* the carrier protein and the two enzymes have molecular weights of 23,000, 98,000 (two subunits of 49,000 each), and 130,000, respectively.

Acetyl CoA carboxylase of eukaryotes is a dimer of two identical subunits (MW = 260,000 each). Each subunit contains one molecule of biotin. Both enzymatic activities (biotin carboxylase and transcarboxylase) and the biotin-binding function are located on a single polypeptide chain. Biotin carboxylase catalyzes the following reaction:

$$\text{BCCP-biotin} + \text{ATP}^{4-} + \text{CO}_2 + \text{H}_2\text{O}$$
$$\downarrow$$
$$\text{BCCP-carboxybiotin}^- + \text{ADP}^{3-} + \text{P}_i^{2-} + 2\text{H}^+$$

Transcarboxylase catalyzes a transfer of the activated carboxyl group from BCCP-carboxybiotin to acetyl CoA (Figure 13.15), yielding malonyl CoA and BCCP-biotin:

$$\text{BCCP-carboxybiotin}^- + \text{acetyl CoA}^{4-} \rightarrow$$
$$\text{Malonyl CoA}^{5-} + \text{BCCP-biotin}$$

13.5.2. Fatty Acid Synthase

The remaining reactions of fatty acid synthesis, beginning with acetyl CoA and malonyl CoA, are catalyzed by the **fatty acid synthase complex.** In animal cells, fatty acid synthase consists of seven enzymatic activities and an acyl carrier protein.

Acyl carrier protein (ACP) is a small protein that functions in fatty acid synthesis much as coenzyme A does in fatty acid degradation. In β-oxidation, fatty acid derivatives become linked to the SH group of a *phosphopantetheine group* that forms part of the structure of coenzyme A. In fatty acid synthesis, fatty acid derivatives become linked to the SH group of a phosphopantetheine group that is covalently attached to an acyl carrier protein (Figure 13.16). ACP, with its attached phosphopantetheine group, carries the acyl group in fatty acid synthesis much as coenzyme A carries it in fatty acid degradation. Scientists believe that the long phosphopantetheine group serves as a "swinging arm" (like the lipoyllysine arm of the pyruvate dehydrogenase complex) that moves the substrate from one catalytic site to another on the enzyme complex (Figure 13.17).

Acyl carrier protein was first isolated from *E. coli.* The 4'-phosphopantetheine group is esterified to the hydroxyl group of serine 36 of the protein. The protein itself

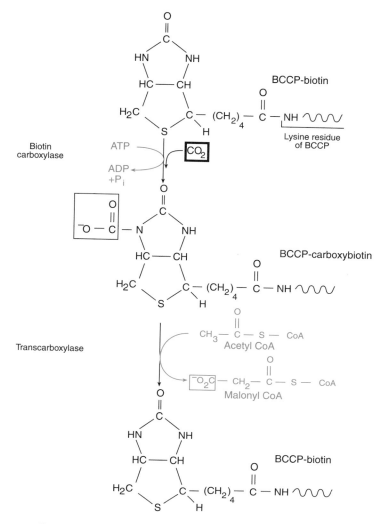

Figure 13.15.   Formation of malonyl CoA by acetyl CoA carboxylase.

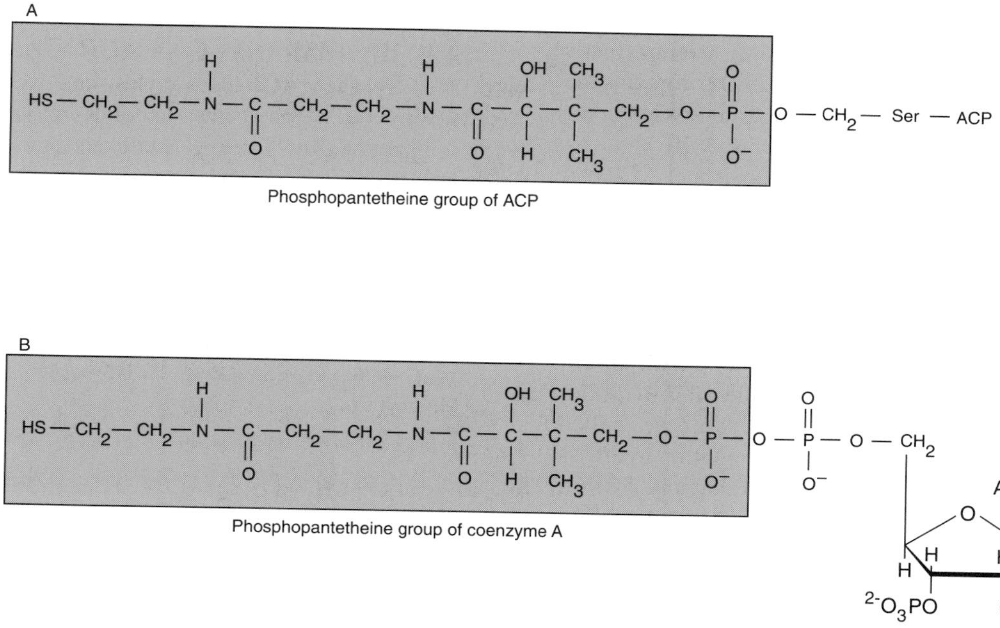

A

HS—CH₂—CH₂—N—C—CH₂—CH₂—N—C—C—C—CH₂—O—P—O—CH₂— Ser —ACP

Phosphopantetheine group of ACP

B

HS—CH₂—CH₂—N—C—CH₂—CH₂—N—C—C—C—CH₂—O—P—O—P—O—CH₂

Adenine

²⁻O₃PO   OH

Phosphopantetheine group of coenzyme A

Figure 13.16.   The phosphopantetheine group: (A) as a prosthetic group of acyl carrier protein (ACP), where it is esterified to the hydroxyl group of a serine residue; (B) as a structural component of coenzyme A.

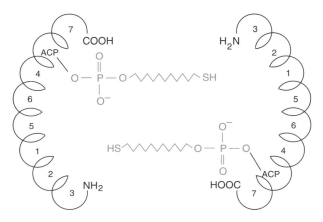

Figure 13.17. The dimeric structure of eukaryotic fatty acid synthase. Each subunit contains an acyl carrier protein (ACP) and catalytic sites of seven enzymatic activities: 1, Acetyl CoA:ACP transacylase; 2, malonyl CoA:ACP transacylase; 3, β-ketoacyl-ACP synthase; 4, β-ketoacyl-ACP reductase; 5, β-hydroxyacyl-ACP dehydratase; 6, enoyl-ACP reductase; 7, palmitoyl-ACP thioesterase. [Adapted, with permission, from S. J. Wakil, J. K. Stoops, and V. C. Joshi, *Annu. Rev. Biochem.* 52:537–579 (1983). © 1983 by Annual Reviews, Inc.]

consists of 77 amino acid residues (MW = 10,000). Assembly of ACP and the enzymes of fatty acid synthesis depends on the type of organism involved. In *E. coli* and plants, fatty acid synthase comprises a multienzyme system, an aggregate of *seven separate polypeptide chains.* One polypeptide chain constitutes ACP, and the remaining *six* represent enzymes.

In yeast, fatty acid synthase also consists of ACP and *six* enzymes except that these are located on *two multifunctional polypeptide chains.* One chain (MW = 185,000) contains the ACP function and two enzymatic activities. The other chain (MW = 175,000) contains the remaining four enzymatic activities. Six dimers associate to form a very large complex (MW ≈ 2.4×10^6).

In animals, fatty acid synthase consists of ACP and *seven* enzymes, all of which are located on *a single multifunctional polypeptide chain* (Figure 13.17). Contiguous regions of the polypeptide chain fold in unique fashion to generate different enzymatic activities and the ACP function. The enzyme is a dimer, containing two identical subunits (MW ≈ 260,000 each) with an antiparallel head-to-tail arrangement. The additional enzymatic activiy of animal fatty acid synthase, called **palmitoyl-ACP thioesterase,** catalyzes the hydrolysis of the final palmitoyl-ACP to palmitate and ACP; it becomes functional only after a fatty acid chain of 16 carbons has been synthesized. Other organisms lack palmitoyl-ACP thioesterase and use palmitoyl-ACP directly.

13.5.3. Individual Reactions

Fatty acid synthesis in animals consists of the following seven reactions, the first six of which are shown in Figure 13.18:

1 Priming: Acetyl CoA:ACP transacylase
2 Loading: Malonyl CoA:ACP transacylase
3 Condensation: β-Ketoacyl-ACP synthase
4 Reduction: β-Ketoacyl-ACP reductase
5 Dehydration: β-Hydroxyacyl-ACP dehydratase
6 Reduction: Enoyl-ACP reductase
7 Release: Palmitoyl-ACP thioesterase

13.5.3A. Acetyl CoA:ACP Transacylase.

Acetyl CoA:ACP transacylase catalyzes a "priming" reaction whereby acetyl CoA becomes transferred, first to ACP and then to an SH group of β-ketoacyl-ACP synthase. We describe the two-step priming process as follows (in mammals, the acetyl-ACP intermediate does not form):

$$\text{Acetyl CoA}^{4-} + \text{HS}-\text{ACP}^- \rightarrow$$

$$\underset{\text{Acetyl-ACP}}{\text{CH}_3-\overset{\displaystyle O}{\overset{\|}{C}}-S-\text{ACP}^-} + \text{CoA}-\text{SH}^{4-} \tag{13.1a}$$

$$\text{CH}_3-\overset{\displaystyle O}{\overset{\|}{C}}-S-\text{ACP}^- + \text{HS}-\text{synthase} \rightarrow$$

$$\underset{\text{Acetyl-synthase}}{\text{CH}_3-\overset{\displaystyle O}{\overset{\|}{C}}-S-\text{synthase}} + \text{HS}-\text{ACP}^- \tag{13.1b}$$

13.5.3B. Malonyl CoA:ACP Transacylase.

Malonyl CoA:ACP transacylase catalyzes a reaction whereby the second substrate of β-ketoacyl-ACP synthase is readied by "loading" malonate onto ACP. In this reaction, the free SH group of ACP attacks the carbonyl group of malonyl CoA to form malonyl-ACP. The priming and loading reactions prepare the two substrates for subsequent condensation.

$$\underset{\text{Malonyl CoA}}{^-\text{O}_2\text{C}-\text{CH}_2-\overset{\displaystyle O}{\overset{\|}{C}}-S-\text{CoA}^{4-}} + \text{HS}-\text{ACP}^- \rightarrow$$

$$\underset{\text{Malonyl-ACP}}{^-\text{O}_2\text{C}-\text{CH}_2-\overset{\displaystyle O}{\overset{\|}{C}}-S-\text{ACP}^-} + \text{CoA}-\text{SH}^{4-} \tag{13.2}$$

13.5.3C. β-Ketoacyl-ACP Synthase. The

next reaction involves a *condensation* between the acetyl

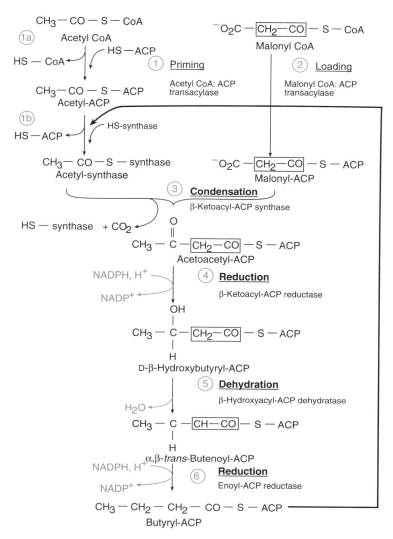

Figure 13.18. Fatty acid synthesis in *E. coli*. In the first round of elongation, an acetyl group becomes extended by a two-carbon fragment (boxed) derived from malonyl CoA. In the second round, butyryl-ACP substitutes for acetyl-ACP in reaction 1b. Subsequent steps lead to hexanoyl-ACP, which then substitutes for acetyl-ACP in reaction 1b, and so on. In mammals, acetyl-ACP is not formed in reaction 1, and elongation stops with palmitoyl-ACP, from which palmitate is then released by palmitoyl thioesterase (reaction 7, not shown).

group (linked to an SH group on the enzyme) and the malonyl group (linked to ACP). Decarboxylation activates the methylene carbon of malonyl CoA and makes it a better nucleophile for attacking the carbonyl carbon of the acetyl group. Loss of CO_2 also helps to make this reaction thermodynamically favorable and irreversible. The condensation is catalyzed by **β-ketoacyl-ACP synthase**, and the product formed consists of an acetoacetyl group linked to ACP.

$$CH_3\overset{O}{\underset{\|}{C}}-S-synthase + {}^-O_2C-CH_2-\overset{O}{\underset{\|}{C}}-S-ACP^- +H^+ \rightarrow$$
Acetyl-synthase   Malonyl-ACP

$$CH_3-\overset{O}{\underset{\|}{\underset{\beta}{C}}}-CH_2-\overset{O}{\underset{\|}{\underset{\alpha}{C}}}-S-ACP^- + HS-synthase + CO_2$$
Acetoacetyl-ACP
(13.3)

13.5.3D. β-Ketoacyl-ACP Reductase.

The reaction catalyzed by **β-ketoacyl-ACP reductase** constitutes the first *reduction* step of fatty acid synthesis. Note that reduction, like oxidation in β-oxidation, takes place at the β-carbon of the molecule. NADPH serves as reducing agent, and the product has the D-configuration.

$$CH_3-\overset{\overset{O}{\|}}{C}-\overset{\beta}{C}H_2-\overset{\overset{O}{\|}}{\underset{\alpha}{C}}-S-ACP^- + NADPH + H^+ \rightarrow$$
Acetoacetyl-ACP

$$CH_3-\overset{\overset{OH}{|}}{\underset{\underset{H}{|}}{C}}-\overset{\beta}{C}H_2-\overset{\overset{O}{\|}}{\underset{\alpha}{C}}-S-ACP^- + NADP^+$$

D-β-Hydroxybutyryl-ACP (13.4)

13.5.3E. β-Hydroxyacyl-ACP Dehydratase.

The next step represents a *dehydration* reaction, catalyzed by **β-hydroxyacyl-ACP dehydratase**. The reaction yields an α, β-unsaturated compound with a *trans* configuration:

$$CH_3-\overset{\overset{OH}{|}}{\underset{\underset{H}{|}}{C}}-\overset{\beta}{C}H_2-\overset{\overset{O}{\|}}{\underset{\alpha}{C}}-S-ACP^- \curvearrowright CH_3-\overset{\beta}{C}=\overset{\overset{H}{|}}{\underset{\underset{H}{|}}{C}}-\overset{\overset{O}{\|}}{\underset{\alpha}{C}}-S-ACP^-$$
$$\qquad\qquad\qquad\qquad\qquad H_2O$$

D-β-Hydroxybutyryl-ACP   α, β-*trans*-butenoyl-ACP

(13.5)

13.5.3F. Enoyl-ACP Reductase.

This step represents the end of the first round of fatty acid synthesis, whereby a four-carbon fatty acid forms from two two-carbon fragments. The step involves a second *reduction,* again at the β-carbon and mediated by NADPH. The enzyme **enoyl-ACP reductase** catalyzes this reaction, which yields a four-carbon fatty acid attached to ACP (butyryl-ACP):

$$CH_3-\overset{\beta}{C}=\overset{\overset{H}{|}}{\underset{\underset{H}{|}}{C}}-\overset{\overset{O}{\|}}{\underset{\alpha}{C}}-S-ACP^- + NADPH + H^+ \rightarrow$$

α, β-*trans*-Butenoyl-ACP

$$CH_3-\overset{\beta}{C}H_2-\overset{\alpha}{C}H_2-\overset{\overset{O}{\|}}{C}-S-ACP^- + NADP^+$$
Butyryl-ACP   (13.6)

At this point, the acetyl group with which the system was originally primed has been elongated by a two-carbon fragment derived from malonate. Butyryl-ACP now becomes the substrate for a second round of elongation. It substitutes for acetyl-ACP, and its acyl group is transferred from butyryl-ACP to the SH group of β-ketoacyl-ACP synthase (reaction 13.1b). The synthase then catalyzes a condensation between the butyryl group (linked to an SH group of the enzyme) and a malonyl group (linked to ACP) according to reaction (13.3).

The condensation yields a six-carbon product, attached to ACP. The condensation product is converted to hexanoyl-ACP, a six-carbon fatty acid attached to ACP, by means of reactions (13.4) through (13.6). In a third round of elongation [reactions (13.1b), (13.3), and (13.4)–(13.6)], another two-carbon fragment is added, forming an eight-carbon fatty acid attached to ACP, and so on.

13.5.3G. Palmitoyl-ACP   Thioesterase.

During each round of elongation, fatty acid chains are elongated by two carbons. In animal cells, elongation stops with synthesis of a 16-carbon fatty acid in the form of *palmitoyl-ACP. Palmitoyl-ACP thioesterase* then catalyzes a reaction whereby palmitate is *released* from the fatty acid synthase complex. The enzyme catalyzes the hydrolysis of palmitoyl-ACP to palmitate and ACP:

$$Palmitoyl\text{-}ACP^- + H_2O \rightarrow$$
$$Palmitate^- + HS-ACP^- + H^+ \qquad (13.7)$$

Elongation of palmitate to form longer chain fatty acids, as well as introduction of double bonds, must be carried out by other enzyme systems (covered below).

Because the first round of synthesis generates a four-carbon, rather than a two-carbon, fatty acyl-ACP, formation of palmitate (16 carbons) requires that reactions (13.3) through (13.6) be carried out only seven times, using seven malonyl CoA molecules. Synthesis of palmitoyl-ACP yields seven molecules of water, but hydrolysis of palmitoyl-ACP to palmitate and ACP requires one molecule of water, so that the net number of water molecules formed is six. Accordingly, the synthesis of palmitate, beginning with acetyl CoA and malonyl CoA, is described by the following equation:

$$Acetyl\ CoA^{4-} + 7malonyl\ CoA^{5-} + 14NADPH + 20H^+$$
$$\downarrow$$
$$Palmitate^- + 7CO_2 + 14NADP^+ + 8CoA-SH^{4-} + 6H_2O$$

To form the seven molecules of malonyl CoA used in this equation requires an input of seven molecules of acetyl CoA:

$$7Acetyl\ CoA^{4-} + 7CO_2 + 7ATP^{4-} + 7H_2O$$
$$\downarrow$$
$$7Malonyl\ CoA^{5-} + 7ADP^{3-} + 7P_i^{2-} + 14H^+$$

Combining the two equations provides the overall stoichiometry for *palmitate synthesis beginning with acetyl CoA:*

$$8Acetyl\ CoA^{4-} + 7ATP^{4-} + 14NADPH + 6H^+$$
$$\downarrow$$
$$Palmitate^- + 14NADP^+ + 8CoA-SH^{4-} + 6H_2O$$
$$+ 7ADP^{3-} + 7P_i^{2-}$$

NADPH and H^+ for the reductive reactions come

from the malic enzyme reaction (see Figure 13.14) and the pentose phosphate pathway.

13.5.4. Comparison with β-Oxidation

We can summarize similarities and differences between the pathways of fatty acid synthesis and fatty acid degradation (β-oxidation) as follows:

- The two pathways differ in their compartmentation. Fatty acid synthesis takes place in the cytosol, whereas fatty acid degradation takes place in the mitochondria.
- Both pathways involve intermediates linked to a carrier. In fatty acid synthesis, the carrier is ACP; in fatty acid degradation, it is coenzyme A.
- Four reactions in one pathway are chemically the reverse of four reactions in the other pathway but use different enzymes and cofactors. These are the condensation, reduction, dehydration, and reduction steps in fatty acid synthesis and the oxidation, hydration, oxidation, and cleavage steps in fatty acid degradation.
- Both pathways require a transport mechanism linking the mitochondrial and cytosolic compartments. The tricarboxylate transport system for acetyl CoA functions in fatty acid synthesis, and the carnitine carrier system for fatty acyl CoA functions in fatty acid degradation.
- Both pathways feature progressive alterations of hydrocarbon chains of fatty acids. In fatty acid synthesis, chains become extended by successive additions of two-carbon fragments derived from acetyl CoA but condensed in the form of malonyl CoA. In fatty acid degradation, chains become shortened by successive removal of two-carbon fragments in the form of acetyl CoA.
- Fatty acids are synthesized from the methyl to the carboxyl end of the molecule (the COOH end is synthesized last). Fatty acids are degraded in the opposite direction (the COOH end is removed first).
- The hydroxyacyl intermediate has a D-configuration in fatty acid synthesis but an L-configuration in fatty acid degradation.
- Fatty acid synthesis constitutes a reductive pathway requiring NADPH; fatty acid degradation constitutes an oxidative pathway requiring FAD and NAD^+.
- Even though the "spiral" or "cyclic" portion (the repeated steps) of each pathway requires addition or removal of a two-carbon fragment, this portion is traversed only seven times for either synthesis or degradation of a 16-carbon fatty acid.
- In animals, enzymes of fatty acid synthesis are located on a single polypeptide chain as part of the fatty acid synthase complex. To what degree corresponding enzymes of fatty acid degradation are associated is still unsettled.

13.5.5. Elongases and Desaturases

You saw that in animals fatty acid synthesis stops with the formation of palmitate, a 16-carbon fatty acid. To produce fatty acids having longer carbon chains requires additional reactions, using palmitate as starting material. These reactions are catalyzed by specific **elongases** and occur both in the mitochondria and in the endoplasmic reticulum. In mitochondria, elongation involves a reversal of β-oxidation, with some modifications. In the endoplasmic reticulum, elongation resembles ordinary fatty acid synthesis except that the fatty acid occurs as a coenzyme A derivative rather than as an ACP derivative.

Unsaturated fatty acids form from saturated ones by action of **desaturases.** These enzymes contain nonheme iron, and in eukaryotes they catalyze the following general reaction:

$$CH_3-(CH_2)_x-CH_2-CH_2-(CH_2)_y-CO-S-CoA$$
$$+ NADH + H^+ + O_2$$
$$\downarrow$$
$$CH_3-(CH_2)_x-CH=CH-(CH_2)_y-CO-S-CoA$$
$$+ NAD^+ + 2H_2O$$

In mammals, introducing double bonds requires a mini electron transport system composed of two enzymes and a cytochrome (Figure 13.19). Mammals lack the enzymes capable of introducing double bonds at carbons beyond C(9). For this reason, linoleic acid and linolenic acid cannot be synthesized by mammals and must be obtained through the diet; they constitute *essential fatty acids.*

Odd-numbered fatty acids are synthesized by priming fatty acid synthase with propionyl CoA rather than with acetyl CoA.

13.5.6. Regulation

The reaction catalyzed by acetyl CoA carboxylase represents a major control point of fatty acid biosynthesis. In eukaryotes, the monomeric form of this enzyme (composed of two subunits) is inactive. However, the monomer polymerizes to form long filamentous structures that are active. Activity of acetyl CoA carboxylase is controlled by shifts in the equilibrium between its two forms:

$$\text{Monomer} \rightleftharpoons \text{polymer}$$
$$\text{(inactive)} \quad \text{(active)}$$

Citrate shifts the equilibrium toward the active polymer and stimulates fatty acid synthesis, but researchers

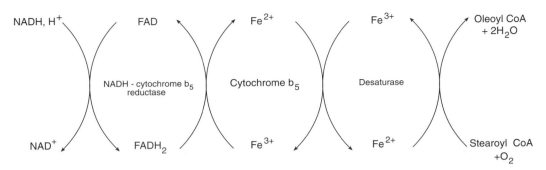

Figure 13.19. The mini electron transport system of fatty acid desaturases in mammals, located on the cytosolic face of the endoplasmic reticulum. Formation of two water molecules constitutes a four-electron reaction; two electrons come from NADH, and two come from the fatty acid bond being reduced.

have not yet decided whether this constitutes a physiological role of citrate. Palmitoyl CoA shifts the equilibrium toward the inactive monomer and inhibits synthesis. Since palmitoyl CoA is the product of fatty acid synthesis, its effect constitutes feedback inhibition.

The equilibrium is also affected by phosphorylation/dephosphorylation of the enzyme in response to certain hormones. Glucagon and epinephrine promote phosphorylation by stimulating cAMP-dependent protein kinase. Phosphorylation shifts the equilibrium toward the inactive monomer (Figure 13.20) and results in inhibition of fatty acid synthesis. Insulin has the opposite effect. Insulin decreases the level of cAMP and promotes dephosphorylation and formation of the active polymer, thus stimulating fatty acid synthesis.

In prokaryotes, acetyl CoA carboxylase is regulated through intracellular levels of guanine nucleotides.

Guanosine 5′-triphosphate (GTP) serves as the major source of energy for protein synthesis, where its hydrolysis produces GDP. Thus, in the course of protein synthesis, intracellular levels of GTP and GDP undergo change. The extent of protein synthesis varies, depending on the state of growth of the cells, so that fatty acid synthesis becomes regulated by the growth requirements of cells.

13.6. BIOSYNTHESIS OF OTHER LIPIDS

Free fatty acids do not occur in cells to any great extent. Most fatty acids are present in esterified form as *triacylglycerols* or *glycerophospholipids*. The biosynthesis of these two types of complex lipids occurs primarily on the endoplasmic reticulum of liver cells or fat cells.

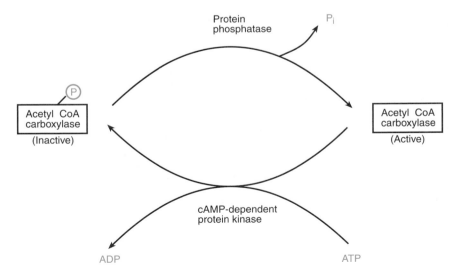

Figure 13.20. Equilibrium between inactive (monomeric) and active (polymeric) acetyl CoA carboxylase. Hormonally controlled phosphorylation and dephosphorylation shift the equilibrium.

13.6.1. Acylglycerols

Acylglycerols are synthesized from two precursors, *fatty acyl CoA* and *glycerol 3-phosphate*. Fatty acyl CoA is the product of fatty acid activation. Glycerol 3-phosphate can be formed either from the glycolytic intermediate dihydroxyacetone phosphate or by phosphorylation of glycerol, a product of the degradation of acylglycerols:

$$\text{Dihydroxyacetone phosphate}^{2-} + \text{NADH} + \text{H}^+ \rightarrow$$
$$\text{glycerol 3-phosphate}^{2-} + \text{NAD}^+$$
$$\text{Glycerol} + \text{ATP}^{4-} \rightarrow \text{glycerol 3-phosphate}^{2-} + \text{ADP}^{3-} + \text{H}^+$$

Mono- and diacylglycerols form from glycerol 3-phosphate by succesive esterifications with fatty acyl

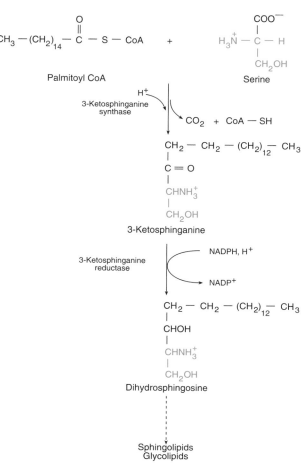

Figure 13.22. Biosynthesis of dihydrosphingosine, the precursor of sphingolipids and glycolipids.

CoA (Figure 13.21). Triacylglycerols form from 1,2-diacylglycerol 3-phosphate (phosphatidic acid) in a two-step process. First, the phosphate group is removed by hydrolysis, yielding 1,2-diacylglycerol. The diacylglycerol then reacts with another molecule of fatty acyl CoA.

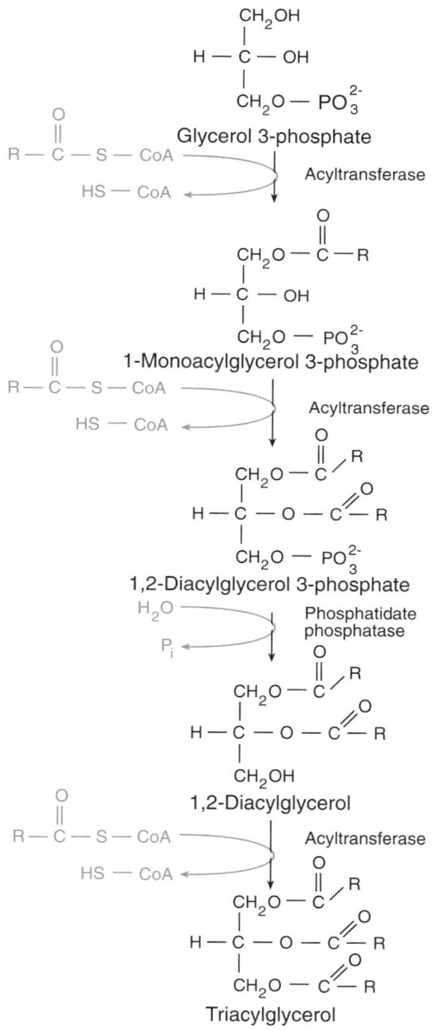

13.6.2. Phospholipids

13.6.2A. Glycerophospholipids. Glycerophospholipids represent important constituents of biological membranes. Pathways for their biosynthesis vary with the type of organism and differ in their details. Some synthetic reactions start with 1,2-diacylglycerol, and others with 1,2-diacylglycerol 3-phosphate. Frequently, CDP (5′-cytidine diphosphate) serves as a carrier for a lipid component much as UDP serves as a carrier for carbohydrates. Some pathways for the biosynthesis of phosphatidyl choline and phosphatidyl ethanolamine require CDP-choline and CDP-ethanolamine, respective-

Figure 13.21. Biosynthesis of acylglycerols from glycerol 3-phosphate. Different fatty acyl groups may be transferred at the three acyltransferase steps.

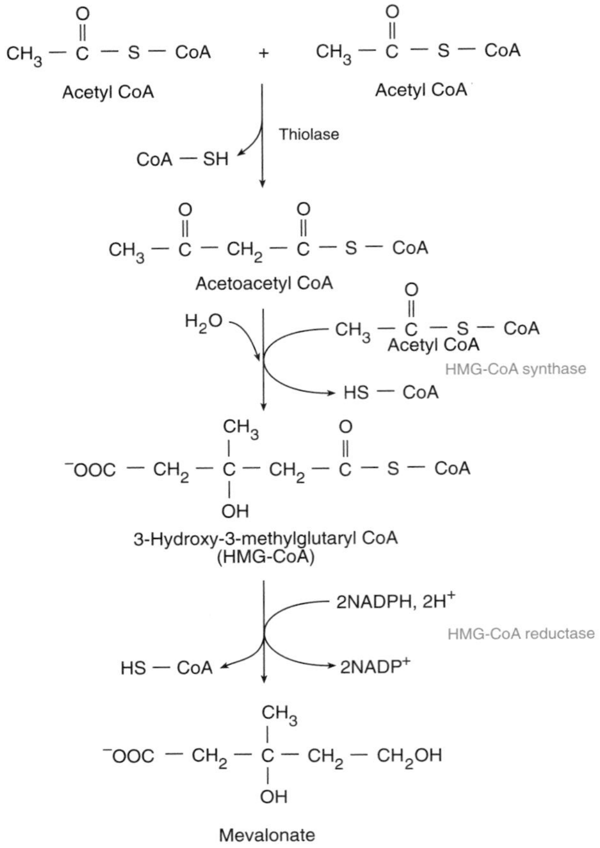

Figure 13.23. Stage I of cholesterol biosynthesis: Formation of mevalonate from three molecules of acetyl CoA.

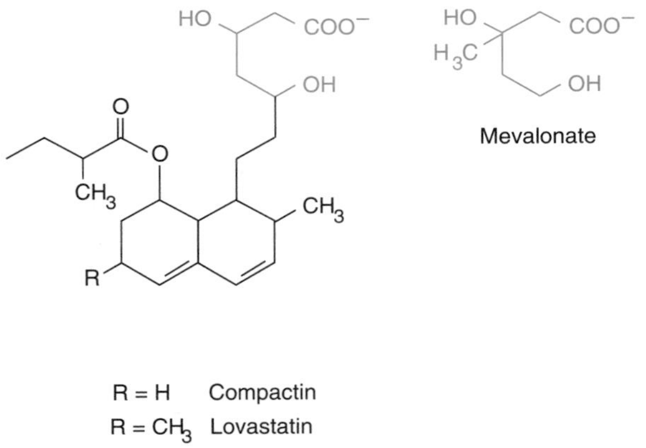

ly. In these compounds, choline and ethanolamine are esterified via a hydroxyl group to the β-phosphate group of CDP.

13.6.2B. Sphingolipids.

Sphingolipids also occur in plant and animal membranes. In mammals, they are particularly abundant in the brain and nervous tissue. Sphingolipids have sphingosine as their structural backbone, but their biosynthetic precursor is dihydrosphingosine, formed from palmitoyl CoA and serine (Figure 13.22). Dihydrosphingosine can be converted to *sphingomyelin, cerebrosides,* or *gangliosides.* A number of hereditary disorders of sphingolipid metabolism occur. These are known as **lipid storage diseases,** and three (gangliosidosis, Gaucher's disease, and Niemann–Pick disease) were listed in Table 8.3.

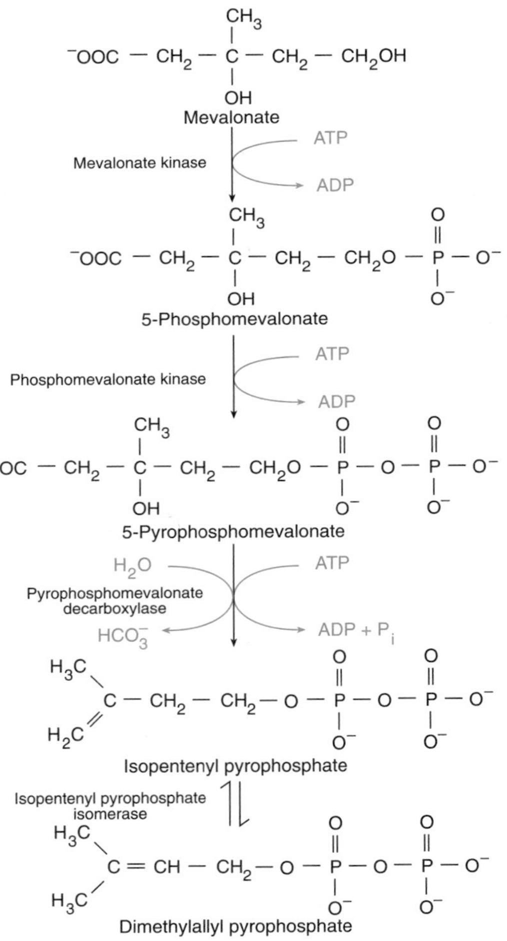

Figure 13.25. Stage II of cholesterol biosynthesis: ATP-dependent conversion of mevalonate to dimethylallyl pyrophosphate.

R = H   Compactin
R = CH₃ Lovastatin

Figure 13.24. Examples of drugs used to lower blood cholesterol levels. Structural similarity to mevalonate accounts for the competitive inhibition of HMG-CoA reductase by these compounds.

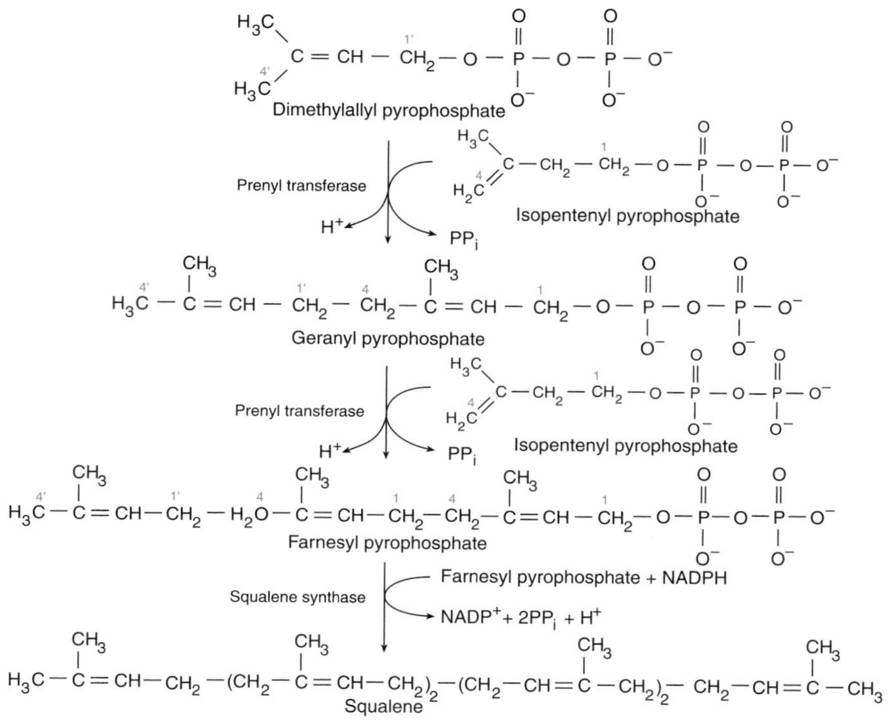

H_3C
$\quad$ C=CH—CH_2—O—P(=O)(O⁻)—O—P(=O)(O⁻)—O⁻
H_3C
Dimethylallyl pyrophosphate

Prenyl transferase

H_3C
$\quad$ C—CH_2—CH_2—O—P(=O)(O⁻)—O—P(=O)(O⁻)—O⁻
H_2C
Isopentenyl pyrophosphate

H⁺ → PP_i

H_3C—C(CH₃)=CH—CH_2—CH_2—C(CH₃)=CH—CH_2—O—P(=O)(O⁻)—O—P(=O)(O⁻)—O⁻
Geranyl pyrophosphate

Prenyl transferase

H_3C
$\quad$ C—CH_2—CH_2—O—P(=O)(O⁻)—O—P(=O)(O⁻)—O⁻
H_2C
Isopentenyl pyrophosphate

H⁺ → PP_i

H_3C—C(CH₃)=CH—CH_2—H_2O—C(CH₃)=CH—CH_2—CH_2—C(CH₃)=CH—CH_2—O—P(=O)(O⁻)—O—P(=O)(O⁻)—O⁻
Farnesyl pyrophosphate

Squalene synthase

Farnesyl pyrophosphate + NADPH → $NADP^+$+ $2PP_i$ + H⁺

H_3C—C(CH₃)=CH—CH_2—(CH_2—C(CH₃)=CH—CH_2)₂—(CH_2—CH=C(CH₃)—CH_2)₂—CH_2—CH=C(CH₃)—CH_3
Squalene

Figure 13.26. Stage III of cholesterol biosynthesis: Condensation reactions from dimethylallyl pyrophosphate to squalene. Loss of PP_i converts dimethylallyl pyrophosphate to an allylic carbonium ion [$(H_3C)_2C=CH-CH_2^+$] that is attacked by the electrons of the double bond in isopentenyl pyrophosphate, resulting in a head-to-tail (1′–4) condensation. The second condensation proceeds likewise, following conversion of geranyl pyrophosphate to an allylic carbonium ion by loss of PP_i.

13.6.3. Cholesterol

The structure of cholesterol was elucidated in the 1930s, but it was not clear how such a complex molecule could be assembled from smaller molecules. The pathway for cholesterol biosynthesis was only unraveled after radioactive tracers became available. In 1941, Rudolf Schoenheimer and Konrad Bloch (who was awarded the Nobel Prize in 1964) showed that labeled acetate was a precursor of cholesterol in rats and mice. In 1952, Bloch proposed a scheme for cholesterol biosynthesis in which the molecule was assembled in a number of stages. The scheme required formation of an unknown precursor in the early part of the pathway. This precursor was identified in 1956 as *mevalonic acid* and provided the missing link for the mechanism of cholesterol biosynthesis. Since then, the individual steps of this lengthy pathway have been defined in detail.

All of the carbons of cholesterol derive from acetyl CoA. Both the methyl and the carboxyl carbons of the acetyl group become incorporated into the steroid nucleus. The entire pathway can be divided into five stages:

$$\begin{array}{ccccccc}
\text{I} & & \text{II} & & \text{III} & & \text{IV} \\
\text{Acetate} & \rightarrow & \text{mevalonate} & \rightarrow & \text{isoprene derivatives} & \rightarrow & \text{squalene} \rightarrow \\
C_2 & & C_6 & & C_5 & & C_{30}
\end{array}$$

$$\begin{array}{ccc}
& \text{V} & \\
\text{lanosterol} & \rightarrow & \text{cholesterol} \\
C_{27} & & C_{27}
\end{array}$$

13.6.3A. Stage I. The first stage (Stage I) leads to synthesis of **mevalonate,** a six-carbon compound, from three molecules of acetyl CoA (Figure 13.23). The initial step of this stage produces *acetoacetyl CoA* from acetyl CoA by reversal of the thiolase reaction. The last step of Stage I involves the reduction of *3-hydroxy-3-methylglutaryl CoA (HMG-CoA)* to mevalonate. This reaction requires two molecules of NADPH and is catalyzed by **HMG-CoA reductase;** it constitutes both the committed and the rate-determining step of cholesterol biosynthesis. The reaction also represents the primary control site of cholesterol biosynthesis. The enzyme is subject to both covalent modification and allosteric control. HMG-CoA reductase exists in two interconvertible forms, a phosphorylated and a dephosphorylated form. Phosphorylation inactivates the enzyme and involves a cAMP-dependent bicyclic enzyme cascade similar to those that function in the control of glycogen phosphorylase and glycogen synthase (see Figures 10.26 and 10.30). The enzyme cascade is hormonally controlled: insulin, which decreases [cAMP], stimulates cholesterol biosynthesis; glucagon, which increases [cAMP], inhibits it.

HMG-CoA reductase is inhibited by long-chain fatty acyl CoA molecules. The inhibition may be due to both a direct allosteric effect on HMG-CoA reductase and an indirect effect produced by an activation of the kinase that catalyzes the phosphorylation of the reductase. Cholesterol also affects the activity of HMG-CoA reductase. High concentrations of cholesterol lead to formation of cholesterol derivatives that inhibit the enzyme allosterically and lead to increased degradation and decreased synthesis of HMG-CoA reductase.

Because of its pivotal role in the regulation of cholesterol biosynthesis, HMG-CoA reductase has been a target of attempts to lower cholesterol levels by means of drug therapy. Two examples of such drugs are the fungal products *compactin* and *lovastatin* (Figure 13.24). These compounds bear a structural similarity to mevalonate and function as competitive inhibitors of HMG-CoA reductase. Lovastatin was approved in 1987 for the treatment of patients with very high cholesterol levels. Drug therapy represents one of two lines of therapy for lowering high levels of serum cholesterol. The other line involves dietary modifications to decrease the intake of dietary cholesterol.

13.6.3B. Stage II.
Stage II of cholesterol biosynthesis involves a phosphorylation and decarboxylation of mevalonate. The product, *isopentenyl pyrophosphate,* is a derivative of *isoprene* (see Figure 6.16). Isopentenyl pyrophosphate isomerizes to a second pyrophosphate derivative of isoprene, *dimethylallyl pyrophosphate* (Figure 13.25). Stage II requires the expenditure of energy; three molecules of ATP are cleaved to ADP and P_i.

13.6.3C. Stage III.
Stage III consists of a series of condensation reactions involving a total of six pyrophosphate derivatives of isoprene. First, dimethylallyl pyrophosphate condenses head-to-tail (1'–4) with isopentenyl pyrophosphate to yield *geranyl pyrophosphate* (C_{10}). A second head-to-tail condensation of geranyl pyrophosphate with isopentenyl pyrophosphate produces *farnesyl pyrophosphate* (C_{15}). Lastly, two molecule of farnesyl pyrophosphate condense head-to-head (1'–1) to form the 30-carbon, openchain, unsaturated hydrocarbon *squalene* (Figure 13.26). Stage III requires one molecule of NADPH and leads to the release of four pyrophosphate groups (PP_i). The exergonic hydrolysis of pyrophosphate, catalyzed by pyrophosphatase, provides a driving force for cholesterol biosynthesis.

13.6.3D. Stage IV.
Squalene is the immediate precursor of the sterols. Its two-step cyclization (Stage IV) leads to *lanosterol.* The first step, catalyzed by *squalene monooxygenase,* requires molecular oxygen and NADPH and forms a reactive intermediate, called *squalene epoxide.* The second step, catalyzed by *squalene cyclase,* produces the remarkable closure of the ring system (Figure 13.27). This step requires a *concerted* movement of electrons through four double bonds and migration of two methyl groups. The term concerted means that each part of the reaction is essential for any other part to take place; all parts occur simultaneously. The total number of enzymatic steps from mevalonate to lanosterol is about 10.

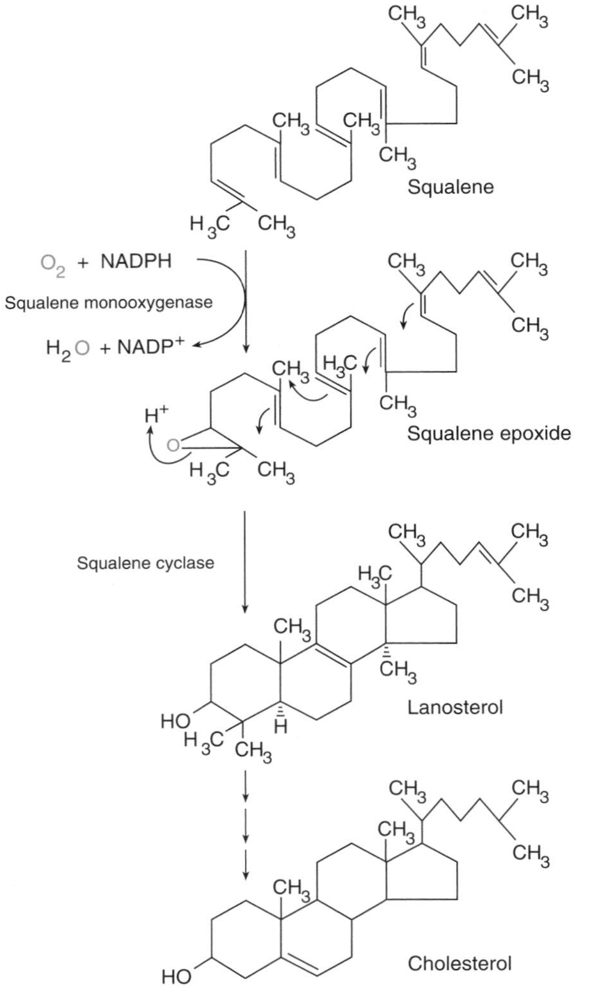

Figure 13.27. The last two stages of cholesterol biosynthesis. Stage IV: Cyclization of squalene to lanosterol by a two-step process. Stage V. Conversion of lanosterol to cholesterol by approximately 20 steps.

13.6.3E. Stage V. The last stage of cholesterol biosynthesis (Stage V) consists of the conversion of lanosterol to cholesterol and includes approximately 20 additional steps, most of which require both NADH (or NADPH) and molecular oxygen. You can see that cholesterol biosynthesis requires expenditure of energy in the form of ATP, is driven in part by hydrolysis of pyrophosphate, and requires a large amount of reducing power in the form of NADH and NADPH. The NADPH required for the reductive reactions comes from the malic enzyme reaction and the pentose phosphate pathway.

SUMMARY

Fats are stored in adipose tissue and transported via the blood as lipoproteins, as fatty acids bound to serum albumin, and as ketone bodies. A hormonally controlled enzyme cascade regulates the release of fatty acids from adipose tissue (mobilization). Execssive mobilization may occur in diabetics and may lead to development of fatty liver.

Fat catabolism begins with hydrolysis, producing glycerol and fatty acids. Glycerol is converted to dihydroxyacetone phosphate, an intermediate of glycolysis. Fatty acids are degraded by β-oxidation, in which chains are shortened by successive removal of two-carbon fragments in the form of acetyl CoA. The acetyl CoA enters the citric acid cycle, yielding NADH and $FADH_2$, which lead to production of ATP via oxidative phosphorylation. Fatty acid activation, located in the cytosol, converts a fatty acid to a fatty acyl CoA that is transported into the mitochondrial matrix by the carnitine carrier system. The remaining four reactions of β-oxidation take place in the matrix. Even-numbered fatty acids are degraded completely to acetyl CoA; odd-numbered fatty acids are degraded to acetyl CoA and propionyl CoA, which is converted to succinyl CoA, an intermediate of the citric acid cycle.

Ketone bodies include acetone, acetoacetate, and β-hydroxybutyrate. Their concentration builds up (ketosis) when acetyl CoA accumulates, as may occur in starvation or diabetes. Ketosis, if not treated, leads to destruction of the alkaline reserve of the body ($NaHCO_3$) and to severe dehydration.

In fatty acid synthesis, located in the cytosol, chains are elongated by two carbons at a time, with the carbons provided in the form of malonyl CoA that is produced from acetyl CoA. Acetyl CoA forms in the mitochondria and is transported to the cytosol by means of the tricarboxylate transport system. Fatty acid synthesis in animals is catalyzed by fatty acid synthase, a dimer of two multifunctional polypeptide chains. Each chain comprises seven enzymatic activities and an acyl carrier protein. Fatty acid synthesis requires both ATP and NADPH.

Acylglycerols are synthesized from fatty acyl CoA and glycerol 3-phosphate. Sphingolipids are produced from dihydrosphingosine. Cholesterol is synthesized entirely from acetyl CoA by a series of some 30 reactions.

SELECTED READINGS

Barron, J. T., Kopp, S. J., Tow, J., and Parrillo, J. E., Fatty acid, tricarboxylic acid cycle metabolites, and energy metabolism in vascular smooth muscle, *Am. J. Physiol.* 267:H764–H769 (1994).

Bieber, L. L., Carnitine, *Annu. Rev. Biochem.* 88:261–283 (1988).

Bradley, W. A., Gianturco, S. H., and Segrest, J. P. (eds.), Plasma lipoproteins, Part C, *Methods in Enzymology,* Vol. 263, Academic Press, San Diego (1996).

Carman, G. M., and Zeimetz, G. M., Regulation of phospholipid biosynthesis in the yeast *Saccharomyces cerevisiae, J. Biol. Chem.* 271:13293–13296 (1996).

Cohen, B. I., Mikami, T., Ayyad, N., Mikami, Y., and Mosbach, E. H., Dietary fat alters the distribution of cholesterol between vesicles and micelles in hamster bile, *Lipids* 30:299–305 (1995).

Hardie, D. G., Regulation of fatty acid synthesis via phosphorylation of acetyl-SCoA carboxylase, *Prog. Lipid. Res.* 28:117–146 (1989).

Kent, C., Eukaryotic phospholipid biosynthesis, *Annu. Rev. Biochem.* 64:315–344 (1995).

Moore, T. S., Jr., *Lipid Metabolism in Plants,* CRC Press, Boca Raton, Florida (1993).

Müller-Newen, G., Janssen, U., and Stoffel, W., Enoyl-CoA hydratase and isomerase form a superfamily with a common active-site glutamate residue, *Eur. J. Biochem.* 228:68–73 (1995).

Salvayre, R., Douste-Blazy, L., and Gatt, S. (eds.), *Lipid Storage Disorders: Biological and Medical Aspects,* Plenum, New York (1988).

Sevanian, A., *Lipid Peroxidation in Biological Systems,* AOCS Press, Champaign, Illinois (1988).

Tuomanen, E., Breaching the blood–brain barrier, *Sci. Am.* 268:80–85 (1993).

Vance, D. E., and Vance, J. E. (eds.), *Biochemistry of Lipids, Lipoproteins, and Membranes,* Elsevier, Amsterdam (1991).

Wakil, S. J., Fatty acid synthase, a proficient multifunctional enzyme, *Biochemistry* 28:4523–4530 (1989).

REVIEW QUESTIONS

A. Define each of the following terms:

Depot fat	Ketone bodies
Lipotropic agents	*S*-Adenosylmethionine
Fatty liver	Ketosis
Alkaline reserve	Blood–brain barrier
β-Oxidation	ACP

B. Differentiate between the two terms in each of the following pairs:

Thiokinase/thiolase	Elongase/desaturase
Tricarboxylate transport system/carnitine carrier system	β-Ketoacyl-ACP synthase/ β-ketoacyl-ACP reductase
Acetyl CoA:ACP transacylase/malonyl CoA: ACP transacylase	Transcarboxylase/biotin carboxylase
Acyl CoA dehydrogenase/ L-3-hydroxyacyl CoA dehydrogenase	Fatty acid mobilization/ fatty acid activation

C. (1) Outline the major steps in β-oxidation of saturated, even- and odd-numbered fatty acids. Write formulas for the intermediates in the β-oxidation of myristic acid (14 carbons, saturated).

(2) How does the overall metabolism of a diabetic differ from that of a normal individual? Why are ketosis and fatty liver often associated with diabetes?

(3) Describe Knoop's experiments. What was the significance of these experiments and how were the results interpreted?

(4) What are the metabolic fates of acetyl CoA?

(5) Compare and contrast the pathways of fatty acid synthesis and fatty acid degradation.

PROBLEMS

13.1. How many molecules of acetyl CoA, $FADH_2$, and NADH are produced in β-oxidation per molecule of a saturated fatty acid that has: (a) 24 carbons; (b) 21 carbons?

13.2. Which carbons in cholesterol will become radioactively labeled when it is synthesized from acetyl CoA in which both carbons of the acetyl group are ^{14}C?

13.3. How many carbons are there in the acyl group of fatty acyl CoA formed from a 17-carbon fatty acid that has been subjected to three cycles of β-oxidation?

13.4. Energetically speaking, how many molecules of glucose can be converted to glyceraldehyde 3-phosphate for every molecule of palmitic acid (16 carbons) that is completely degraded to CO_2 and H_2O?

13.5. You are told that some animals have adapted to life in arid environments by using their lipid stores to produce metabolic water. How can this be?

13.6. Write a balanced equation for the overall β-oxidation to acetyl CoA of: (a) arachidic acid (see Table 6.1); (b) a 15-carbon, saturated fatty acid.

13.7. What is the similarity among the mechanisms of hydrolysis, phosphorolysis, and thiolysis?

13.8. Why is it advantageous that fatty acid mobilization is regulated by means of a hormonally controlled enzyme cascade?

13.9.* Calculate the net yield of ATP molecules per molecule of glycerol for the following catabolic sequences. Assume that the reactions occur under aerobic conditions with participation of the glycerol phosphate shuttle.

(a) Glycerol to pyruvate

(b) Glycerol to acetyl CoA

(c) Glycerol to CO_2 and H_2O

13.10.* Calculate the net yield of ATP molecules per molecule of acylglycerol for the following catabolic sequences. The names refer to the mono- and tri-acylglycerols, respectively, of palmitic and stearic acid. Assume aerobic conditions and participation of the glycerol phosphate shuttle.

(a) Monopalmitin to acetyl CoA
(b) Tristearin to CO_2 and H_2O

13.11.* Assume that 15% of the body mass of a 70.0-kg adult consists of adipose tissue composed entirely of tristearin (MW = 892). Tristearin is oxidized as indicated in the previous problem.

(a) What is the fuel reserve of this adipose tissue in terms of kilojoules?
(b) Using a daily energy requirement of 12,134 kJ (Table 8.6), calculate the number of days this adult could survive using only depot fat as a source of energy.
(c) What is the daily weight loss of the individual under these starvation conditions?

13.12.* According to a popular misconception, camels store water in their humps for long journeys through the desert. In actuality, these humps consist of large fat deposits that serve as a source of metabolic water. Assuming that such a deposit consists entirely of tristearin (the triacylglycerol of stearic acid; MW = 892), calculate the amount of water (in milliliters) that could be produced from β-oxidation of the stearic acid contained in 1.00 kg of fat. Ignore the water required for β-oxidation, and use 1.00 g/ml for the density of water.

13.13.* How much glycogen would an animal have to store in order to have an amount of stored energy equivalent to that contained in 1.00 g of tristearin (see Problem 13.10)? Given that 1.00 g of glycogen sequesters 2.80 g of bound water, what is the total weight that would have to be stored? (MW of any glucose residue = 162.) Assume aerobic conditions and operation of the glycerol phosphate shuttle.

13.14. Which of the following constitute energy-rich compounds? (a) Fatty acyl adenylate; (b) succinyl CoA; (c) pyrophosphate; (d) carnitine; (e) propionyl CoA; (f) acetoacetic acid; (g) acetyl CoA; (h) β-hydroxybutyric acid; (i) trans-Δ^2-enoyl CoA; (j) β-ketoacyl CoA; (k) acetone; (l) acetoacetyl CoA

13.15. A diabetic suffers from a severe case of ketosis. When acetyl CoA (labeled with ^{14}C in both carbons of the acetyl group) is administered to the diabetic, is it likely that her breath will contain labeled acetone? Why or why not?

13.16.* Consider an even-numbered, saturated fatty acid in which all of the methylene (CH_2) groups are doubly labeled (^{14}C and ^3H). When this fatty acid is ca-

tabolized under aerobic conditions, will labeled CO_2 be formed when (a) the first and (b) the last acetyl CoA produced from the fatty acid passes through one turn of the citric acid cycle? Will labeled water be produced under the same conditions?

13.17. Write balanced equations for the conversion of: (a) glycerol to pyruvate; (b) propionyl CoA to succinyl CoA; (c) acetyl CoA to acetoacetate.

13.18. A person subsists on a diet that is entirely devoid of carbohydrates. Would he be better off consuming odd- or even-numbered fatty acids? Why?

13.19. Some bacteria can grow using hydrocarbons as their sole nutrients. The hydrocarbons serve as a source of both carbon atoms and energy and are oxidized to their corresponding carboxylic acids. For example,

$$NAD^+ + CH_3(CH_2)_6CH_3 + O_2 \rightarrow$$
$$\text{octane}$$
$$CH_3(CH_2)_6COOH + NADH + H^+$$
$$\text{octanoic acid}$$

How might such organisms be used to clean up oil spills?

13.20.* What is the net yield of ATP per molecule of acetoacetate when this ketone body is used as a source of energy in the brain?

13.21. Palmitic acid is labeled with ^{14}C at the following specific sites: (a) C(1); (b) C(2); (c) C(7); (d) C(1, 4); (e) C(16). During which cycles in β-oxidation will each of these molecules yield labeled acetyl CoA?

13.22. Write a balanced equation for the complete oxidation of octanoic acid (Problem 13.19) to CO_2 and H_2O.

13.23. Which of the following is not likely to lead to the development of fatty liver?

(a) Excessive fatty acid mobilization
(b) A high-starch diet
(c) Diabetes
(d) Starvation
(e) Extensive exposure to pyridine

13.24. Consider the complete oxidation of glucose to CO_2 and H_2O via glycolysis/glycerol phosphate shuttle and the citric acid cycle/electron transport system. Under these conditions, how many moles of palmitate can be synthesized from acetyl CoA per mole of glucose catabolized? Base your calculation on the number of moles of ATP involved.

13.25. Calculate the number of moles of ATP (a) obtainable per mole of myristic acid (14 carbons, saturated) when it is degraded via β-oxidation to acetyl CoA and (b) required to synthesize a mole of myristic acid from acetyl CoA. How do you explain the difference between the two numbers?

13.26. A cell-free system for fatty acid synthesis is exposed to ^{14}C-labeled CO_2. Will the palmitate produced be labeled and, if so, at what position(s)?

13.27.* How many deuterium atoms (D, heavy isotope of hydrogen) are incorporated into palmitate, and what are their locations, when a fatty acid-synthesizing system is provided with the following? (a) Labeled acetyl CoA (CD_3—CO-S-CoA) and unlabeled malonyl CoA; (b) labeled malonyl CoA ($^-$OOC-CD_2-CO-S-CoA) and unlabeled acetyl CoA.

13.28. On the basis of this chapter, predict the effects on lipid metabolism that are likely to arise from insulin deficiency.

Amino Acid and Nucleotide Metabolism

14

Amino acid and nucleotide metabolism, like that of carbohydrates and lipids, can be divided into five major parts—*digestion, transport, storage, degradation,* and *biosynthesis.*

Digestion of proteins (see Section 8.4) produces free amino acids that are absorbed across the intestinal wall into the bloodstream. Proteins and amino acids are transported in biological fluids primarily as plasma proteins, lipoproteins, nucleoproteins, peptides, and free amino acids. In metabolism, nitrogenous compounds are degraded to, and synthesized from, intermediates of carbohydrate and lipid metabolism.

Animals do not store proteins as they do carbohydrates and lipids. The muscles of adult organisms are a functional tissue, not a storage form of energy. Likewise, nucleic acids and their components function in genetic information transfer and not as deposits of metabolic energy. If it becomes necessary for the body to draw on protein tissues, as during fasting or starvation, a "sparing effect" ensures use of carbohydrate and lipid deposits before protein tissues are attacked. Because of the absence of protein stores, humans cannot tolerate a protein-deficient diet, especially one lacking in essential amino acids, for prolonged times. Nutritional deficiency of protein produces *kwashiorkor* (an African word, meaning "weaning disease"), a disease that frequently strikes children who, after having been weaned, exist on a diet containing insufficient protein. The severity of kwashiorkor, particularly in growing children, is due not only to lack of dietary protein but also to breakdown of the body's own proteins.

14.1. NITROGEN UTILIZATION

All living organisms require nitrogen to synthesize building blocks like amino acids, purines, and pyrimidines and to synthesize other nitrogen-containing compounds. All of the nitrogen found in biological systems ultimately derives from gaseous N_2, which constitutes about 80% of the atmosphere. We can divide the overall process of nitrogen utilization by living organisms into three stages:

- Stage I: *Formation of ammonia.* Ammonia forms in living organisms via two major mechanisms. Some bacteria are capable of **nitrogen fixation,** involving reduction of nitrogen gas (N_2) to am-

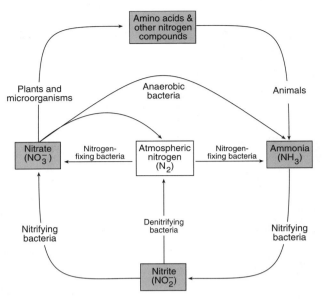

Figure 14.1.   The nitrogen cycle.

• Stage III: *Nitrogen transfer.* Nitrogen-containing compounds formed by ammonia fixation participate in metabolic reactions involving intermediates of carbohydrate and lipid metabolism. Such reactions (for example, *transamination*) generate a variety of other nitrogen-containing biomolecules.

14.1.1. Nitrogen Fixation

Some organisms can use atmospheric nitrogen directly and reduce it to ammonia. Nitrogen-fixing bacteria, sometimes in symbiosis with plants, carry out this conversion. Nitrogen fixation is catalyzed by **nitrogenase,** an enzyme complex composed of two parts. One part, *MoFe-protein,* is a tetrameric protein ($\alpha_2\beta_2$) containing both molybdenum and iron; it has a molecular weight of about 220,000. The second part, *Fe-protein,* is a dimer of two identical subunits, contains only iron, and has a molecular weight of about 64,000. Nitrogenase has several oxidation–reduction centers, and most of its iron occurs in the form of [4Fe-4S] complexes (see Figure 12.5). The overall reaction is:

$$N_2 \xrightarrow[\text{$6e^-$, $6H^+$}]{\text{nitrogenase}} 2NH^3$$

$$:N:::N: \qquad\qquad 2\,H:\overset{\displaystyle \cdot\cdot}{\underset{\displaystyle \cdot\cdot}{N}}:H$$

(10 electrons)   (16 electrons)

Nitrogen fixation is initated by generating electrons through photosynthesis or from oxidation reactions. These electrons are transferred to *ferredoxin,* an electron carrier containing [4Fe-4S] complexes, which transfers the electrons to the Fe-protein component of nitrogenase (Figure 14.2).

monia (NH_3). Other bacteria, plants, and fungi carry out **nitrate assimilation,** in which they convert nitrite (NO_2^-) and nitrate (NO_3^-) to ammonia. The interconversions of nitrogen gas, nitrate, nitrite, and ammonia constitute part of the **nitrogen cycle** (Figure 14.1).

• Stage II: *Utilization of ammonia.* All living organisms can use ammonia and convert it to nitrogen-containing organic compounds. Conversion of ammonia to organic compounds, **ammonia fixation,** is accomplished by three reactions, one or more of which occur in all levels of organisms. Animals require ammonia as starting material for using nitrogen.

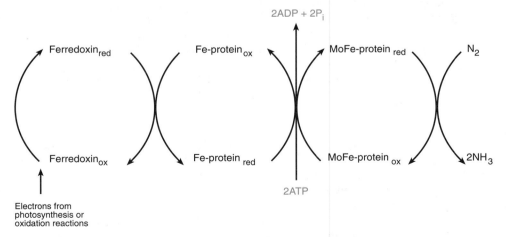

Figure 14.2.   The electron transport system of nitrogenase.

Two molecules of ATP bind to reduced Fe-protein and are hydrolyzed as the electron passes from Fe-protein to MoFe-protein. Chemists think that ATP hydrolysis produces a conformational change in Fe-protein that results in the redox potential of Fe-protein becoming significantly more negative; $E^{\circ\prime}$ changes from -0.29 to -0.40 V. The change in $E^{\circ\prime}$ permits the electron to reduce N_2 to NH_3, a reaction that has an $E^{\circ\prime}$ of -0.34 V.

Actual reduction of nitrogen occurs on the MoFe-protein and proceeds in three discrete steps, each requiring two electrons:

$$N_2 + 2H^+ + 2e^- \rightarrow HN{=}NH$$
Diimine
$$HN{=}NH + 2H^+ + 2e^- \rightarrow H_2N{-}NH_2$$
Hydrazine
$$H_2N{-}NH_2 + 2H^+ + 2e^- \rightarrow 2NH_3$$
Ammonia

Because iron–sulfur proteins undergo oxidation–reduction reactions involving a single electron at a time, the electron transfer from ferredoxin to MoFe-protein must occur six times per N_2 and, therefore, requires hydrolysis of 12 molecules of ATP. This leads to the following overall stoichiometry:

$$N_2 + 6e^- + 6H^+ + 12ATP^{4-} + 12H_2O \rightarrow$$
$$2NH_3 + 12ADP^{3-} + 12P_i^{2-} + 12H^+$$

so that the *net* equation is

$$N_2 + 6e^- + 12ATP^{4-} + 12H_2O \rightarrow$$
$$2NH_3 + 12ADP^{3-} + 12P_i^{2-} + 6H^+$$

You can see that these reactions represent an energetically costly process. Actually, the energy price tag is even higher than indicated. Nitrogenase also reduces H^+ to H_2, which reacts with diimine to re-form the original substrate (N_2):

$$HN{=}NH + H_2 \rightarrow N_2 + 2H_2$$

When you combine this equation with that of diimine formation, you obtain an overall reaction

$$NH{=}NH + N_2 + 2H^+ + 2e^- \longrightarrow NH{=}NH + N_2 + H_2$$

that constitutes a *futile cycle* (see Figure 10.35). Although such cycles appear to achieve nothing—except dissipate free energy or possibly generate some heat—they may provide regulatory mechanisms if the two reactions are subject to different degrees of activation and inhibition. The nitrogenase futile cycle usually occurs to some extent. Because its operation involves a reduction by means of two electrons, four additional molecules of ATP must be hydrolyzed. If the cycle were to occur with *every* passage of N_2 through the pathway (which generally is not the case), the overall *net* equation for nitrogen fixation would become

$$N_2 + 8e^- + 16ATP^{4-} + 16H_2O \rightarrow$$
$$2NH_3 + H_2 + 16ADP^{3-} + 16P_i^{2-} + 8H^+$$

14.1.2. Nitrate Assimilation

Nitrate assimilation constitutes a second major ammonia-forming mechanism in living systems. In contrast to nitrogen fixation, which is limited to nitrogen-fixing bacteria and some plants, nitrate assimilation can be carried out by virtually all plants, fungi, and bacteria. Nitrate assimilation consists of a two-stage reduction of nitrate to ammonia. In the first stage, nitrate undergoes reduction to nitrite:

$$NO_3^- + 2H^+ + 2e^- \rightarrow NO_2^- + H_2O$$

This reaction is catalyzed by **nitrate reductase,** a large (MW $\approx$ 800,000) multisubunit enzyme that contains several electron carriers: FAD, molybdenum, and cytochrome$_{557}$ (contains a [4Fe-4S] complex). The electron donor is NADH in plants and NADPH in fungi and bacteria. The reduction comprises the following flow of electrons:

$$NAD(P)H \rightarrow FAD \rightarrow Cyt_{557} \rightarrow Mo \rightarrow NO_3^-$$

and results in the overall reaction

$$NO_3^- + NAD(P)H + H^+ \rightarrow NO_2^- + NAD(P)^+ + H_2O$$

The second stage of nitrate assimilation consists of three steps that accomplish the reduction of nitrite to ammonia. The same enzyme, **nitrite reductase,** catalyzes all three reactions. Nitrite reductase contains a number of electron carriers, including an iron–sulfur complex and an iron porphyrin, and leads to the following reaction sequence:

$$NO_2^- + 2H^+ + 2e^- \rightarrow NO^- + H_2O$$
$$NO^- + 3H^+ + 2e^- \rightarrow NH_2OH$$
$$NH_2OH + 2H^+ + 2e^- \rightarrow NH_3 + H_2O$$

Each step requires two electrons, donated by ferredoxin. NADPH represents the ultimate source of electrons. Three molecules of NADPH are required for every nitrite ion. The overall reaction is

$$NO_2^- + 3NADPH + 4H^+ \rightarrow NH_3 + 3NADP^+ + 2H_2O$$

14.1.3. Ammonia Fixation

Living organisms use three reactions to convert ammonia to organic compounds, which can then can be used in metabolism. In one reaction, catalyzed by **carbamoyl phosphate synthase,** ammonia, CO_2 (as HCO_3^-), and ATP serve as reactants for the synthesis of *carbamoyl phosphate:*

$$NH_4^+ + HCO_3^- + 2ATP^{4-} + H_2O \rightarrow$$

$$\underset{\text{Carbamoyl phosphate}^{2-}}{H_2N-\overset{\displaystyle O}{\overset{\|}{C}}-OPO_3^{2-}} + 2ADP^{3-} + P_i^{2-} + 2H^+$$

Carbamoyl phosphate is an important metabolite, not only because it represents a "fixed" form of ammonia, but also because it constitutes an energy-rich compound owing to its mixed anhydride structure. Synthesis of carbamoyl phosphate requires the expenditure of two molecules of ATP even though only one phosphate group is incorporated into the newly made compound (Figure 14.3).

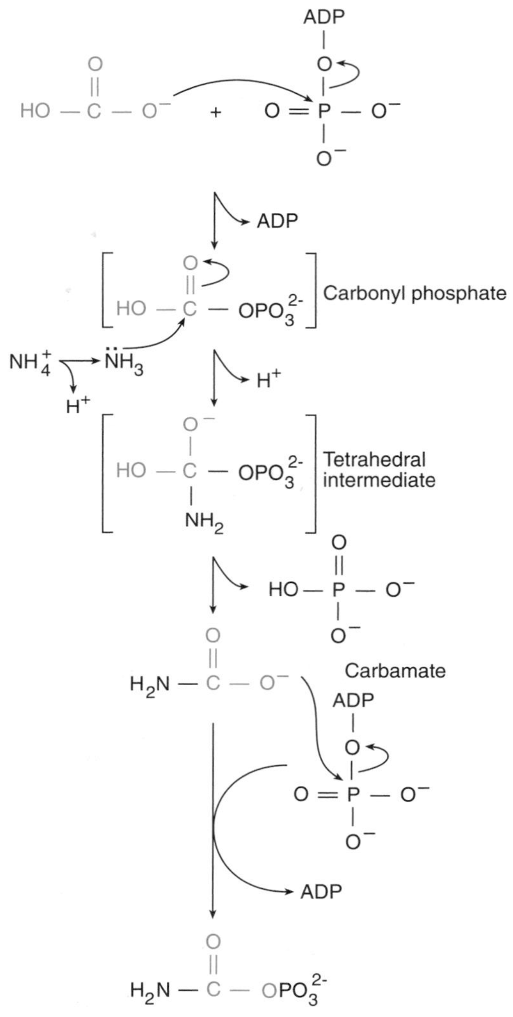

Carbamoyl phosphate participates in arginine synthesis in the urea cycle (Section 14.3) and in pyrimidine biosynthesis (Section 14.4). In eukaryotes, synthesis of the carbamoyl phosphate used in the urea cycle is catalyzed by a mitochondrial enzyme (*carbamoyl phosphate synthase I*) that uses ammonia as nitrogen source. Synthesis of the carbamoyl phosphate used in pyrimidine biosynthesis is catalyzed by a different cytosolic enzyme (*carbamoyl phosphate synthase II*) that uses glutamine as nitrogen source. Prokaryotes have only one carbamoyl phosphate synthase, which catalyzes both arginine and pyrimidine synthesis and uses glutamine for both synthetic processes.

A second reaction for converting ammonia to nitrogen-containing compounds is catalyzed by **glutamate dehydrogenase** and reduces α-ketoglutarate to *glutamate.* Glutamate dehydrogenase is widespread in animal, plant, and microbial cells. In plants and animals, the enzyme is located in the mitochondria. Glutamate dehydrogenase of some species uses NADH as coenzyme, and that of other species uses NADPH. Some organisms can use either NADH or NADPH as coenzyme for glutamate dehydrogenase:

$$NH_4^+ + NAD(P)H + H^+ + \underset{\text{α-Ketoglutarate}^{2-}}{^-OOC-\overset{\displaystyle O}{\overset{\|}{C}}-CH_2-CH_2-COO^-} \rightleftharpoons$$

$$\underset{\text{Glutamate}^-}{^-OOC-\overset{\displaystyle NH_3^+}{\overset{|}{C}H}-CH_2-CH_2-COO^-} + NAD(P)^+ + H_2O$$

The third reaction of ammonia fixation involves a conversion of glutamate to *glutamine,* catalyzed by **glutamine synthase:**

$$\underset{\text{Glutamate}^-}{^-OOC-\overset{\displaystyle NH_3^+}{\overset{|}{C}H}-CH_2-CH_2-COO^-} + NH_4^+ + ATP^{4-} \rightarrow$$

$$\underset{\text{Glutamine}}{^-OOC-\overset{\displaystyle NH_3^+}{\overset{|}{C}H}-CH_2-CH_2-\overset{\displaystyle O}{\overset{\|}{C}}-NH_2} + ADP^{3-} + P_i^{2-} + H^+$$

Glutamine occupies a pivotal position in nitrogen metabolism since it serves as a precursor for many metabolites. This makes glutamine synthase a critical enzyme. *E. coli* glutamine synthase has been studied extensively. It is an allosteric enzyme, composed of 12 identical subunits, and

Figure 14.3.　Synthesis of carbamoyl phosphate by carbamoyl phosphate synthase I. Activation of bicarbonate to carbonyl phosphate prepares the carbon for nucleophilic attack by ammonia. The mechanism's final step closely resembles the initial step.

subject to multiple regulation by allosteric effectors, feed-back inhibition, and covalent modification. The covalent modification consists of an *adenylylation* whereby an AMP group of ATP (splitting out PP_i) becomes linked to a specific tyrosine residue in the enzyme. The bond forms by esterification of the phosphate of AMP and the phenolic hydroxyl of tyrosine. Adenylylation decreases glutamine synthase activity.

The mammalian enzyme differs from the bacterial enzyme in both structure and properties. It consists of eight identical subunits but also exists in the form of tetramers. Brain and liver glutamine synthase have been most thoroughly characterized. The liver enzyme is not subject to covalent modification. Glycine, serine, alanine, and carbamoyl phosphate inhibit the enzyme, and α-ketoglutarate activates it. In mammals, glutamine is synthesized primarily in muscle tissue and transported from there via the blood to other tissues. Glutamine concentration in the blood is about 0.6mM, higher than that of any other amino acid.

14.1.4. Essential Amino Acids

Based on their nutritional value, we classify amino acids as essential or nonessential ones (see Table 2.4). *Essential amino acids* cannot be synthesized by an organism or are not synthesized in sufficient quantity and must be obtained through the diet. Essential amino acids in humans represent a good cross section of the different types of amino acids that occur in nature—aromatic, aliphatic, sulfur-containing, and hydroxy amino acids. W. C. Rose identified them in the 1940s, using measurements of **nitrogen balance,** the difference between the amounts of nitrogen ingested and excreted:

Nitrogen balance = N ingested − N excreted

Normally, an animal's nitrogen balance equals zero; the amount ingested equals that excreted, and the animal is in **nitrogen equilibrium.** In this state, any added nitrogen intake (e.g., a high-protein meal) leads to an equivalent increase in nitrogen excretion so that the difference between intake and excretion remains zero.

A nitrogen balance may, however, be greater or smaller than zero. In growing organisms, dietary nitrogen is continuously assimilated, and muscle tissue is constantly produced. Hence, nitrogen ingestion exceeds nitrogen excretion, resulting in a positive nitrogen balance. Conversely, in organisms suffering from certain wasting diseases, muscle tissue breakdown continues unabated regardless of nitrogen intake. Accordingly, nitrogen excretion exceeds nitrogen ingestion, resulting in a negative nitrogen balance.

To study nitrogen balance, Rose designed a synthet-

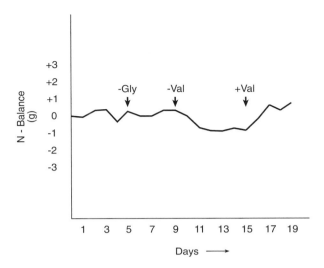

Figure 14.4. Determining nitrogen balance in humans. Omitting an essential amino acid, such as valine, from the diet leads to a negative nitrogen balance. Restoring the amino acid to the diet reestablishes a zero nitrogen balance.

ic diet consisting of a mixture of purified amino acids, starch, sucrose, butter, inorganic salts, and vitamins. He put individuals on this diet and controlled the daily amount of nitrogen ingested. He measured daily nitrogen excretion and determined the nitrogen balance as a function of time (Figure 14.4).

So long as all 20 amino acids were present in the diet, the nitrogen balance was essentially zero. After several days, Rose omitted *one amino acid at a time* from the diet. Omission of some amino acids (e.g., glycine) had no effect on the nitrogen balance, and Rose judged these amino acids to be nonessential. Evidently, the body could compensate for lack of these amino acids in the diet by synthesizing them from other metabolites. Omission of other amino acids (e.g., valine) led to a negative nitrogen balance, and Rose judged these amino acids to be essential. Absence of such amino acids results in impaired protein synthesis but does not decrease protein degradation. Therefore, nitrogen excretion exceeds nitrogen intake, producing a negative nitrogen balance. Restoring an essential amino acid to the diet reestablishes nitrogen equilibrium. (Rose's studies also led to the identification of threonine as one of the 20 amino acids occurring in proteins.)

14.2. PATHWAYS OF AMINO ACID METABOLISM

Amino acid metabolism consists of many different reactions and a variety of pathways. Some pathways are limited to a few amino acids, whereas others are more general in nature. **Transamination** represents a general pathway that applies to most amino acids and constitutes an important mechanism for interconversion of nitrogen-containing compounds.

14.2.1. Transamination

Transamination reactions are catalyzed by **transaminases (aminotransferases)** and involve the transfer of an amino group from an *amino acid* to a *keto acid*. The coenzyme of transaminases contains *vitamin B_6* as a structural component. Vitamin B_6 is widely distributed in nature and occurs in three forms, having an alcoholic ($-CH_2OH$), an amine ($-CH_2NH_2$), or an aldehyde ($-CHO$) group attached to the benzene ring (Figure 14.5). *Pyridoxal phosphate (PLP)* and *pyridoxamine phosphate (PMP)* constitute phosphorylated forms of vitamin B_6 and serve as coenzymes of transaminases. Thus, the vitamin and the coenzyme differ only by a phosphate group.

Figure 14.6 shows the mechanism of transamination. An incoming amino acid (AA_1) displaces *pyridoxal phosphate,* which is linked to a lysine residue of the enzyme in the form of a *Schiff base* (an *aldimine*). This *transimination* produces another aldimine (also a Schiff base), which

tautomerizes to a *ketimine* (also a Schiff base) via a *quinonoid intermediate* (resonance-stabilized carbanion). Hydrolysis of the ketimine yields a keto acid (KA_1), derived from the original amino acid, and *pyridoxamine phosphate*. Thus, AA_1 has been converted to KA_1 while the coenzyme has been altered from the aldehyde to the amine form.

The entire sequence of steps then proceeds in the opposite direction, beginning with an incoming *keto acid* (KA_2) and *pyridoxamine phosphate*. The incoming keto acid differs from that produced in the forward sequence. The reverse sequence yields an *amino acid* (AA_2), derived from the incoming keto acid, and *pyridoxal phosphate*. The entering keto acid has been converted to an amino acid, while the coenzyme has been altered from the amine to the aldehyde form. The reverse sequence regenerates the original form of the coenzyme, pyridoxal phosphate.

Transamination brings about an interconversion of two amino acids and two keto acids. The coenzyme carries out the actual chemical reaction by having its vitamin component shuttle back and forth between the aldehyde and the amine forms.

Transamination reactions exemplify Stage III of nitrogen utilization. In this stage, compounds formed by ammonia fixation interact with intermediates of carbohy-

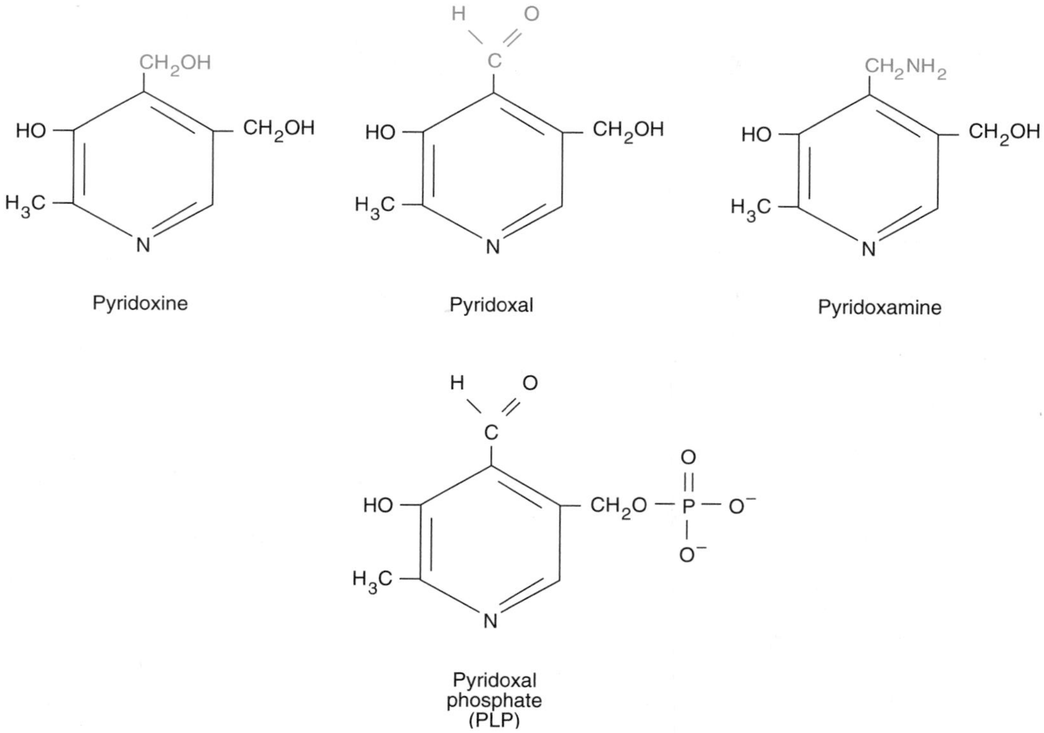

Figure 14.5.   Vitamin B_6. The vitamin occurs as an alcohol, an aldehyde, and an amine. Phosphorylated forms are coenzymes.

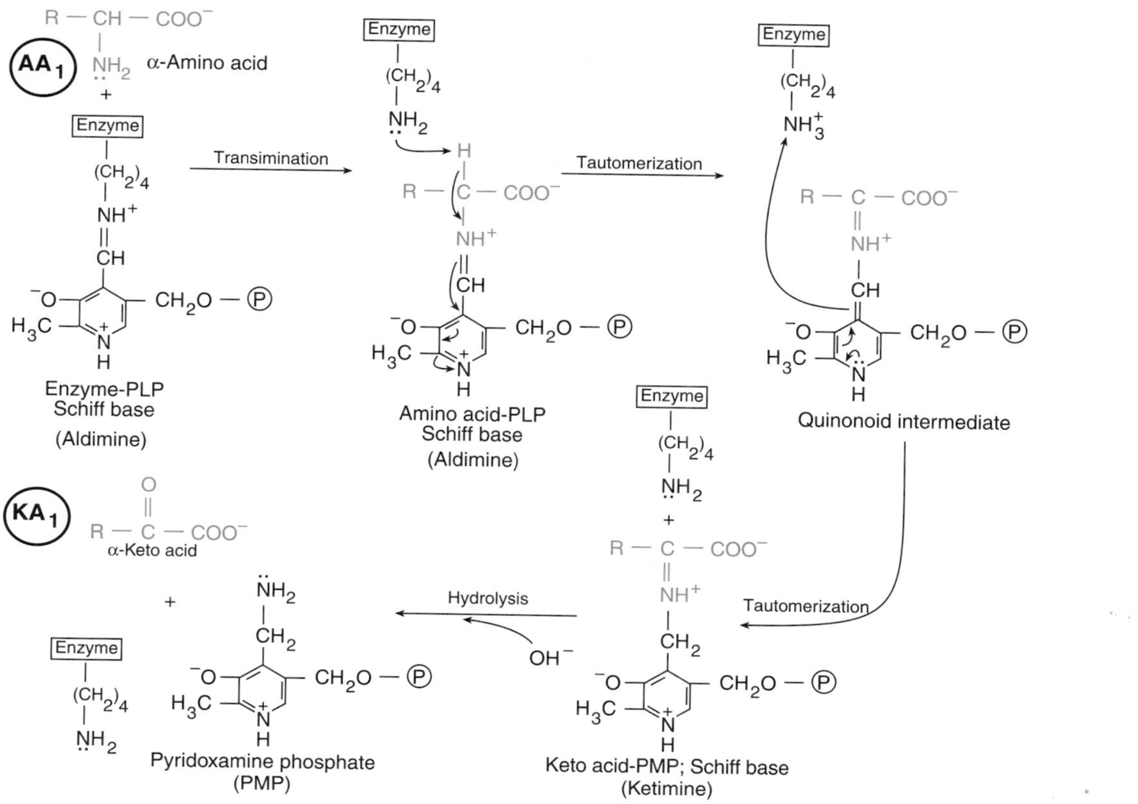

Figure 14.6. The mechanism of transamination. The reaction sequence shown then proceeds in the reverse direction, converting a different keto acid (KA_2) to the corresponding amino acid (AA_2). In the process, PLP is regenerated from PMP. The overall reaction is:

$$AA_1 + PLP \rightleftharpoons KA_1 + PMP$$
$$KA_2 + PMP \rightleftharpoons AA_2 + PLP$$
$$\overline{AA_1 + KA_2 \rightleftharpoons KA_1 + AA_2}$$

drate and lipid metabolism. Transamination provides a means of using the amino group of glutamate, a key product of ammonia fixation, to form other amino acids by means of pyruvate and oxaloacetate, key intermediates of carbohydrate and lipid metabolism (Figure 14.7). The amino acids formed can subsequently be converted to other nitrogenous compounds. Transamination also provides simple routes for synthesis of nonessential amino acids. Alanine and aspartate are nonessential amino acids because they are readily synthesized via transamination from the common metabolites pyruvate and oxaloacetate. Lastly, transamination has a regulatory function. By varying rates of transamination, organisms can maintain a steady-state level of nonessential amino acids regardless of their concentrations in the diet.

Two transaminases, **serum glutamate-oxaloacetate transaminase (SGOT)** and **serum glutamate-pyruvate transaminase (SGPT)**, were once of great clinical interest since their plasma levels were used to diagnose heart and liver damage. Both the heart and the liver normally contain high concentrations of SGOT and SGPT. In the event of a myocardial infarct (heart attack), a portion of the heart is damaged, causing SGOT and SGPT to leak into the bloodstream from the injured cells. By measuring the activity of these enzymes in the blood, physicians can estimate the extent of heart damage. Liver damage can be assessed similarly since injured liver cells also lead to leakage of SGOT and SGPT into the bloodstream.

In recent times, use of these enzymes has been supplanted by measurements of the isozymes of *lactate dehydrogenase* and *creatine kinase,* which have greater specificity and predictive accuracy. Creatine kinase catalyzes the phosphorylation of creatine to phosphocreatine.

14.2.2. Oxidative Deamination

With few exceptions, the first step in amino acid degradation involves a removal of the amino group by either

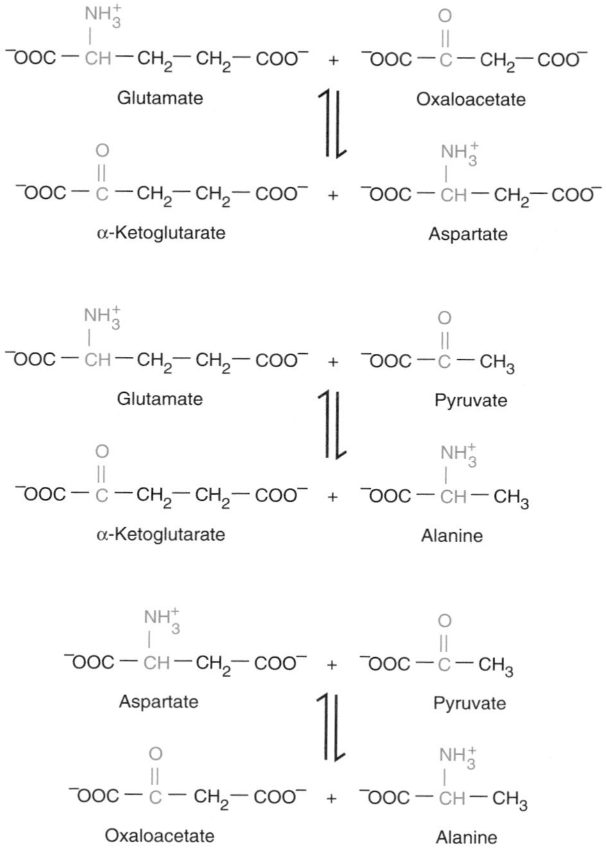

$$\overset{\overset{NH_3^+}{|}}{^-OOC-CH-CH_2-CH_2-COO^-} + {^-OOC-\overset{\overset{O}{\|}}{C}-CH_2-COO^-}$$

Glutamate   Oxaloacetate

$$\rightleftarrows$$

$$^-OOC-\overset{\overset{O}{\|}}{C}-CH_2-CH_2-COO^- + {^-OOC-\overset{\overset{NH_3^+}{|}}{CH}-CH_2-COO^-}$$

α-Ketoglutarate   Aspartate

$$\overset{\overset{NH_3^+}{|}}{^-OOC-CH-CH_2-CH_2-COO^-} + {^-OOC-\overset{\overset{O}{\|}}{C}-CH_3}$$

Glutamate   Pyruvate

$$\rightleftarrows$$

$$^-OOC-\overset{\overset{O}{\|}}{C}-CH_2-CH_2-COO^- + {^-OOC-\overset{\overset{NH_3^+}{|}}{CH}-CH_3}$$

α-Ketoglutarate   Alanine

$$\overset{\overset{NH_3^+}{|}}{^-OOC-CH-CH_2-COO^-} + {^-OOC-\overset{\overset{O}{\|}}{C}-CH_3}$$

Aspartate   Pyruvate

$$\rightleftarrows$$

$$^-OOC-\overset{\overset{O}{\|}}{C}-CH_2-COO^- + {^-OOC-\overset{\overset{NH_3^+}{|}}{CH}-CH_3}$$

Oxaloacetate   Alanine

Figure 14.7.   Examples of transamination reactions.

transamination or **deamination.** In deamination, the amino group is removed as ammonia. Deamination usually occurs coupled to an oxidation, resulting in conversion of an amino acid to a keto acid. This makes the overall reaction one of **oxidative deamination.** The ammonia enters the urea cycle, where it is converted to urea and excreted. The keto acid enters carbohydrate or lipid metabolism. Oxidative deamination occurs via two mechanisms using either glutamate dehydrogenase or amino acid oxidases.

14.2.2A. Glutamate Dehydrogenase. Most
amino acids can donate their amino groups and form glutamate via transamination. Amino acids undergoing transamination have their amino groups concentrated into glutamate, which can subsequently be oxidatively deaminated at an appreciable rate by glutamate dehydrogenase to provide ammonia for the urea cycle. Both the transaminase and glutamate dehydrogenase reactions are reversible.

Transaminase reaction:

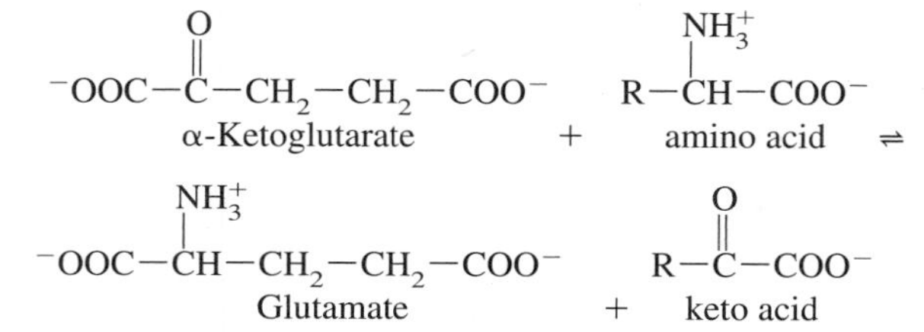

$$^-OOC-\overset{\overset{O}{\|}}{C}-CH_2-CH_2-COO^- \quad {^-R-\overset{\overset{NH_3^+}{|}}{CH}-COO^-}$$

α-Ketoglutarate   +   amino acid   ⇌

$$\overset{\overset{NH_3^+}{|}}{^-OOC-CH-CH_2-CH_2-COO^-} \quad R-\overset{\overset{O}{\|}}{C}-COO^-$$

Glutamate   +   keto acid

Glutamate dehydrogenase reaction:

$$\overset{\overset{NH_3^+}{|}}{^-OOC-CH-CH_2-CH_2-COO^-}$$

Glutamate $+ NAD^+ + H_2O \rightleftarrows$

$$^-OOC-\overset{\overset{O}{\|}}{C}-CH_2-CH_2-COO^-$$

α-Ketoglutarate $+ NADH + H^+ + NH_4^+$

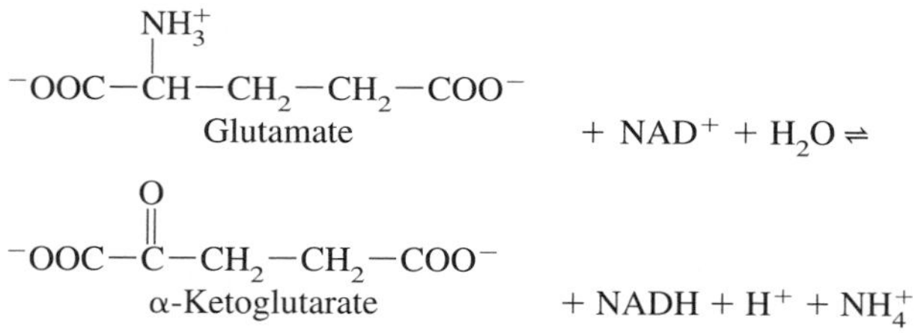

The overall reaction is:

Amino acid $+ NAD^+ + H_2O \rightleftarrows$ keto acid $+ NADH + H^+ + NH_4^+$

14.2.2B. Amino Acid Oxidases. Oxidative
deamination of amino acids when catalyzed by amino acid oxidases proceeds in a single step. Amino acid oxidases occur predominantly in the liver and the kidney. *L-Amino acid oxidases* are specific for L-amino acids and have FMN as coenzyme. *D-Amino acid oxidases* are specific for D-amino acids and have FAD as coenzyme. The function of D-amino acid oxidases is enigmatic since D-amino acids are rare in animals, being associated primarily with bacterial cell walls. The general reaction with an amino acid oxidase takes the form:

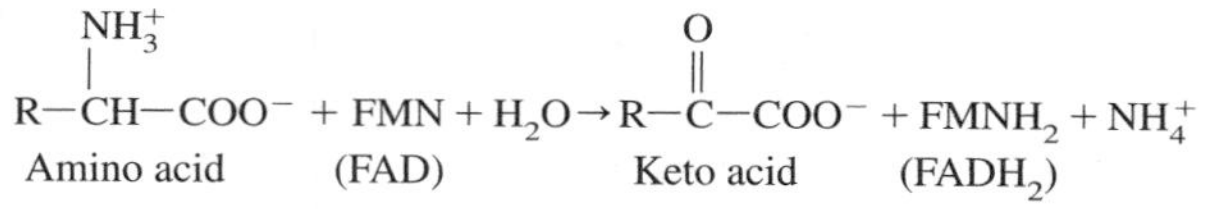

$$R-\overset{\overset{NH_3^+}{|}}{CH}-COO^- + FMN + H_2O \rightarrow R-\overset{\overset{O}{\|}}{C}-COO^- + FMNH_2 + NH_4^+$$

Amino acid   (FAD)   Keto acid   (FADH$_2$)

Amino acid oxidases also catalyze reoxidation of the reduced coenzyme FMNH$_2$ (FADH$_2$) by oxygen. Recall that any oxidase catalyzes the direct interaction of a substrate with molecular oxygen:

$$FMNH_2 + O_2 \rightarrow FMN + H_2O_2$$
$$(FADH_2) \qquad\quad (FAD)$$

The toxic hydrogen peroxide (H$_2$O$_2$) formed in this reaction is decomposed by action of the enzyme *catalase:*

$$2H_2O_2 \longrightarrow 2H_2O + O_2$$

Table 14.1. Glucogenic and Ketogenic Amino Acids

Amino acid	Glucogenic	Ketogenic
Alanine	×	
Arginine	×	
Asparagine	×	
Aspartate	×	
Cysteine	×	
Glutamate	×	
Glutamine	×	
Glycine	×	
Histidine	×	
Isoleucine	×	×
Leucine		×
Lysine		×
Methionine	×	
Phenylalanine	×	×
Proline	×	
Serine	×	
Threonine	×	×
Tryptophan	×	×
Tyrosine	×	×
Valine	×	

14.2.3. Metabolic Fates of Amino Acids

Following oxidative deamination, carbon skeletons of all amino acids undergo degradation to one of seven common intermediates. Based on these end products of their catabolism, we can place amino acids in two large groups (Table 14.1). **Glucogenic (glycogenic) amino acids** are catabolized to *precursors of carbohydrates.* Catabolism of these amino acids yields pyruvate, oxaloacetate, α-ketoglutarate, succinyl CoA, and fumarate—all compounds that lead to synthesis of glucose and glycogen. **Ketogenic amino acids** are catabolized to *precursors of lipids.* Catabolism of these amino acids yields acetyl CoA and acetoacetyl CoA—compounds that lead to synthesis of ketone bodies. Most of the amino acids (13) are strictly glucogenic; leucine and lysine represent the only amino acids that are strictly ketogenic. Five amino acids turn out to be both glucogenic and ketogenic; their degradation yields two products, one of which is glucogenic, and the other ketogenic.

Both amino acid catabolism and anabolism are intimately linked to operation of the citric acid cycle (Figure

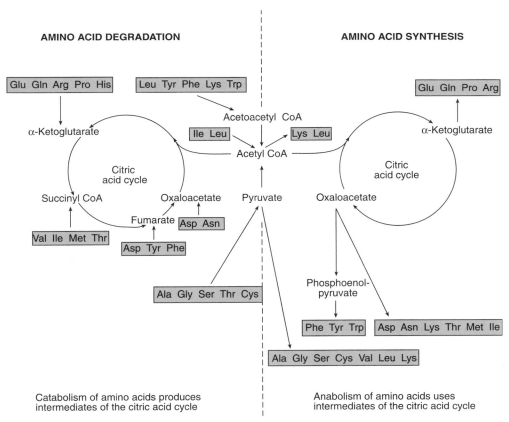

Figure 14.8.   The relationships between amino acid metabolism and the citric acid cycle.

14.8). When degraded, amino acids yield intermediates of the cycle or precursors of such intermediates. When synthesized, amino acids form from intermediates of the cycle or from precursors of such intermediates.

14.2.4. Catabolism of Phenylalanine and Tyrosine

As an illustration of amino acid degradation, consider the specific catabolic pathways of phenylalanine and tyrosine (Figure 14.9). The major catabolic pathway of these two aromatic amino acids proceeds from phenylalanine through tyrosine to fumarate and acetoacetate. Because fumarate is glucogenic whereas acetoacetate is ketogenic, we classify phenylalanine and tyrosine as being both glucogenic and ketogenic. As Figure 14.9 indicates, tyrosine is normally synthesized by hydroxylation of phenylalanine, a reaction catalyzed by *phenylalanine hydroxylase.* Because of this reaction, we consider tyrosine to be a nonessential amino acid. The first step in tyrosine catabolism, as in the catabolism of most amino acids, is an oxidative deamination.

Phenylalanine hydroxylase catalyzes the conversion of phenylalanine to tyrosine in a single step. The enzyme requires molecular oxygen and the cofactor *tetrahydro-*

biopterin. Phenylalanine hydroxylase is a *monooxygenase,* or *mixed-function oxygenase,* since one atom of O_2 appears in the product (tyrosine) and the other is converted to water (Figure 14.10). The quinonoid form of dihydrobiopterin produced in the hydroxylation of phenylalanine is reduced back to tetrahydrobiopterin by NADH in a reaction catalyzed by *dihydropteridine reductase.*

In addition to initiating the major catabolic route, phenylalanine can undergo a second set of reactions. This pathway functions much as the "overflow pathway" of ketone body formation. Normally of minor significance, the pathway becomes accentuated when phenylalanine accumulates because of a block in the major pathway. In some individuals such a block occurs at the phenylalanine hydroxylase step.

We refer to the genetic disease associated with this metabolic defect as **phenylketonuria,** or **PKU.** It is caused by an autosomal recessive gene. In phenylketonuria, phenylalanine cannot be hydroxylated to tyrosine and is, instead, converted to phenylpyruvate and phenyllactate (Figure 14.9). The first step of this catabolic pathway also involves an oxidative deamination. The PKU pathway leads to accumulation of large quantities of phenylpyruvate and phenyllactate in the blood, followed

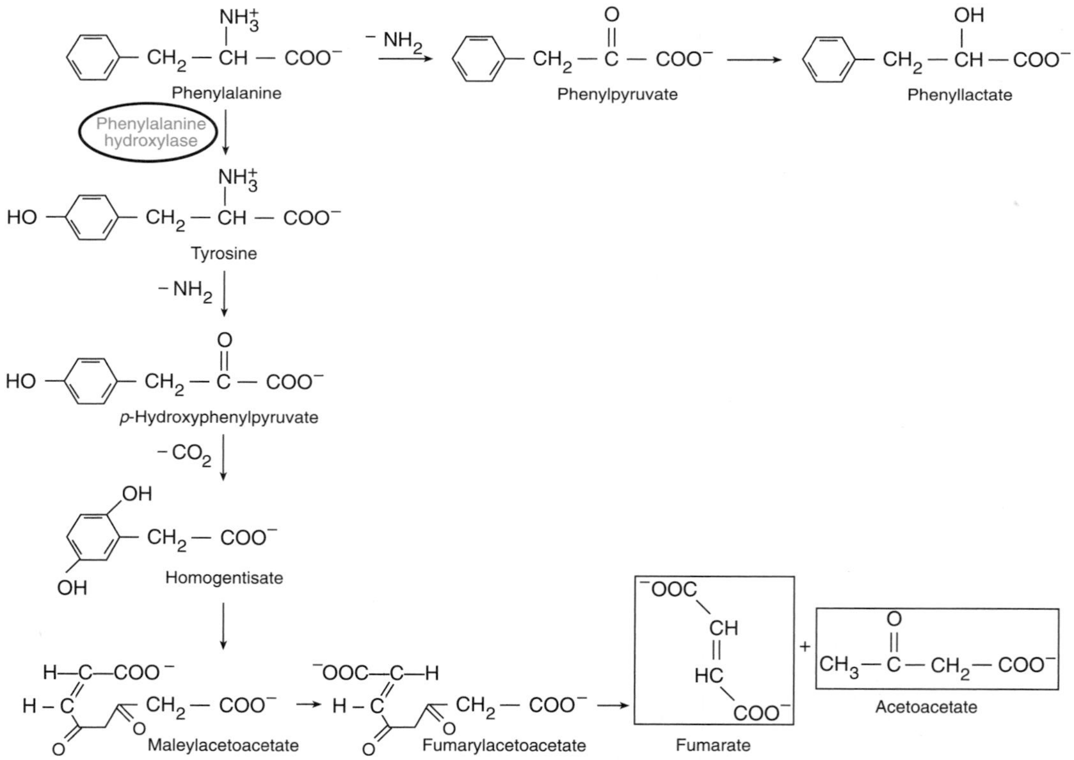

Figure 14.9. Catabolism of phenylalanine and tyrosine. The major pathway leads to fumarate and acetoacetate.

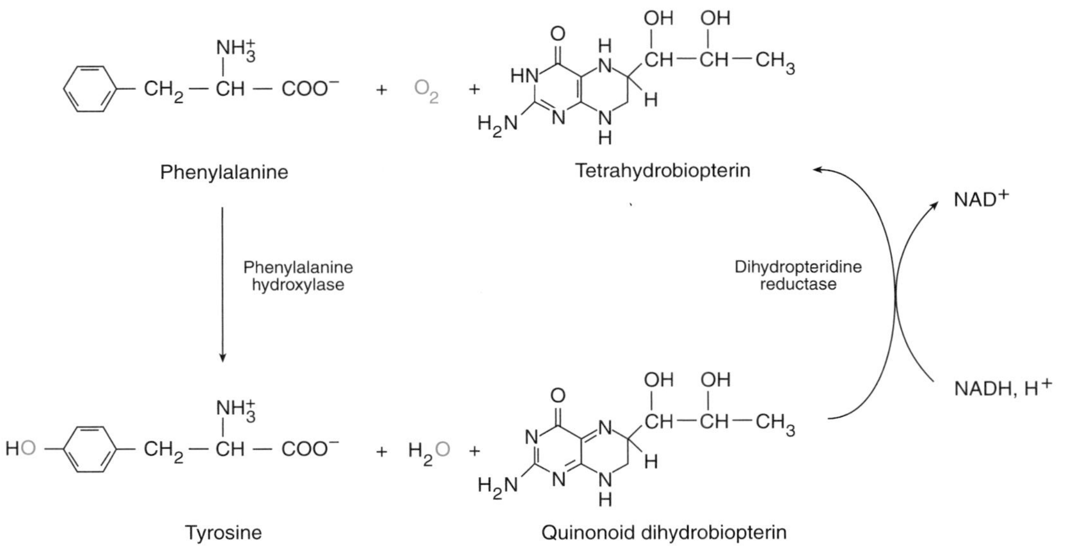

Figure 14.10. Tyrosine conversion to phenylalanine, catalyzed by phenylalanine hydroxylase. The reaction involves an oxidation of tetrahydrobiopterin, which is then reduced by means of NADH in a reaction catalyzed by dihydropteridine reductase.

by their excretion in the urine. Physicians consider a concentration of over 20 mg of phenylalanine per 100 ml of blood to be a positive indicator for PKU.

Someone suffering from PKU in early childhood becomes mentally retarded. Researchers generally attribute mental retardation to toxic effects of phenylalanine, possibly on the transport of other aromatic amino acids in the brain. Because of the severity of the disease, hospitals today routinely check children at birth for PKU.

When laboratory analysis positively identifies the condition, the child is placed on a synthetic diet low in phenylalanine. This regimen is continued at least through early childhood (up to the age of 5). Following that, protein is restricted in the diet for three to five more years. At that age, the deleterious effects of PKU seem to disappear since, by the time a child reaches age 8, a large portion of the brain has been formed. However, whether and when phenylalanine can be added to the diet at normal levels is still a controversial subject.

14.2.5. Amino Acids as Biosynthetic Precursors

In addition to their major function as building blocks of proteins, amino acids serve as precursors for a variety of biomolecules. Glycine, for example, functions as a building block of heme. The entire tetrapyrrole ring system is assembled from glycine and acetate (Figure 14.11). Heme degradation entails an opening of the ring system and its conversion to linear tetrapyrroles termed *bile pigments.*

Tyrosine serves as a precursor of *melanins,* dark-colored substances responsible for skin pigmentation.

A number of physiologically active amines, including epinephrine (Figure 10.3), histamine (Figure 14.12), and serotonin (Figure 14.12), are derivatives of amino acids. Epinephrine is a hormone derived from tyrosine that functions in regulating glycogen metabolism and fatty acid mobilization. Histamine derives from histidine and has multiple functions. It acts as a strong vasodilator and is released during inflammation and allergic reactions. It also promotes increased capillary permeability and contraction of smooth muscle and enhances secretion of hydrochloric acid in the stomach.

Serotonin derives from tryptophan. It is a neurotransmitter and acts as a potent vasoconstrictor. It helps to control blood pressure and regulate peristalsis in the small intestine. A reaction common to the synthesis of epinephrine, histamine, and serotonin consists of a pyridoxal phosphate-dependent decarboxylation of the amino acid from which each derives.

Glutathione is a tripeptide composed of glutamic acid, cysteine, and glycine (see Figure 2.12) that functions in oxidation–reduction reactions of sulfhydryl groups. Glutamic acid residues also occur as structural components of folate coenzymes.

14.2.6. Biosynthesis of Amino Acids

Biosynthesis of amino acids proceeds from simple metabolic precursors (see Figure 14.8). Based on the relationships among their biosynthetic pathways, we group amino

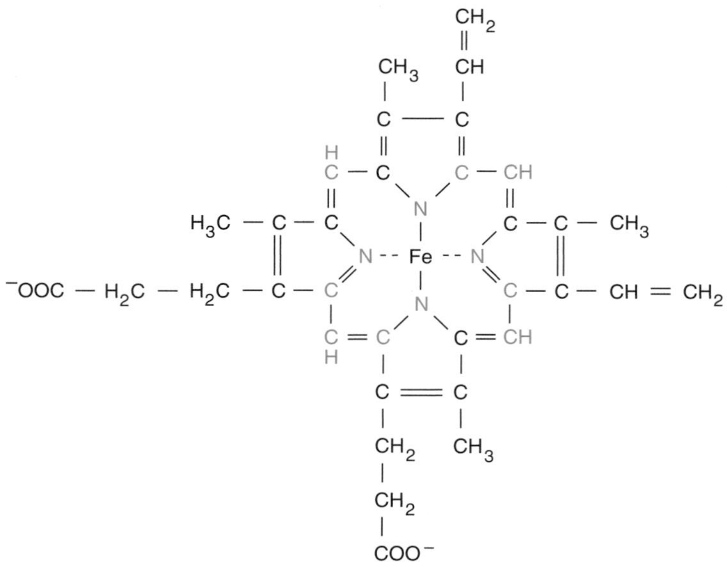

Figure 14.11.   Assembly of the heme nucleus. Except for the iron, all atoms are derived from glycine (red) and acetate (black).

acids into six families (Figure 14.13). Some nonessential amino acids are produced directly from intermediates of carbohydrate and lipid metabolism, and some are formed by way of other amino acids. Hydroxylation of phenyl-alanine yields tyrosine. Serine serves as a precursor of glycine and cysteine, and glutamate serves as a precursor of proline and arginine. Aspartate and glutamate are converted to asparagine and glutamine, respectively, by *asparagine synthase* and *glutamine synthase*. Glutamine synthase, an allosteric enzyme, occupies a key position in regulating nitrogen metabolism. Glutamine represents a storage form of ammonia and functions as an amino group donor in many biosynthetic reactions.

Pathways for the biosynthesis of essential amino acids

occur only in plants and microorganisms and usually require a larger number of steps than pathways for the synthesis of nonessential amino acids. Histidine forms from ribose 5-phosphate by way of phosphoribosyl pyrophosphate.

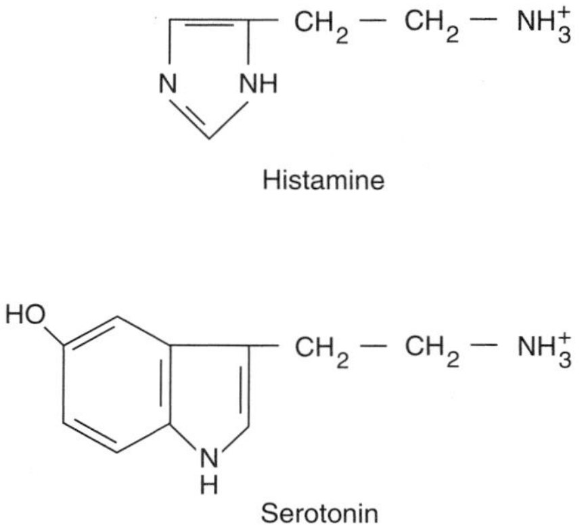

14.3. UREA CYCLE

The **urea cycle** constitutes a general pathway of amino acid catabolism whereby amino groups of amino acids are converted to urea that can be excreted. The cycle was elucidated by Hans Krebs and Kurt Henseleit in 1932, several years before Krebs discovered the citric acid cycle. The urea cycle was the first known metabolic cycle.

14.3.1. Operation of the Cycle

14.3.1A. Individual Reactions. The urea cycle consists of five enzymatic reactions, three of which occur in the cytosol and two of which take place in the mitochondrial matrix (Figure 14.14). Specific carrier mechanisms, located in the inner mitochondrial membrane, provide for transport of ornithine and citrulline into and out of the matrix. The reactions of the cycle are:

Histamine

Serotonin

Figure 14.12.   Some physiologically active amines derived from amino acids.

1. $NH_4^+ + HCO_3^- + 2ATP^{4-} + H_2O \rightarrow$
    carbamoyl phosphate^{2-} + 2ADP^{3-} + P$_i^{2-}$ + 2H$^+$
2. Carbamoyl phosphate^{2-} + ornithine$^+$ $\rightarrow$
    citrulline + P$_i^{2-}$ + H$^+$
3. Citrulline + aspartate$^-$ + ATP^{4-} $\rightarrow$
    AMP^{2-} + PP$_i^{3-}$ + argininosuccinate$^-$ + H$^+$

4. Argininosuccinate$^-$ + H$_2$O →
   arginine$^+$ + fumarate^{2-}
5. Arginine$^+$ + H$_2$O → ornithine$^+$ + urea

Reaction 1 represents one of the ammonia fixation reactions in which ammonia from oxidative deamination of amino acids gives rise to **carbamoyl phosphate.** The reaction requires expenditure of energy in the form of ATP, which is cleaved to ADP and P$_i$. Carbamoyl phosphate formation constitutes the *committed step* of the urea cycle and its major control point. The reaction is catalyzed by *carbamoyl phosphate synthase,* an allosteric enzyme. Recall that in eukaryotes, this reaction is catalyzed by *carbamoyl phosphate synthase I* (see Section 14.1).

Carbamoyl phosphate reacts with **ornithine** to yield **citrulline** (reaction 2). Both ornithine and citrulline are α-amino acids, but do not occur as structural components of proteins. In reaction 3, *aspartate* reacts with citrulline, and ATP is cleaved to AMP and PP$_i$. Because the reaction product can be considered to be an addition compound of arginine and succinate, we call it **argininosuccinate.**

Reaction 4 involves a cleavage of argininosuccinate, yielding *arginine* and *fumarate.* **Arginase** catalyzes the last step of the cycle (reaction 5), which results in a cleavage of arginine to **urea** (H$_2$N—CO—NH$_2$) and ornithine. In mammals, urea synthesis occurs exclusively in the liver. Urea is then transported by the blood from the liver to the kidney, where it is excreted in the urine.

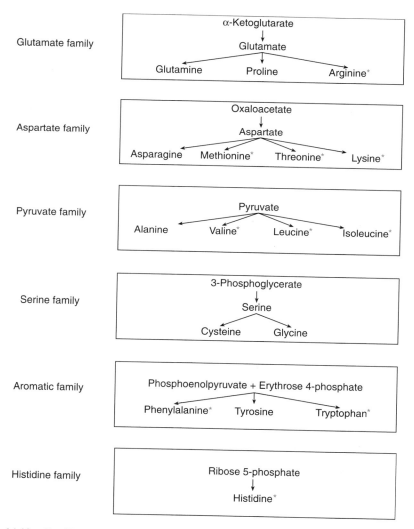

Figure 14.13.   Families of amino acids based on their biosynthetic pathways (* = essential amino acid).

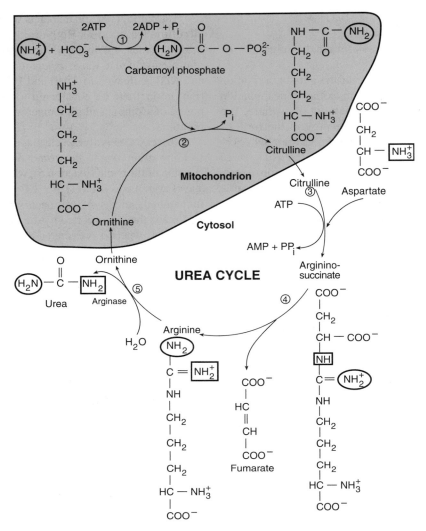

Figure 14.14.   The urea cycle. Numbers designate enzymes: ① Carbamoyl phosphate synthase I; ② ornithine transcarbamoylase; ③ argininosuccinate synthetase; ④ argininosuccinase; ⑤ arginase.

14.3.1B. Overall Reaction.

By adding the above five reactions, you obtain the overall reaction of the urea cycle:

$$NH_4^+ + HCO_3^- + 3ATP^{4-} + 2H_2O + aspartate^-$$
$$\downarrow$$
$$Urea + fumarate^{2-} + 2ADP^{3-} + AMP^{2-} + PP_i^{3-} + 2P_i^{2-} + 4H^+$$

One turn of the cycle achieves the synthesis of one molecule of urea and requires the input of *one carbon atom and two amino groups*. The carbon derives from CO_2 in reaction 1, and the two amino groups derive from two amino acids. One amino group comes from aspartate, which is thereby converted to fumarate (reactions 3 and 4). The second amino group (reaction 1) comes from deamination of some other amino acid, primarily glutamate via the glutamate dehydrogenase reaction.

14.3.1C. Energetics.

Although the urea cycle constitutes a catabolic pathway, it does not yield usable energy. On the contrary, this pathway is endergonic and requires the input of energy. Between them, reactions 1 and 3 require three molecules of ATP. Additionally, as is common in metabolism, when pyrophosphate forms, it undergoes hydrolysis to inorganic phosphate, an exergonic reaction catalyzed by *pyrophosphatase*. Such hydrolysis always helps to drive the overall process that includes pyrophosphate formation. Allowing for pyrophosphate hydrolysis, the energy requirement for urea synthesis becomes equivalent to the hydrolysis of four energy-rich bonds:

Reaction	Energy-rich bonds
$2ATP \rightarrow 2ADP + 2P_i$	2
$ATP \rightarrow AMP + PP_i$	1
$PP_i \rightarrow 2P_i$	1
Total	4

14.3.1D. Metabolic Interrelationships.

Operation of the urea cycle illustrates the interconnections of metabolic pathways (Figure 14.15). The substrates for reaction 1 derive from deamination (NH_3) and from oxidations via the citric acid cycle and the electron transport system (CO_2, H_2O, ATP). Reaction 1 "fixes" ammonia into the organic compound carbamoyl phosphate. The aspartate in reaction 3 comes from transamination of oxaloacetate, a citric acid cycle intermediate. Fumarate, the product of reaction 4, enters the citric acid cycle and is converted back to oxaloacetate. Conversion of fumarate to oxaloacetate generates NADH, which can be oxidized via the electron transport system to produce ATP and thereby furnish energy for operation of the urea cycle. You can see that the cycle has many links to major aspects of metabolism.

14.3.2. Comparative Biochemistry of Nitrogen Excretion

Because ammonia is highly toxic for living cells, its concentration is generally maintained at low levels. We do not fully understand the reasons for ammonia's great toxicity, but we know of two important contributing factors—the high pK_a' of ammonia (9.3 at 25°C) and the free movement of NH_3 across cell and mitochondrial membranes.

Because of its high pK_a' only a small fraction of ammonia exists as NH_3 at physiological pH (7.0); most occurs in the form of the ammonium ion, NH_4^+. However, because of its great permeation capacity, NH_3 passes readily across the membranes of brain cells and the membranes of their mitochondria. Once inside the mitochondria, ammonia reacts with α-ketoglutarate via the glutamate dehydrogenase reaction. Conversion of α-ketoglutarate to glutamate lowers the α-ketoglutarate concentration in the citric acid cycle pool, thereby lowering the cycle's activity. Decreased citric acid cycle activity translates into decreased oxidation of glucose, the principal fuel of the brain.

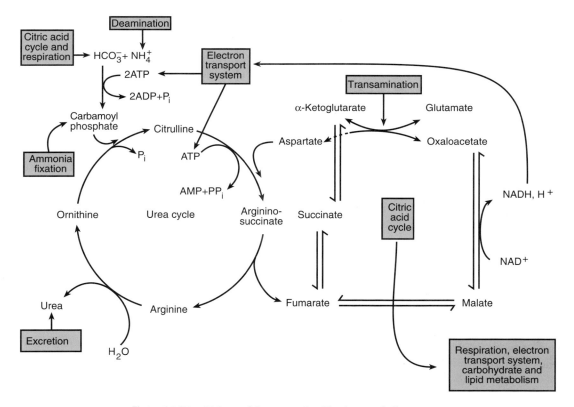

Figure 14.15. Linkage of the urea cycle with other metabolic systems.

Excess ammonia produced by amino acid catabolism must be excreted. The form in which amino acid nitrogen is excreted by living organisms varies. Some organisms excrete ammonia directly; others excrete waste nitrogen in the form of urea or uric acid. We may classify living organisms according to the form in which they excrete their nitrogen. **Ureotelic, uricotelic,** or **ammonotelic organisms** excrete waste nitrogen primarily in the form of *urea, uric acid,* or *ammonia,* respectively. The diversity of nitrogen excretion represents an interesting case of comparative biochemistry. It relates to water intake and availability and to solubility and toxicity of the three excretory nitrogen-containing compounds.

Ammonia represents the most toxic but also the most soluble of the three compounds. Aquatic animals and teleost fish have an unlimited supply of water. These organisms excrete ammonia despite its great toxicity. Because of its great solubility, ammonia becomes effectively diluted by the environment as soon as it is excreted and is thereby rendered harmless.

Birds and land-dwelling reptiles, whose water intake is limited and whose excretion consists largely of semisolid material, excrete nitrogen as uric acid. Uric acid is less toxic than ammonia and is the least soluble of the three compounds. Its low toxicity results in part from its low solubility in water.

Mammals and most terrestrial vertebrates fall somewhere in between as regards their water supply. These organisms excrete nitrogen in the form of urea, a nontoxic compound. Urea is intermediate in its solubility, being less soluble than ammonia but more soluble than uric acid.

Researchers have suggested that the "choice" between excretion of urea and uric acid depends on the conditions under which an embryo develops. In the case of mammals, the early embryonic stage takes place in an *internal environment* and in close contact with the circulatory system of the mother. Urea, a nontoxic compound of good water solubility, can safely be formed by the embryo for ready removal by the maternal circulation.

The early embryonic stage of birds and reptiles, however, takes place in an *external environment* and in an egg surrounded by a hard shell. The egg contains sufficient water to allow hatching but not enough to permit excretion of large quantities of toxic substances. Excretion of ammonia by these organisms would be fatal, and they have evolved to eliminate nitrogen as uric acid, which precipitates out in a fluid-filled sac (amnion) located on the interior surface of the shell.

Scientists believe that these metabolic systems, necessary for development of the embryo, are then carried over to the adult organism. Support for this theory comes from a number of living systems. Amphibia fall in between ureotelic and ammonotelic organisms. The tadpole lives in an aqueous environment and excretes nitrogen as ammonia. However, during metamorphosis, its liver acquires the necessary enzymes for urea synthesis, and the organism begins to excrete urea. By the time that metamorphosis has been completed and the adult frog has formed, nitrogen excretion occurs predominantly in the form of urea.

Nitrogen excretion of the lungfish also varies. So long as the lungfish can exist in plenty of water, its nitrogen excretion consists primarily of ammonia. When the riverbeds and lakes become dry, the fish settles down in the mud and begins to accumulate nitrogen as urea. When the rainy season returns to fill the rivers and lakes, the lungfish first eliminates a large amount of accumulated urea and then reverts to excreting nitrogen as ammonia.

Lastly, the order Chelonia (tortoises and turtles) includes some species that are strictly aquatic, some that are strictly terrestrial, and some that are semiterrestrial. The aquatic species excrete a mixture of urea and ammonia, the terrestrial ones excrete predominantly uric acid, and the semiterrestrial organisms excrete urea.

14.4. PURINE AND PYRIMIDINE METABOLISM

Purines and pyrimidines occur widely in cells in the form of nucleotides and nucleic acids. We will discuss nucleotide metabolism in this chapter and that of nucleic acids in later chapters. In addition to being *building blocks* of nucleic acids, nucleotides function in many other capacities in living cells. Nucleoside triphosphates, such as ATP and GTP, represent *energy-rich compounds* that drive metabolic reactions. Nucleotides form parts of *coenzymes,* like NAD^+, $NADP^+$, FAD, FMN, and coenzyme A. These coenzymes participate in many metabolic systems and pathways, such as the electron transport system, the citric acid cycle, and β-oxidation. Nucleotides constitute versatile *"carriers"* of metabolites. AMP serves as a carrier of methionine (*S*-adenosylmethionine), UDP as a carrier of glucose (UDP-glucose), and CDP as a carrier of choline (CDP-choline). Lastly, nucleotides play an important role in the *regulation* of metabolism. The levels of ATP, ADP, and AMP, the levels of NAD^+ and NADH, and the mediation of hormonal signals by means of cAMP all function to control the network of metabolic reactions.

14.4.1. Purine Biosynthesis

14.4.1A. Synthesis of IMP. Atoms of the purine ring come from five sources: formate, glycine, aspartate (amino group), glutamine (amide group), and CO_2 (Figure 14.16). **Inosine 5′-monophosphate (IMP)** serves

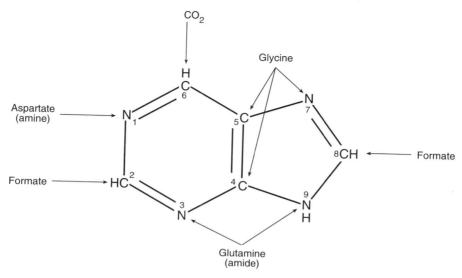

Figure 14.16.   Assembly of the purine nucleus.

as the common precursor of all purine nucleotides. It is synthesized *de novo* ("anew," from simple precursors) from ribose 5-phosphate by a series of 11 reactions (Figure 14.17). Assembly of the purine ring takes place on the sugar phosphate, which serves as an anchor.

Synthesis of IMP begins with a conversion of ribose 5-phosphate to 5-phospho-α-D-ribosyl-1-pyrophosphate, or **phosphoribosyl pyrophosphate (PRPP)**. The reaction is catalyzed by **PRPP synthase,** an unusual kinase that catalyzes the transfer of a pyrophosphate group, rather than a phosphate group, from ATP. PRPP synthase occupies a key position in purine metabolism and is subject to control via a number of activators and inhibitors. PRPP also serves as a precursor in the biosynthesis of pyrimidines.

The second step of purine biosynthesis represents the committed step of the pathway. The reaction, catalyzed by **glutamine-PRPP amidotransferase,** involves displacement of the pyrophosphate group of PRPP by the amide nitrogen of glutamine. The proper anomeric (β) configuration of nucleotides becomes established at this step. Pyrophosphatase-catalyzed hydrolysis of the PP_i produced drives the reaction to completion. Glutamine-PRPP amidotransferase is subject to feedback inhibition by purine nucleotides.

Among the subsequent steps of IMP synthesis occur two that require **tetrahydrofolate (THF),** the coenzyme form of *folic acid* (Figure 14.18). THF functions in the metabolism of *one-carbon fragments.* This area of metabolism includes decarboxylations, methylations, and transfers of other groups containing only one carbon atom. Biotin (see Section 13.5) serves as coenzyme for most carboxylation reactions, and *S*-adenosylmethionine (see Section 13.1) serves as coenzyme for most methyla-

tion reactions. THF serves as coenzyme for other one-carbon fragment reactions. THF is a versatile coenzyme that can carry one-carbon fragments in various oxidation states, except the most oxidized state (CO_2). As such, THF can serve as a carrier for methyl ($-CH_3$), methylene ($-CH_2-$), formyl ($-CH=O$), formimino ($-CH=NH$), and methenyl ($-CH=$) groups. Figure 14.18 illustrates how THF carries some of these groups.

14.4.1B. Interconversions of Purine Nucleotides.
Inosine monophosphate does not accumulate in cells because it is rapidly converted to AMP or GMP. Each conversion involves a two-step mechanism. Synthesis of AMP proceeds via adenylosuccinate, and synthesis of GMP proceeds via xanthosine monophosphate. All other purine nucleotides are derived from AMP and GMP (Figure 14.19).

Regulation of purine biosynthesis occurs via multiple feedback inhibitions (Figure 14.20). Both AMP and GMP serve as negative allosteric effectors for the enzymes at steps 1 and 2 of the purine biosynthetic pathway. In addition, AMP inhibits the first step of its synthesis from IMP, and GMP inhibits the first step of its synthesis from IMP.

Both purine and pyrimidine nucleotides are also synthesized by means of **salvage pathways.** A salvage pathway uses catabolic compounds for biosynthetic purposes even though they do not constitute true intermediates of the corresponding normal biosynthetic pathway. Purine and pyrimidine salvage pathways use bases obtained from nucleotide and nucleoside catabolism for biosynthesis of nucleosides and nucleotides. Salvage pathways constitute "recycling" systems in that they allow for reuse of bases

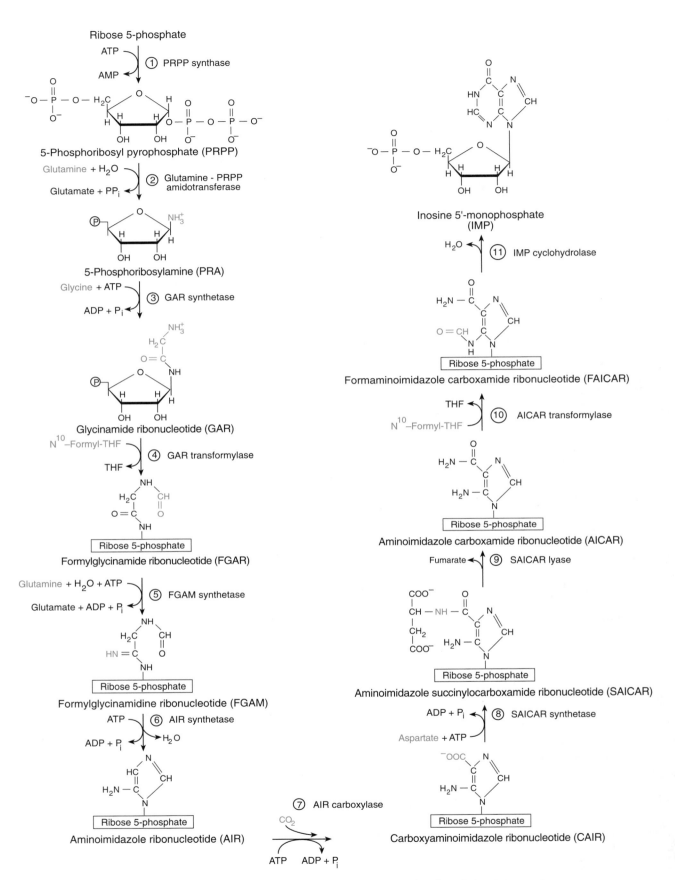

Figure 14.17. Purine biosynthesis. IMP is synthesized *de novo* using ribose 5-phosphate as anchor.

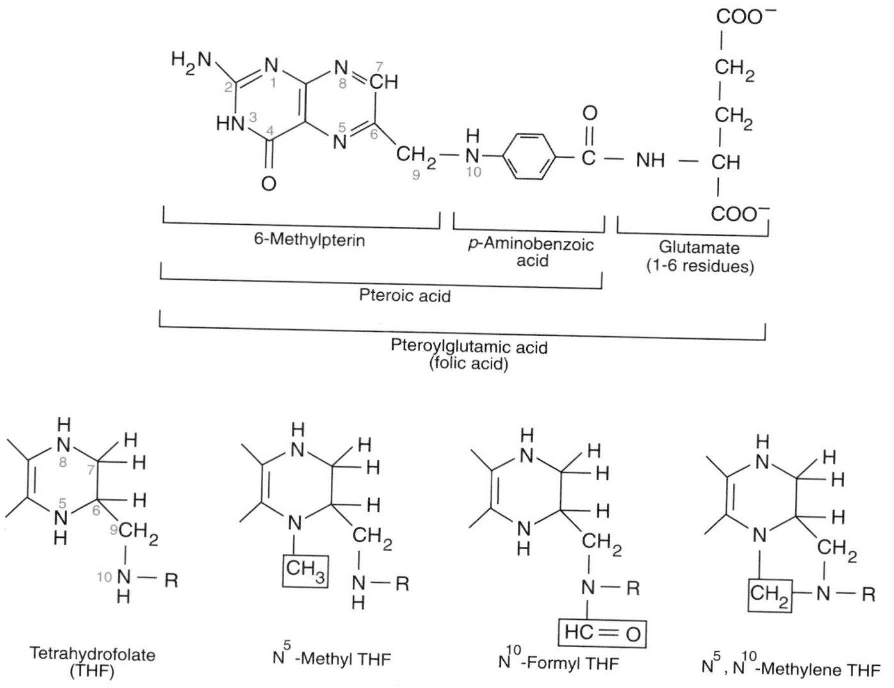

Figure 14.18.   Folic acid and its coenzyme forms.

formed in the more complicated and energy-requiring *de novo* biosynthetic pathways. PRPP plays an important role in salvage of purines and pyrimidines because these pathways involve *phosphoribosyl transferases,* enzymes that catalyze reactions of the type

$$\text{Base} + \text{PRPP} \rightleftarrows \text{ribonucleoside } 5'\text{-phosphate} + \text{PP}_i$$

This readily reversible reaction is pulled to completion by pyrophosphatase-catalyzed hydrolysis of PP_i to inorganic phosphate.

Purine deoxyribonucleotides are synthesized from corresponding ribonucleotides by reduction at C(2'), a re-action catalyzed by **ribonucleotide reductase.** Several forms of the enzyme occur in different species. The most widely distributed form in both prokaryotes and eukaryotes contains a nonheme iron. The enzyme has a reactive disulfide bond at its active site.

In some organisms, ribonucleotide reductase reduces nucleoside triphosphates (NTP) directly to deoxynucleoside triphosphates (dNTP). In most organisms, however, reduction occurs at the nucleoside diphosphate level. Reduction of nucleoside diphosphates (NDP) produces deoxynucleoside diphosphates (dNDP), which are then converted to deoxynucleoside triphosphates (dNTP) by phosphorylation, catalyzed by *nucleoside diphosphate kinase:*

$$\text{NDP} \xrightarrow[\substack{\text{ribonucleotide} \\ \text{reductase}}]{} \text{dNDP} \xrightarrow[\substack{\text{Nucleoside diphosphate} \\ \text{kinase}}]{\substack{\text{ATP} \quad \text{ADP}}} \text{dNTP}$$

NADPH provides the reducing power for reduction of nucleotides to deoxynucleotides by ribonucleotide reductase. Electrons are transferred from NADPH to the reductase by means of a mini electron transport system (Figure 14.21) centered about **thioredoxin,** a small electron carrier (MW = 12,000) that contains a reactive disulfide bond.

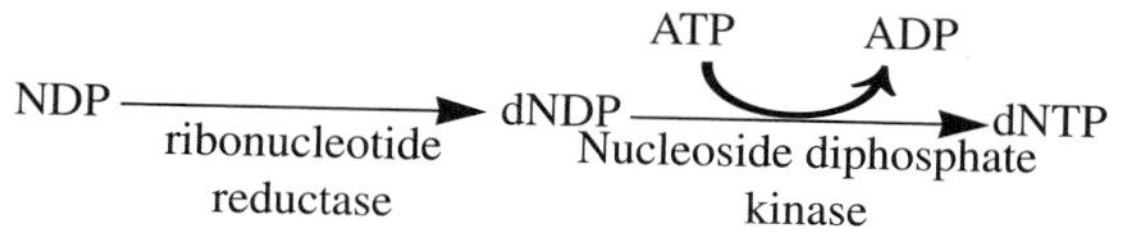

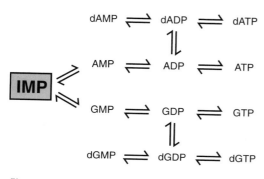

Figure 14.19.   Interconversions of purine nucleotides.

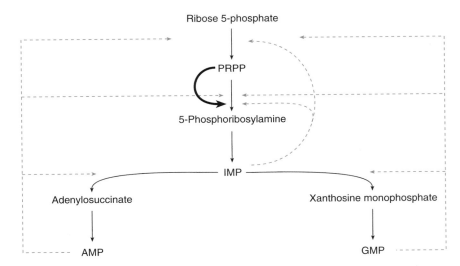

Figure 14.20.   Regulation of purine biosynthesis. Feedback inhibiton is shown in red, and activation is shown in black.

As electrons flow from NADPH, they first reduce *thioredoxin reductase,* an FAD-containing flavoprotein with a reactive disulfide bond. FAD is reduced to $FADH_2$ and then regenerated by coupling of the oxidation of $FADH_2$ to reduction of the enzyme's reactive disulfide bond. From thioredoxin reductase, electrons flow to the reactive disulfide bond of thioredoxin. Reduced thioredoxin then reduces the reactive disulfide bond of ribonucleotide reductase to two sulfhydryl groups. Formation of the two SH groups at the enzyme's active site allows

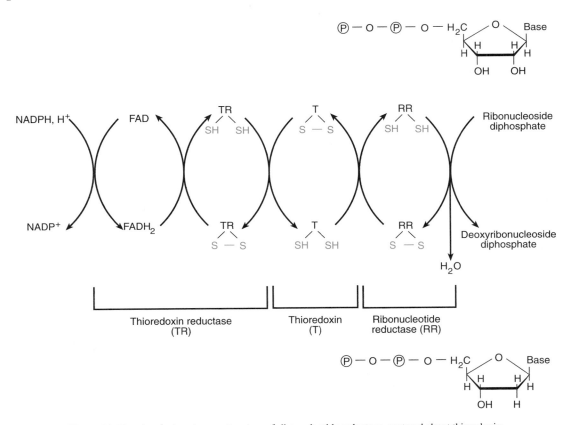

Figure 14.21.   An electron transport system of ribonucleotide reductase, centered about thioredoxin.

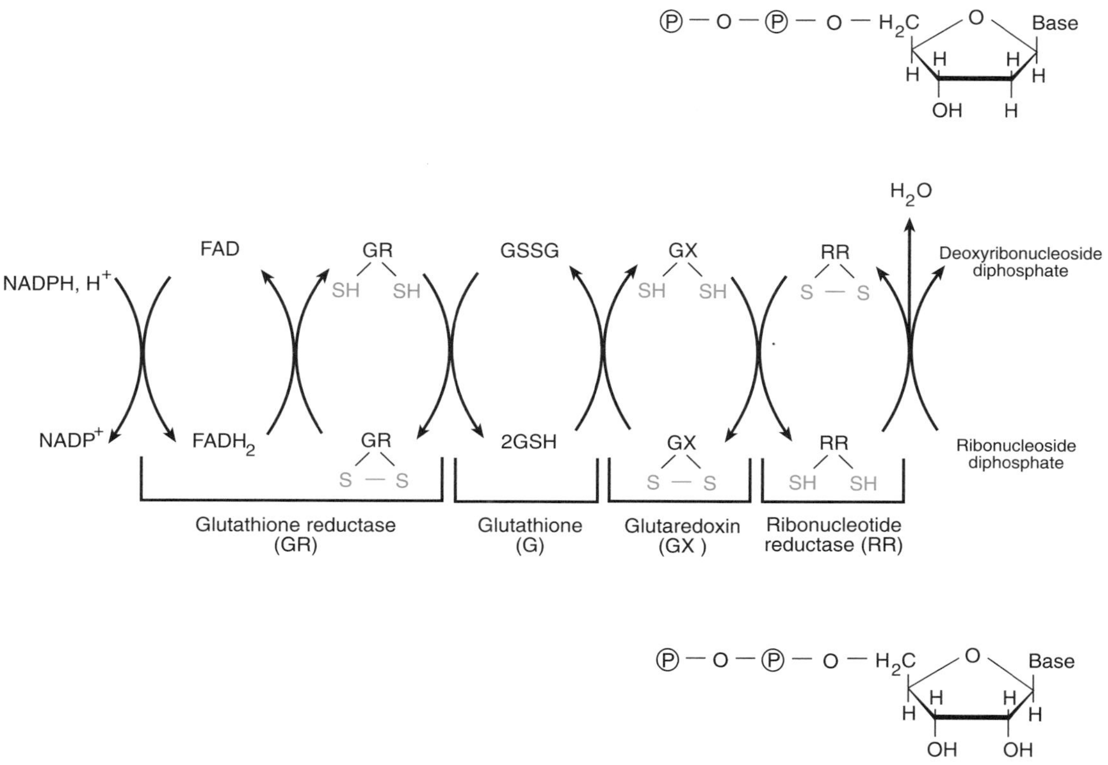

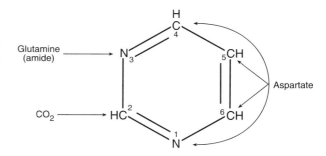

Figure 14.22.   An electron transport system of ribonucleotide reductase, centered about glutaredoxin.

reduction of ribose to deoxyribose by a complex free-radical mechanism.

In the absence of thioredoxin (as in *E. coli* mutants lacking this protein), electron flow from NADPH to ribonucleotide reductase proceeds via a second electron transport system, centered about *glutathione* and also containing glutathione reductase and **glutaredoxin** (Figure 14.22). *Glutathione reductase,* like thioredoxin reductase, is an FAD-containing flavoprotein with a reactive disulfide bond. Glutaredoxin resembles thioredoxin in its structure, particularly in the segment containing the two reactive SH groups.

14.4.2. Pyrimidine Biosynthesis

14.4.2A. Synthesis of UMP. Atoms of the pyrimidine ring derive from aspartate, glutamine (amide group), and CO_2 (Figure 14.23). Pyrimidine synthesis differs from purine synthesis in that assembly of the ring system proceeds without a sugar anchor. Only after the ring has been completed does it become coupled to ribose 5-phosphate. **Uridine 5'-monophosphate (UMP)** serves as the common precursor of all the pyrimidines. UMP is synthesized *de novo* in six steps that proceed through *orotate,*

a pyrimidine that does not normally occur in nucleic acids (Figure 14.24). The first step of UMP synthesis, catalyzed by carbamoyl phosphate synthase, leads to formation of carbamoyl phosphate. Recall that in eukaryotes the enzyme that catalyzes carbamoyl phosphate synthesis in this pathway is a cytosolic enzyme, *carbamoyl phosphate synthase II.* This enzyme uses glutamine and differs from the mitochondrial carbamoyl phosphate synthase I, which uses ammonia and functions in arginine synthesis in the urea cycle (Section 14.3). Prokaryotes have a single carbamoyl phosphate synthase that functions in both pyrim-

Figure 14.23.   Assembly of the pyrimidine nucleus.

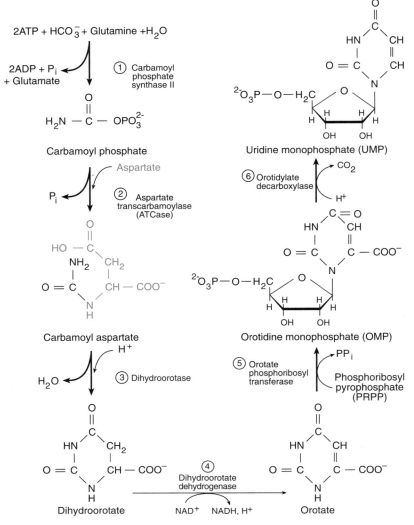

Figure 14.24.   Pyrimidine biosynthesis. The pyrimidine ring is assembled first (as orotate) and then linked to ribose 5-phosphate.

idine and arginine biosynthesis and uses glutamine for both processes.

The second reaction of the UMP biosynthetic pathway involves catalysis by **aspartate transcarbamoylase (ATCase).** Bacterial ATCase is an allosteric enzyme that constitutes a major control point for purine biosynthesis. *E. coli* ATCase is one of the most thoroughly studied allosteric enzymes. Note that phosphoribosyl pyrophosphate (PRPP), which serves as a precursor in purine biosynthesis, also serves as a precursor in pyrimidine biosynthesis.

14.4.2B. Interconversions of Pyrimidine Nucleotides.
All of the pyrimidine nucleotides are derived from UMP (Figure 14.25). Thymine nucleotides are derived from dUMP. Methylation of dUMP yields dTMP in a reaction catalyzed by *thymidylate synthase* (Figure 14.26). The enzyme uses N^5,N^{10}-methyl-

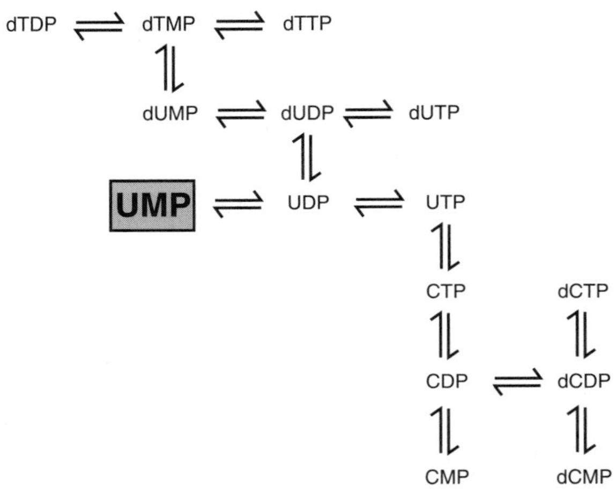

Figure 14.25.   Interconversions of pyrimidine nucleotides.

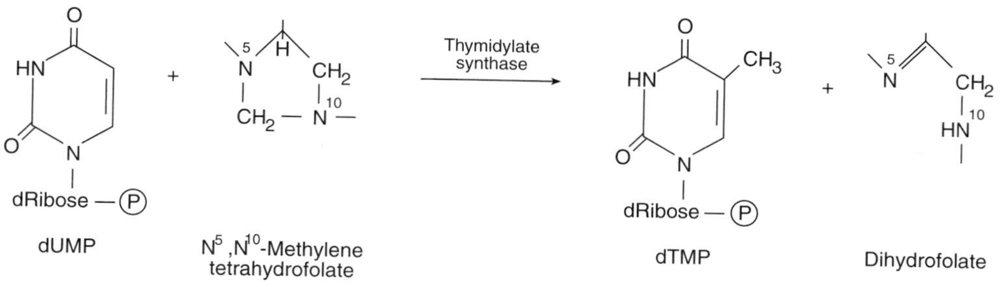

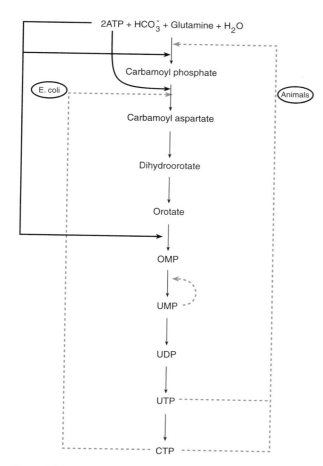

Figure 14.26.   Synthesis of dTMP from dUMP by thymidylate synthase.

enetetrahydrofolate as methyl-group donor. The folate coenzyme has a dual function in this reaction, serving as both a carrier of a one-carbon fragment and as a reducing agent. Tetrahydrofolate carries and donates a methylene group ($-CH_2-$) during the reaction. However, since the methylene group becomes converted to a methyl group ($-CH_3-$) in thymidine monophosphate, the coenzyme also serves as a reducing agent. In the process, it is oxidized from N^5,N^{10}-methylenetetrahydrofolate to 7,8-dihydrofolate. The active tetrahydrofolate carrier must be regenerated, since only it serves as a methylene group carrier for this reaction. Reduction of dihydrofolate to tetrahydrofolate is catalyzed by *dihydrofolate reductase,* an enzyme that requires NADPH as coenzyme. Because of this reduction, the overall conversion of dUMP to dTMP requires reducing power in the form of NADPH.

Major control points in the biosynthesis of pyrimidines occur at the first two steps of the UMP biosynthetic pathway (Figure 14.27). *E. coli* ATCase is activated by ATP and is subject to feedback inhibition by CTP. In many other bacterial systems, UTP is the major inhibitor of ATCase. The ATCase of animals is not an allosteric enzyme. In animals, control of pyrimidine biosynthesis occurs via feedback inhibition of UTP and CTP on carbamoyl phosphate synthase II.

Synthesis of pyrimidine nucleotides and deoxynucleotides by means of salvage pathways proceeds via the same reactions described for purines. Deoxynucleotides are synthesized from the corresponding nucleoside diphosphates by ribonucleotide reductases. As in the case of purines, pyrimidine salvage pathways involve catalysis by phosphoribosyl transferases. A special salvage pathway exists for dTMP, which can be synthesized from thymidine (deoxythmidine) in a reaction catalyzed by an ATP-dependent *thymidine kinase:*

$$\text{Thymidine} \xrightarrow[\text{thymidine kinase}]{\text{ATP}\quad\text{ADP}} \text{dTMP}$$

This pathway has a practical application for experiments requiring the labeling of DNA by means of radioactive thymidine. Thymidine enters cells readily and is largely metabolized to dTMP by the above reaction. Subsequent conversion of dTMP to dTTP produces a substrate for DNA polymerase and results in incorporation of labeled thymidine into newly synthesized DNA.

Figure 14.27.   Regulation of pyrimidine biosynthesis in *E.coli* and animals. Feedback inhibiton is shown in red, and activation is shown in black.

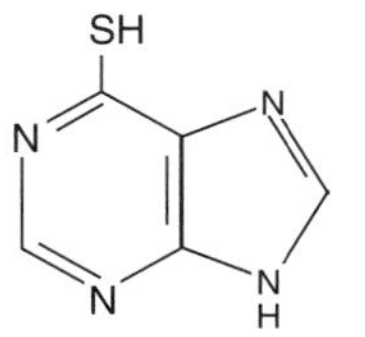

6-Mercaptopurine

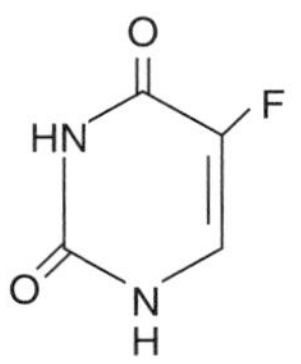

5-Fluorouracil

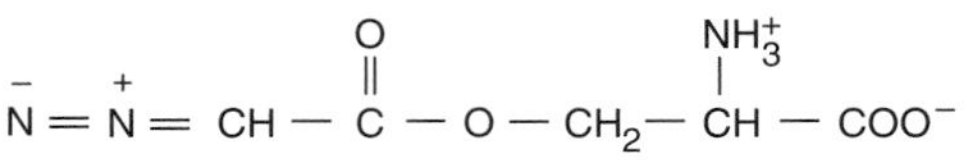

Azaserine

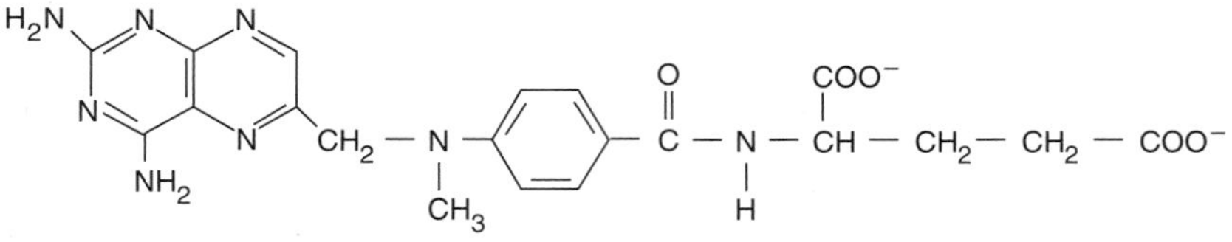

Methotrexate

Figure 14.28. Some drugs used in cancer chemotherapy.

Uric acid

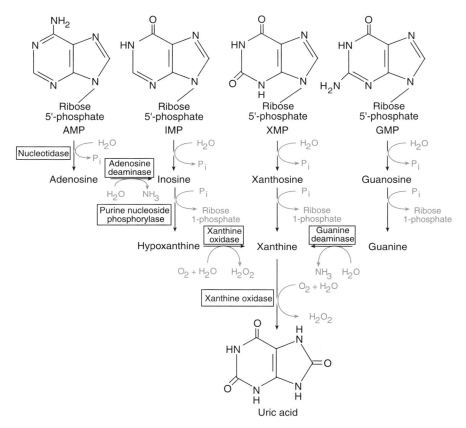

Figure 14.29. Catabolism of purines in animals. XMP, Xanthosine 5′-monophosphate. Xanthine oxidase catalyzes both the conversion of hypoxanthine to xanthine and that of xanthine to uric acid. Deoxyribonucleotides are degraded likewise.

14.4.2C. Cancer Chemotherapy.

Our understanding of the pathways for *de novo* synthesis of purines and pyrimidines has been of great benefit in the design of drugs for chemotherapy of cancer. Knowledge of the biosynthetic pathways allows for synthesis of drugs that inhibit purine and pyrimidine biosynthesis at specific steps. Such inhibitors are very toxic to all cells, since all cells require an adequate supply of nucleotides to synthesize DNA and RNA for chromosome duplication and protein synthesis, respectively. However, inhibitors have particularly great toxicity for cells that grow rapidly and require an above-average supply of nucleotides. Because many tumors contain cells that grow and proliferate rapidly, they are affected by the anticancer drugs to a greater extent than normal cells. Consequently, destruction of cancer cells exceeds that of normal cells.

6-Mercaptopurine and **5-fluorouracil** (Figure 14.28) are two useful drugs in cancer chemotherapy. Neither compound by itself possesses cytotoxicity, but their conversion in humans leads to formation of potent nucleotide metabolism inhibitors. 6-Mercaptopurine undergoes conversion to the corresponding purine nucleotide, which acts as a powerful inhibitor of glutamine-PRPP amidotransferase in purine biosynthesis (step 2). 5-Fluorouracil undergoes conversion via a salvage pathway to its deoxyribonucleotide, 5-fluorodeoxyuridylate, which serves as an irreversible inhibitor of thymidylate synthase. 5-Fluorodeoxyuridylate constitutes a *suicide substrate* of the enzyme (see Section 11.3) because it forms an inactive ternary complex composed of thymidylate synthase, 5-fluorodeoxyuridylate, and tetrahydrofolate.

Azaserine and **methotrexate (amethopterin)** are two other drugs that have proven effective in treating some cancers (Figure 14.28). Azaserine is an analog of glutamine, and methotrexate is an analog of folic acid. These drugs belong to two groups of compounds called *glutamine antagonists* and *antifolates,* respectively. Azaserine inhibits glutamine amidotransferases and therefore inhibits purine biosynthesis. Methotrexate is a potent inhibitor of dihydrofolate reductase, which functions in the thymidylate synthase reaction. Methotrexate binds at least 1000 times more tightly to the enzyme than the normal substrate. Administration of methotrexate results in decreased levels of N^5,N^{10}-methylenetetrahydrofolate and greatly decreased dTMP synthesis by thymidylate synthase. The effect of methotrexate on thymidylate synthase is so pronounced because the enzyme is particularly sensitive to depletion of tetrahydrofolate compared to other tetrahydrofolate-requiring enzymes. The high sensitivity of thymidylate synthase results from the dual role that tetrahydrofolate plays in the reaction.

Compounds that interfere with purine and pyrimidine metabolism may also be useful used as antiviral drugs, since viruses also constitute rapidly replicating systems critically dependent on a supply of nucleotides. Antiviral drugs, like cancer drugs, must be such that the destruction of the rapidly replicating viruses exceeds the destruction of the more slowly replicating host cells. An example of an antiviral drug is *AZT* (Figure 7.8), used for treatment of AIDS, caused by a virus termed HIV (human immunodeficiency virus). AZT inhibits DNA polymerase. The viral enzyme (an RNA-dependent DNA polymerase) is at least 100 times more sensitive to the drug than the host cell enzyme (a DNA-dependent DNA polymerase).

14.4.3. Purine Catabolism

Purine catabolism begins with hydrolysis of nucleotides to nucleosides, a reaction catalyzed by *nucleotidases.* Cleavage of nucleosides by *phosphorolysis* (see Section 10.5) yields ribose 1-phosphate (or deoxyribose 1-phos-

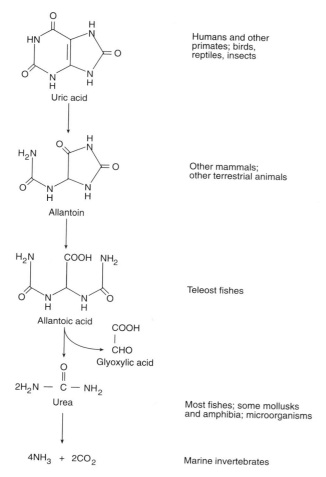

Humans and other primates; birds, reptiles, insects

Other mammals; other terrestrial animals

Teleost fishes

Most fishes; some mollusks and amphibia; microorganisms

Marine invertebrates

Figure 14.30. End products of purine catabolism. The nitrogenous compounds shown are excreted by the animals listed on the right.

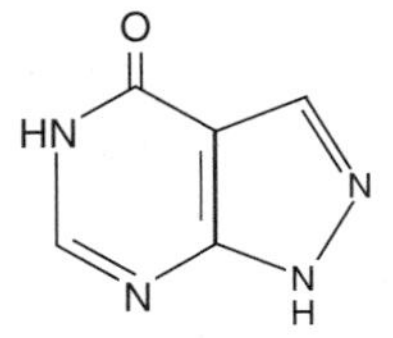

Figure 14.31.   Allopurinol, an inhibitor of xanthine oxidase.

phate) and free purines. Phosphorolysis is catalyzed by *purine nucleoside phosphorylase,* an enzyme widely distributed in both mammalian tissues and microbial cells. All of the purines ultimately undergo conversion to **xanthine,** a purine not normally found in nucleic acids. Xan-

thine is converted to **uric acid** by means of **xanthine oxidase** (Figure 14.29).

The subsequent fate of uric acid varies with the organism (Figure 14.30). In some organisms, uric acid constitutes the ultimate breakdown product of purines and is excreted as such. In others, uric acid undergoes further degradation to **allantoin, allantoic acid,** urea, or ammonia.

Gout is a disease caused by abnormal uric acid metabolism. The disease is fairly common, afflicting about 3 out of every 1000 people, mostly males. It is associated with an elevated plasma level of uric acid. The high concentration of uric acid causes formation of painful deposits of sodium urate in the cartilage of joints, especially that of the big toe,

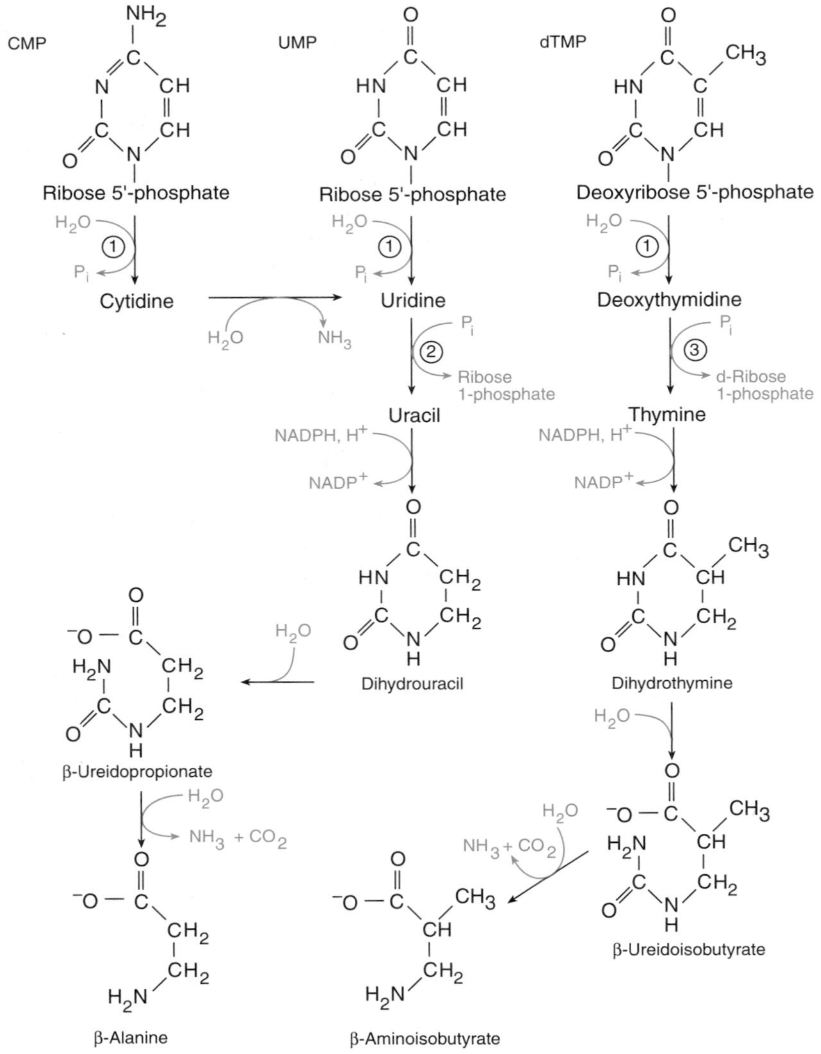

Figure 14.32.   Catabolism of pyrimidines in animals. β-Alanine and β-aminoisobutyrate are subsequently metabolized to malonyl CoA and methylmalonyl CoA, respectively. Deoxyribonucleotides are degraded likewise. Numbers designate enzymes: ① Nucleotidase; ② uridine phosphorylase; ③ thymidine phosphorylase.

and can also form kidney stones. Many cases of gout can be treated successfully with the antimetabolite **allopurinol** (Figure 14.31), an inhibitor of xanthine oxidase.

14.4.4. Pyrimidine Catabolism

Pyrimidine catabolism also begins with nucleotidase-catalyzed hydrolysis of nucleotides to nucleosides, followed by phosphorolysis of nucleosides. The latter reaction, catalyzed by *pyrimidine nucleoside phosphorylase,* yields ribose 1-phosphate (or deoxyribose 1-phosphate) and free pyrimidines. Cytosine and uracil are subsequently degraded to β-alanine; thymine is degraded to β-aminoisobutyrate (Figure 14.32).

SUMMARY

Some bacteria can reduce atmospheric nitrogen to ammonia (nitrogen fixation) by means of nitrogenase. Other bacteria, plants, and fungi can reduce nitrate and nitrite to ammonia (nitrate assimilation) using nitrate reductase. All living organisms can use ammonia and convert it to nitrogen-containing organic compounds. Three key reactions lead to conversion of ammonia to carbamoyl phosphate, glutamate, or glutamine. These nitrogenous compounds react with intermediates of carbohydrate and lipid metabolism to form other nitrogen-containing compounds. One such reaction, catalyzed by transaminases, results in conversion of an amino acid to a keto acid while a different keto acid undergoes conversion to an amino acid.

Essential amino acids are those that an organism cannot synthesize or cannot synthesize in sufficient quantity; they must be obtained through the diet. Essential amino acids in humans were determined by nitrogen balance studies in which the difference between the amount of nitrogen ingested and that excreted was measured. Glucogenic and ketogenic amino acids are catabolized to precursors of carbohydrates and lipids, respectively.

The first step in amino acid catabolism usually consists of an oxidative deamination. The ammonia removed enters the urea cycle, where it is converted to urea that can be excreted in the urine. The urea cycle constitutes a catabolic pathway that requires input of energy and has close links to many other metabolic pathways. Mammals and terrestrial vertebrates excrete waste nitrogen as urea. Other organisms excrete waste nitrogen as uric acid or ammonia.

Individuals suffering from the genetic disease phenylketonuria have a deficiency of phenylalanine hydroxylase. This enzyme catalyzes the hydroxylation of phenylalanine to tyrosine in the major pathway of phenylalanine catabolism. In afflicted individuals, a minor pathway becomes accentuated and leads to conversion of phenylalanine to phenylpyruvate and phenyllactate.

Purine and pyrimidine ring systems are synthesized from small metabolic precursors. The purine ring is assembled on a sugar phosphate molecule. The pyrimidine ring is assembled by itself and subsequently linked to a sugar phosphate. A number of compounds function as anticancer drugs by inhibiting specific steps in purine or pyrimidine biosynthesis. Hydrolysis of purine and pyrimidine nucleotides to nucleosides is catalyzed by nucleotidases. The nucleosides are then cleaved by phosphorolysis to yield free bases.

SELECTED READINGS

Bender, D., *Amino Acid Metabolim,* 2nd ed., Wiley, Chichester, England (1985).

Carreras, C. W., and Santi, D. V., The catalytic mechanism and structure of thymidylate synthase, *Annu. Rev. Biochem.* 64:721–762 (1995).

Cavenee, W. K., and White, R. L., The genetic basis of cancer, *Sci. Am.* 272:72–79 (1995).

Dou, Q. P., and Pardee, A. B., Transcriptional activation of thymidine kinase, a marker for cell cycle control, *Prog. Nucleic Acid Res. Mol. Biol.* 53:197–217 (1996).

Fontecave, M., Nordlund, P., Eklund, H., and Reichard, P. The redox centers of ribonucleotide reductase of *E. coli, Adv. Enzymol.* 65:147–183 (1992).

Greenberg, G. R., and Hilfinger, J. M., Regulation of ribonucleotide reductase and relationship to DNA replication in various systems, *Prog. Nucleic Acid Res. Mol. Biol.* 53:345–395 (1996).

Howard, J. B., and Rees, D. C., Nitrogenase: A nucleotide-dependent molecular switch, *Annu. Rev. Biochem.* 63:235–264 (1994).

Kaufman, S., The phenylalanine hydroxylating system, *Adv. Enzymol.* 67:77–264 (1993).

Lipscomb, W. N., Aspartate transcarbamylase from *E. coli:* Activity and regeneration, *Adv. Enzymol.* 68:67–151 (1994).

Peters, J. W., Fisher, K., and Dean, D. R., Nitrogenase structure and function: A biochemical-genetic perspective, *Annu. Rev. Microbiol.* 49:335–366 (1995).

Smith, J. L, Zaluzec, E. J., Wery, J. P., Niu, L., Switzer, R. L., Zalkin, H., and Satow, Y., Structure of the allosteric regulatory enzyme of purine biosynthesis, *Science* 264:1427–1433 (1994).

Traut, T. W., and Jones, M. E., Uracil metabolism—UMP synthesis from orotic acid or uridine and conversion of uracil to β-alanine: Enzymes and cDNAs, *Prog. Nucleic Acid Res. Mol. Biol.* 53:1–78 (1996).

REVIEW QUESTIONS

A. Define each of the following terms:

Transamination	Nitrogen balance
Oxidative deamination	Phenylketonuria
Salvage pathway	Phosphoribosyl pyrophosphate
Nitrate assimilation	Nitrogenase
Ribonucleotide reductase	Tetrahydrofolate

B. Differentiate between the two terms in each of the following pairs:

UMP/IMP	SGOT/SGPT
Ureotelic organisms/ uricotelic organisms	Nitrogen fixation/ ammonia fixation
Nitrate reductase/ nitrite reductase	Glucogenic amino acids/ ketogenic amino acids
Glutamine synthase/ glutamate dehydrogenase	Aspartate transcarbamoylase/carbamoyl phosphate synthase

C. (1) Outline the steps whereby living organisms convert atmospheric nitrogen to nitrogen-containing organic compounds.

(2) Write two balanced equations showing the action of adenosine phosphorylase and uridine phosphorylase.

(3) Complete the following transamination reactions, using structural formulas for reactants and products:

$$\text{Phenylalanine} + \text{pyruvate} \rightleftarrows$$
$$\text{Glycine} + \text{oxaloacetate} \rightleftarrows$$

(4) Outline the operation of the urea cycle. Write balanced equations for the individual steps. What are the energy requirements of the cycle and how is the cycle linked to other metabolic pathways?

(5) What are some of the similarities and differences between purine and pyrimidine biosynthesis and between purine and pyrimidine catabolism?

PROBLEMS

14.1. Which of the following conditions is likely to lead to a positive nitrogen balance in humans?

    (a). Inability to digest dietary protein
    (b). Chronic deterioration of muscle tissue
    (c). Inhibition of amino acid deamination
    (d). Inhibition of protein biosynthesis
    (e). Deficiency of amino acid absorption from the intestine

14.2. We classify arginine as an essential amino acid for young organisms but *not* for adults. How do you explain this, considering that arginine is synthesized, even in young organisms, via the urea cycle?

14.3. In terms of energy-rich bonds, how many molecules of urea could be synthesized from the energy released by the complete aerobic oxidation to CO_2 and H_2O of one molecule of: (a) glucose; (b) palmitic acid?

14.4. Name the labeled compound formed by transamination of α-ketoglutarate labeled with ^{14}C in its keto group. Where is the label located in the compound formed?

14.5. You place a rat on a diet including [^{15}N]alanine. Will urea excreted by the rat become labeled with ^{15}N and, if so, will it become labeled in one or both of its amino groups?

14.6. Write the overall, balanced equation describing the oxidative deamination of D-phenylalanine with D-amino acid oxidase and catalase.

14.7. Name the enzyme that catalyzes each of the following reactions:

(a) $HN=NH + 2H^+ + 2e^- \rightleftarrows H_2N\text{-}NH_2$
(b) $NO_3^- + 2H^+ + 2e^- \rightleftarrows NO_2^- + H_2O$
(c) $NO^- + 3H^+ + 2e^- \rightleftarrows NH_2OH$
(d) $NH_4^+ + HCO_3^- + 2ATP^{4-} + H_2O \rightarrow$ carbamoyl phosphate^{2-} + 2ADP^{3-} + P$_i^{2-}$ + 2H$^+$

14.8.* We define the *oxidation number* of an atom in a polyatomic ion or in a compound as the actual or assigned charge that the atom would have if all the electrons of each bond were assigned exclusively to the more electronegative of the bonded atoms. The sum of the oxidation numbers in a polyatomic ion equals the charge of the ion; that in a compound equals zero. On that basis, determine the oxidation number of nitrogen in each of the following: (a) N_2; (b) NO^-; (c) NO_2^-; (d) NO_3^-; (e) $H_2N\text{-}NH_2$; (f) $HN=NH$; (g) NH_2OH.

14.9. Write a balanced equation for the synthesis of alanine from glucose using the reactions of glycolysis and transamination.

14.10. What is the energy requirement, in terms of the number of molecules of ATP used, for the biosynthesis of IMP and UMP?

14.11. Consider two human populations—one subsisting on a diet rich in meat, and the other subsisting on a diet rich in rice. Which of the two populations is likely to have a higher incidence of gout? Why?

14.12. Write a balanced equation for the reaction catalyzed by each of the following enzymes:

(a) Glutamate dehydrogenase
(b) Glutamine synthase
(c) Phenylalanine hydroxylase
(d) Arginase
(e) Adenine phosphoribosyl transferase
(f) Cytidine 5'-triphosphate reductase

14.13.* A student isolates ribonucleotide reductase from bacteria. The cell-free extract needs to be stored in the refrigerator for some time. Would you suggest adding some glutathione to the preparation to stabilize the enzyme and minimize the loss of activity upon storage? Why or why not? Would you sug-

gest adding some iodoacetamide when assaying the enzyme *in vitro*? Why or why not?

14.14. When rabbits are grown for a long time on a diet that includes acetyl CoA labeled with ^{14}C at both carbons of the acetyl group, will urea excreted by the rabbits become labeled with ^{14}C as soon as the acetyl CoA enters the citric acid cycle (i.e., during the first turn of the cycle)? Why or why not? Will the urea become labeled at a later time? Why or why not?

14.15. Draw the structures, and name the α-keto acid formed, when the following amino acids undergo transamination with α-ketoglutarate: (a) alanine; (b) glutamate; (c) aspartate; (d) phenylalanine; (e) tyrosine.

14.16. A researcher assays the transamination between alanine and α-ketoglutarate spectrophotometrically by including lactate dehydrogenase and NADH in the incubation mixture. What does the investigator measure? Why can this measurement be used as an assay for the transamination?

14.17. Deficient absorption of vitamin B_{12} from the intestine causes the disease *pernicious anemia*. In amino acid catabolism, isoleucine, methionine, and valine produce propionyl CoA, which undergoes conversion to succinyl CoA. From your knowledge of this latter set of reactions, predict the changes you would expect in isoleucine, methionine, and valine catabolism in pernicious anemia.

14.18.* Calculate the net yield of ATP when alanine serves as a fuel and undergoes complete oxidation to CO_2 and H_2O under aerobic conditions.

14.19.* Cats that have fasted are given a single meal containing all of the amino acids except arginine. Would you expect the catabolism of glucogenic amino acids to be accelerated or slowed down following the meal? Would the blood concentration of ammonia increase or decrease during the same time interval? Explain your answers. (Hint: Arginine is synthesized in insufficient quantities to compensate for its absence from the cat's diet.)

14.20.* Patients suffering from acute leukemia have extensive breakdown of nucleic acids. When treated with anticancer drugs, allopurinol is often added as well. What is the reason for this?

14.21.* The enzyme *hypoxanthine-guanine phosphoribosyl transferase (HGPRT)* catalyzes the following two reactions:

Hypoxanthine + PRPP → IMP + PP$_i$
Guanine + PRPP → GMP + PP$_i$

A severe HGPRT deficiency causes an extreme neurological disorder called the *Lesch-Nyhan syndrome*. The syndrome constitutes a sex-linked trait because the gene for HGPRT is located on the sex chromosome. Afflicted individuals exhibit mental retardation, aggressive behavior, and a compulsion for self-mutilation. Biochemically speaking, the disorder is characterized by an overproduction of uric acid (6 times normal) and excessive *de novo* biosynthesis of purines (200 times normal). How are these metabolic changes brought about?

14.22. A physician advises individuals on a high-protein diet to drink large amounts of water. What is the rationale for this?

14.23. What energy saving, in terms of energy-rich bonds, is achieved by producing IMP via the salvage pathway of Problem 14.21 as opposed to its *de novo* synthesis as shown in Figure 14.17?

14.24. How many α-amino acids participate in the operation of the urea cycle? How many of these are used for the biosynthesis of proteins?

14.25.* A nurse constructs a *phenylalanine tolerance curve* (similar to a glucose tolerance curve) by injecting a large dose of phenylalanine into a person's bloodstream and then determining the phenylalanine concentration in the serum as a function of time. What would such a curve look like for (a) a normal person and (b) a person suffering from PKU disease?

14.26. Assume that a person suffering from phenylketonuria is also diet-conscious and wants to avoid sugar. Would you advise this individual to use aspartame (see Section 2.5) or saccharin (see Figure 5.17) as a food sweetener? Why?

14.27. One of the symptoms of children afflicted with kwashiorkor is a depigmentation of the skin and hair. What is the biochemical basis for this?

14.28.* The daily energy requirement for an average young man is 12,134 kJ (see Table 8.6). You know that proteins contain 16.0% (by weight) of nitrogen. If a man excretes 40.0 g of urea per day, what is his protein intake in grams per day? Given that the energy derivable from proteins is 17.0 kJ/g, calculate the percentage of the man's energy requirement contributed by protein.

14.29. Oxidation of 1.0 g of protein in humans generally yields a smaller number of ATP than oxidation of 1.0 g of carbohydrate or lipid. Why is this so?

14.30. Why might a potent inhibitor of purine or pyrimidine biosynthesis be useful as an anticancer and/or an antiviral drug?

Photosynthesis

15

Photosynthesis is *the reaction by which algae, bacteria, and plants (photosynthetic organisms), in the presence of certain pigments (especially chlorophyll) and radiant energy from the Sun, synthesize carbohydrates from carbon dioxide and water.* Solar energy is the driving force of photosynthesis and the ultimate source of all the energy living systems require.

Photosynthesis constitutes an oxidation–reduction reaction in which water is oxidized to oxygen and carbon dioxide is reduced to carbohydrate (*CO_2 fixation*):

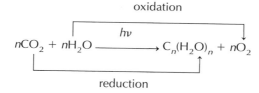

$$\text{oxidation}$$

$$n\text{CO}_2 + n\text{H}_2\text{O} \xrightarrow{\ h\nu\ } \text{C}_n(\text{H}_2\text{O})_n + n\text{O}_2$$

$$\text{reduction}$$

where $h\nu$ is the energy of a photon of light, and $\text{C}_n(\text{H}_2\text{O})_n$ is a carbohydrate.

The reactions of photosynthesis lead to production of fructose 6-phosphate. Other monosaccharides (e.g., glucose), oligosaccharides (e.g., sucrose), and polysaccharides (e.g., starch) are then made from fructose 6-phosphate. Because the monomeric unit in these oligomers and polymers is a six-carbon sugar, we usually use a coefficient of $n = 6$ in the above equation, corresponding to the synthesis of glucose:

$$6\text{CO}_2 \ + \ 6\text{H}_2\text{O} \xrightarrow[\text{chlorophyll}]{\text{sunlight}} \text{C}_6\text{H}_{12}\text{O}_6 \ + \ 6\text{O}_2$$

| From the air | From the soil | Monosaccharide (glucose) | Returned to the air |

This reaction constitutes the reverse of carbohydrate metabolism in plants and animals:

$$\underset{\substack{\text{Sugar} \\ \text{ingested}}}{C_6H_{12}O_6} \; + \; \underset{\substack{\text{From} \\ \text{the air}}}{6O_2} \; \xrightarrow{\text{respiration}} \; \underset{\substack{\text{Expired} \\ \text{to the air}}}{6CO_2} \; + \; \underset{\substack{\text{Returned} \\ \text{to the soil}}}{6H_2O} \; + \; \text{energy}$$

Photosynthesis and respiration provide a balance between plant and animal life, between photosynthetic and nonphotosynthetic organisms (Figure 15.1). Photosynthetic organisms convert sunlight energy to chemical energy in the form of ATP and NADPH, both of which then serve to synthesize glucose and other organic compounds from carbon dioxide and water. Simultaneously, photosynthetic organisms release oxygen into the atmosphere.

Both photosynthetic and nonphotosynthetic organisms use the oxygen produced to degrade energy-containing products of photosynthesis to carbon dioxide and water. In the process, energy is released and conserved in the form of ATP. The stored energy of ATP later drives endergonic metabolic reactions.

15.1. THE SCOPE OF PHOTOSYNTHESIS

15.1.1. Brief Historical Perspective

The concept of photosynthesis—plants synthesizing carbohydrates from light and air—took almost two centuries to become firmly established. In 1727, Stephen Hales proposed that plants obtain some of their matter from air. Joseph Priestley, an English clergyman and chemist, investigated the nature of this gaseous component (1770–1780). He showed that a mouse died rapidly if placed in a closed container in which the air had been "depleted" by burning a candle until it went out. However, if the con-

tainer was connected to a second one containing a green plant exposed to light, the air regained its ability to support a mouse's life.

This classic experiment provided the first indication that plants produce a gas when they are irradiated with light. We now know that this gas consists of oxygen produced by photosynthesis. Priestley later discovered oxygen and called it "dephlogisticated air," but it was Antoine Lavoisier who elucidated oxygen's role in combustion and respiration.

In 1779, Jan Ingen-Housz, a Dutch court physician to the Austrian empress, made another important discovery. He showed that light is essential for photosynthesis

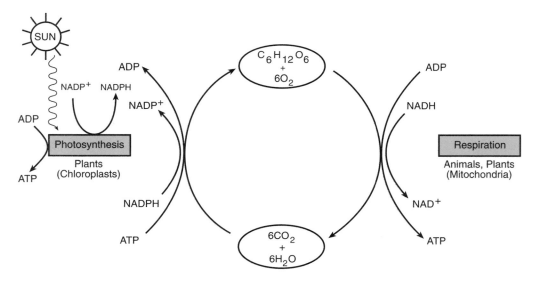

Figure 15.1. The balance between photosynthesis and respiration.

and that only the green parts of a plant produce oxygen. Shortly thereafter, Jean Senebier, a Swiss pastor, showed that CO_2 was taken up during photosynthesis (1782).

By the early 19th century, researchers initiated quantitative measurements of photosynthesis. Some made attempts to determine the amount of CO_2 assimilated, the amount of oxygen evolved, and the amount of plant material produced. Some also considered the energetics of photosynthesis. Robert Mayer, a German surgeon and one of the formulators of the first law of thermodynamics, proposed that plants convert light energy to chemical energy (1842).

At the beginning of the 20th century (1905), F. Blackman proposed that photosynthesis consists of two stages: (1) a *light reaction* that is light dependent but temperature independent, like typical photochemical reactions, and (2) a *dark reaction* that is temperature dependent, like typical enzymatic reactions.

In 1931, Cornelius van Niel showed that photosynthetic sulfur bacteria use H_2S to generate sulfur and proposed a general equation to describe photosynthesis in both plants and sulfur bacteria:

$$CO_2 + 2H_2A \xrightarrow{\text{light}} C(H_2O) + H_2O + 2A$$

where H_2A and 2A are H_2O and O_2 in green plants but H_2S and 2S in photosynthetic sulfur bacteria. Van Niel proposed that the first stage of photosynthesis consists of the oxidation of H_2A to 2A and the formation of a reducing agent,

$$2H_2A \xrightarrow{\text{light}} 2A + \text{reducing agent}$$

followed by a second stage in which the reducing agent converts CO_2 to carbohydrate:

$$\text{Reducing agent} + CO_2 \longrightarrow C(H_2O) + H_2O$$

Based on this reasoning, van Niel predicted that the oxygen evolved in photosynthesis comes from water. His prediction was confirmed when ^{18}O became available. By using ^{18}O-labeled H_2O, Samuel Ruben and Martin Kamen showed that the evolved oxygen was derived from water (1941). The oxygen atoms of carbon dioxide appear in the other two products of the reaction, carbohydrate and water:

$$CO_2^* + 2H_2{}^{18}O \xrightarrow{\text{light}} C(H_2O^*) + H_2O^* + {}^{18}O_2$$

In 1932, Robert Emerson made a crucial discovery while measuring photosynthesis as a function of the wave-length of light. He observed that photosynthesis decreased when he exposed plants to monochromatic light of longer wavelengths (about 700 nm), even though the plants still absorbed light significantly at these longer wavelengths. We refer to this decrease in photosynthetic activity as **red drop** (Figure 15.2). When the same plants were supplemented with light of shorter wavelengths (about 650 nm), photosynthetic activity increased, a phenomenon called the **Emerson enhancement effect.** Emerson's observations suggested that there are *two photosystems* in photosynthesis: one operating at about 700 nm, and one at about 650 nm.

In the late 1930s, the work of Robert Hill led to another milestone in the study of photosynthesis. Hill showed that isolated *chloroplasts* (first described by A. Meyer in 1883) could produce oxygen in response to light and that ferric ions could substitute for CO_2 in the reaction. The photochemical evolution of oxygen was coupled to the reduction of ferric to ferrous ions, resulting in what became known as the **Hill reaction:**

$$4Fe^{3+} + 2H_2O \xrightarrow[\text{chloroplasts}]{\text{light}} 4Fe^{2+} + 4H^+ + O_2$$

Hill's discovery ushered in an era of cell-free studies of photosynthesis. In 1954, D. I. Arnon and his co-workers showed that isolated chloroplasts could carry out complete photosynthesis—that is, produce oxygen *and* fix carbon dioxide. They also discovered *photophosphorylation,* the synthesis of ATP coupled to operation of a pho-

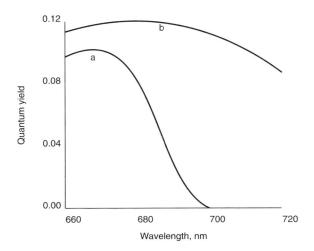

Figure 15.2. The "red drop" of photosynthesis. The quantum yield (defined in Section 15.2) decreases (a) when the illumination consists of monochromatic light of longer wavelengths but increases (b) when the illumination is supplemented with light of shorter wavelengths (<650 nm).

tosynthetic electron transport system. Detailed steps of the dark reaction were not unraveled until the advent of radioactive isotopes. Melvin Calvin and his associates finally elucidated the pathway of carbon dioxide fixation in the 1950s.

15.1.2. Evolution of Photosynthesis

The scenario outlined below is based on the assumption that the Earth's gaseous cover evolved from a reducing to an oxidizing atmosphere. As you read in Section 1.1, recent findings have led many researchers to postulate that the primordial atmosphere was oxidizing. If this view prevails, the scenario must be modified accordingly.

Assuming that photosynthesis arose when the Earth had a reducing atmosphere and an abundance of reducing agents, anaerobic organisms are thought to have been the first to carry out photosynthesis. Modern photosynthetic sulfur bacteria (Table 15.1), which use H_2S to reduce CO_2, are believed to resemble these early photosynthetic organisms.

As anaerobic photosynthetic organisms multiplied and became established, they began sealing their own doom. Their very success at carrying out reduction reactions resulted in continuous depletion of reducing agents in the environment. Some theorists believe that forerunners of modern *cyanobacteria* (Table 15.1) were the first to adapt to these changing environmental conditions and evolved a system capable of extracting electrons from water. This probably occurred some 3×10^9 yr ago and led to accumulation of oxygen, a toxic waste product (see Section 12.5).

The advent of aerobic photosynthesis produced a gradual change in the Earth's environment, converting the reducing atmosphere to an oxidizing one. Conversion of the atmosphere paved the way for development of aerobic metabolism, evolution of animals, and establishment of a balance between photosynthesis and respiration. As the oxidizing atmosphere spread, anaerobic photosynthetic organisms decreased in significance. Today, they constitute only a small fraction of the total number of photosynthetic organisms.

Photosynthesis has evolved to become the single most important chemical reaction in the biosphere. It exceeds any other kind of "manufacturing" reaction in the world in the quantities of reactants and products involved. You can appreciate the magnitude of photosynthesis by considering the following statistics:

- An estimated 4×10^{11} tons of CO_2 (about 10^{11} tons of carbon) are fixed by photosynthesis per year.
- Fixation of 10^{11} tons of carbon represents an energy trapping of about 4×10^{18} kJ/yr. This energy, derived from the Sun, amounts to well over 10 times the total energy generated by all fossil fuels used the world over.
- As immense as the energy trapped is, it constitutes only about 0.1% of the total radiant energy striking the Earth.
- At least one-half of the total photosynthetic activity on Earth takes place in oceans, rivers, and lakes due to algae and microorganisms. The rest is due to terrestrial plants.
- Agricultural crops of the United States alone produce approximately 6×10^6 tons of chlorophyll per year.
- Approximately 3 centuries (for CO_2) and 20 centuries (for O_2) are required to process by photosynthesis an amount of gas equal to that present in the atmosphere.

15.2. LIGHT AND ENERGY

To understand the essence of photosynthesis, we must review some of the properties of light. Recall that light has both wavelike and particulate character. We refer to the particles of light as **photons.** Each photon represents a fixed package of light energy, termed a **quantum.**

Table 15.1. Photosynthesis in Prokaryotes and Eukaryotes

Organisms	Formation of O_2	Presence of chloroplasts	Type of chlorophyll	Number of photosystems
Prokaryotes				
Sulfur bacteria	No	No	Bacteriochlorophyll *a* or *b*	1
Cyanobacteria	Yes	No	Chlorophyll *a*	2
Eukaryotes				
Plants and algae	Yes	Yes	Chlorophyll *a* and *b*	2

15.2.1. Energy of Photons

A photon has no charge and is believed to have no mass. The energy of a photon varies with the wavelength of light as defined by **Planck's law:**

$$E = h\nu \qquad (15.1)$$

where E is the energy of a photon, h is Planck's constant (6.626×10^{-34} J s), and ν is the frequency of the light (cycles/s).

The frequency is equal to the velocity of light in a vacuum ($c = 3.00 \times 10^{10}$ cm s^{-1}) divided by the wavelength (λ):

$$\nu = c/\lambda \qquad (15.2)$$

Because photon energy varies inversely with photon wavelength, the shorter the wavelength, the greater the energy. A photon of blue light (e.g., $\lambda = 500$ nm) has greater energy than a photon of red light (e.g., $\lambda = 700$ nm). You can calculate the energy of a photon by means of Eq. (15.1). For example, a photon of red light having a wavelength of 700 nm has an energy equal to

$$E_{700} = (6.626 \times 10^{-34} \text{ J s}) (3.00 \times 10^{10} \text{ cm s}^{-1}/7.00 \times 10^{-5} \text{ cm})$$
$$= 28.4 \times 10^{-20} \text{ J} = 28.4 \times 10^{-23} \text{ kJ}$$

A mole of photons contains Avogadro's number (6.023×10^{23}) of photons. We call a mole of photons an **einstein.** An einstein of 700-nm light represents a quantity of energy equal to

$$(28.4 \times 10^{-20} \text{ J/photon})(6.023 \times 10^{23} \text{ photons}) = 171 \text{ kJ}$$

Table 15.2 lists the energies of several other photons.

15.2.2. Absorption of Light

A molecule can exist in numerous energy states defined by the distribution of electrons in orbitals of different energies. Associated with each of these electronic states are various vibrational and rotational substates. When a molecule absorbs energy in the form of light, a photon strikes the molecule and boosts an electron from an orbital of lower energy to one of higher energy. Both the electron and the entire molecule progress from the **ground state** to an **excited state.** Energy differences between orbitals are **quantized.** It takes a specific amount of energy (*quantum*) to raise an electron from one orbital to another.

To be effective, the striking photon must have a *minimum* amount of energy equal to the energy difference between the lower and higher electronic orbitals. If the photon does not have that energy (has a smaller quantum of energy; is of longer wavelength), it will be ineffective in promoting the electron to a higher energy level; the molecule will not absorb light of that wavelength. If the photon has more energy than is required for boosting the electron (has a greater quantum of energy; is of shorter wavelength), the extra amount of energy may be used to increase the kinetic energy of the boosted electron and/or the vibrational and rotational energies of the molecule.

Thus, a given molecule can only absorb photons of specific wavelengths, and those must provide at least sufficient energy to bring about the particular electronic transitions between orbitals. Once a molecule has been electronically excited, it can dissipate its excitation energy in a number of ways. Four common ones involve internal conversion, fluorescence, resonance energy transfer, and electron transfer (Figure 15.3).

In many cases, an excited molecule returns to its ground state by a radiationless transfer of energy called **internal conversion.** In this process, excitation energy is converted to kinetic energy of surrounding molecules; the energy is converted to heat. Internal conversion occurs very rapidly and is completed in less than 10^{-11} s.

At times, an excited molecule *loses only part* of its excitation energy by emitting radiation in the form of **fluorescence.** In this case, the excited molecule emits a photon that is *less energetic* than the one that excited the molecule; the emitted photon has a *longer wavelength* than the exciting photon. Fluorescence is slower than internal conversion and requires about 10^{-8} s for completion.

When excitation energy is dissipated by **resonance energy transfer,** the excited molecule directly excites an adjacent molecule. Excitation energy is transferred from one molecule to a neighboring one through interaction of the molecular orbitals of the participating molecules. Resonance energy transfer is a radiationless process, also called *exciton transfer.*

Lastly, excitation energy can be dissipated via **electron transfer.** In this radiationless process, the excited electron itself is transferred from one molecule to a neigh-

Table 15.2. Energy Content of Photons

Color	Wavelength (nm)	Energy/einstein (kJ)
Red	700	171.0
Yellow	600	199.5
Blue	500	239.5
Violet	400	299.3
Ultraviolet	300	399.1

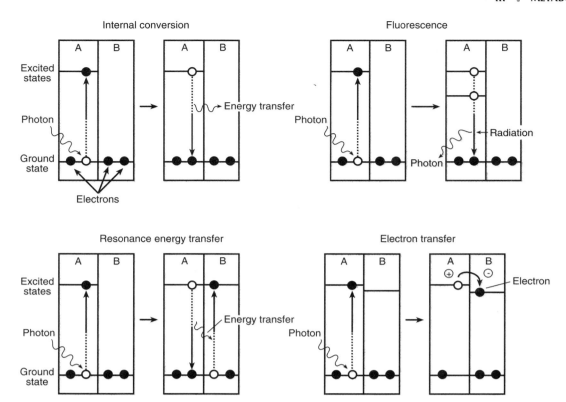

Figure 15.3.   Common modes for the dissipation of excitation energy. A and B denote two molecules.

boring one with a slightly lower excited state. The electron donor molecule undergoes oxidation (*photooxidation*) while the acceptor molecule undergoes reduction (*photoreduction*). Electron transfer occurs because the excited electron is bound less tightly to the donor molecule in its excited state than it is to the same molecule in its ground state. Both resonance energy transfer and electron transfer play major roles in the light reactions of photosynthesis.

15.2.3. Quantum Yield

In a photochemical reaction, not all of the incident photons are absorbed. Moreover, as you just saw, excitation energy can be dissipated in different ways, not all of which lead to a chemical reaction. To allow for these factors, we define the efficiency of a photochemical reaction in terms of its quantum yield.

The **quantum yield** of a photochemical reaction *is a ratio equal to the number of molecules (moles) that react or are formed, divided by the number of photons (einsteins) absorbed*. Recall that the biochemical standard free energy change for the complete oxidation of glucose to CO_2 and water is $\Delta G^{\circ\prime} = -2870$ kJ mol^{-1}. In photosynthesis, six moles of CO_2 are reduced. Hence, the

amount of energy that has to be expended to reduce *one mole* of CO_2 is $+2870/6 = 478$ kJ.

Photosynthesis uses light that has a wavelength of about 700 nm. Photons of such light have an energy of 171 kJ/einstein. Assuming an efficiency of 100%, three moles of these photons are required to reduce one mole of CO_2, or three photons are required per molecule of CO_2. This represents a quantum yield of 0.33. At an efficiency of 50%, the same reaction would require six photons, resulting in a quantum yield of 0.17. As you will see, photosynthesis actually requires *eight photons* for the reduction of *one molecule of CO_2*. This represents a quantum yield of 0.13 and an efficiency of 37%, which is similar to the efficiency of other metabolic pathways. Remember, however, that using $\Delta G^{\circ\prime}$ provides only an approximate answer and that an accurate evaluation of efficiency requires the use of $\Delta G^\prime$.

15.3. PHOTOSYNTHETIC MACHINERY

15.3.1. Prokaryotes and Eukaryotes

Most photosynthetic organisms use water as an electron donor and produce oxygen. Some organisms use other electron donors, such as H_2S, and do not evolve oxygen during photosynthesis. These organisms are generally strict anaerobes for which oxygen is toxic.

Both prokaryotes and eukaryotes carry out photosynthesis. Photosynthetic prokaryotes include cyanobacteria, purple and green sulfur bacteria (Table 15.1), and nonsulfur photosynthetic bacteria. Biologists formerly called cyanobacteria blue-green algae and considered them to be plants because of their content of chlorophyll and their evolution of oxygen. Scientists now consider them a class of bacteria. Cyanobacteria use water as an electron donor. Sulfur bacteria use hydrogen sulfide, sulfite, thiosulfate, or other sulfur compounds as electron donors and do not evolve oxygen. Nonsulfur bacteria use a variety of compounds such as hydrogen gas, lactate, succinate, or acetate as reducing agents. In prokaryotes, photosynthesis takes place in **chromatophores,** vesicular structures formed by invaginations of the cell membrane.

Photosynthetic eukaryotes include plants, multicellular algae, dinoflagellates, and diatoms. These organisms all use water as an electron donor, and photosynthesis takes place in specialized subcellular structures called chloroplasts.

15.3.2. Chloroplasts

Chloroplasts are subcellular, membrane-bounded organelles that resemble mitochondria in several ways. Like the mitochondrion, a chloroplast has an *inner* and an *outer membrane,* separated by an *intermembrane space* (Figure 15.4). The inner chloroplast membrane, like the mitochondrial membrane, is nearly impermeable whereas the outer membrane is highly permeable. **Stroma,** the soluble portion of a chloroplast, is analogous to the matrix of mitochondria. Mitochondria and chloroplasts are *semiautonomous;* they have their own DNA and machinery for replicating and expressing this DNA, but functions of both organelles depend on additional products encoded in nuclear DNA. According to the *endosymbiotic theory,* both mitochondria and chloroplasts have evolved from prokaryotic cells (specifically, cyanobacteria) that assumed symbiotic relationships with nonphotosynthetic eukaryotic cells.

Within the chloroplast occur the **grana,** bodies consisting of flattened membranes termed **thylakoid disks.** The membrane of thylakoid disks has an unusual lipid composition. It contains only small amounts of phospholipids but large amounts of glycolipids (about 10% and 90%, respectively). The membrane also contains proteins. Specific multisubunit protein complexes embedded in the thylakoid membrane are termed **reaction centers.** They play a key role in the photosynthetic process and represent sites at which chemical reactions occur. Thylakoid disks form by invaginations of the inner chloroplast membrane and resemble the cristae of mitochondria.

A network of membranes, called **stroma lamellae** or **intergranal lamellae,** connects the grana. Thylakoid disks, stacked into grana, are embedded in the stroma much as the cristae of mitochondria are embedded in the matrix. Typically, each chloroplast contains about 40–80 grana.

Chloroplasts vary in number, size, and shape. There

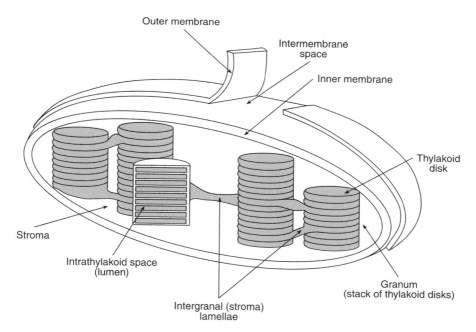

Figure 15.4.   Schematic drawing of a chloroplast.

may be from 1 to 1000 per cell. They usually have an elongated and ellipsoidal shape. They have a length of about 5–10 μm and a diameter of about 0.5–2 μm. Chloroplasts are approximately 2–5 times larger than mitochondria. Light trapping and oxygen production (the light reactions) take place in the thylakoid disks. CO_2 fixation (the dark reactions) occurs in the stroma.

15.3.3. Photosynthetic Pigments

15.3.3A. Chlorophylls. The primary light-absorbing pigment is **chlorophyll (Chl).** Major chlorophylls of eukaryotes are **chlorophylls *a* and *b*,** both of which occur in plants and algae. Cyanobacteria contain only Chl *a*. Other photosynthetic bacteria have **bacteriochlorophylls *a* and *b*.** Chlorophylls are planar, metal-containing tetrapyrroles that resemble heme (Figure 15.5). They differ structurally from heme in four respects:

1. Chlorophylls contain magnesium rather than iron; they are magnesium porphyrins rather than iron porphyrins.
2. Chlorophylls contain a cyclopentanone ring (ring V) fused to pyrrole ring III.
3. Chlorophylls contain **phytol,** a long-chain alcohol esterified to the propionyl group of pyrrole ring IV. The nonpolar phytol serves to anchor chlorophyll molecules in the hydrophobic portion of thylakoid membranes.
4. Chlorophylls are more reduced (saturated) than heme. Chlorophylls *a* and *b* contain a partially reduced pyrrole ring IV; bacteriochlorophylls *a* and *b* contain partially reduced pyrrole rings II and IV.

Because of their extensive conjugated double-bond system, chlorophylls absorb light strongly in the visible region. Chlorophylls *a* and *b* have distinct absorption spectra and complement each other in their absorptive properties (Figure 15.6). The composite effect of light absorption by chlorophylls *a* and *b* is responsible for chlorophyll's green color. Bacteriochlorophylls have fewer conjugated double bonds than chlorophylls and have different absorptive properties. They absorb at longer wavelengths, in some cases up to 1100 nm.

15.3.3B. Accessory Pigments. In addition to chlorophylls, several other pigments, termed **accessory pigments,** function in photosynthesis. Two major classes of accessory pigments are the **carotenoids,** which are purple, red, or yellow pigments, and the **phycobilins,**

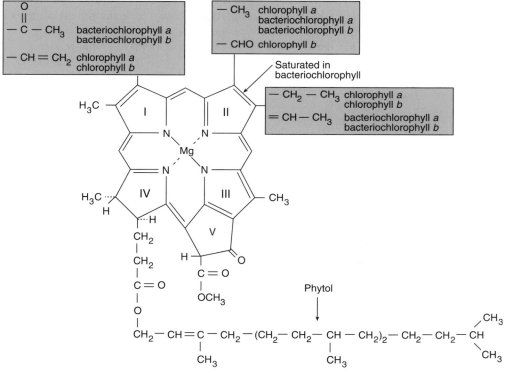

Figure 15.5.   Structure of plant and bacterial chlorophylls.

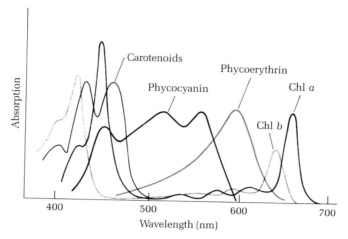

which are blue or red pigments. In chloroplasts, there occur primarily two types of carotenoids, *carotenes* (precursors of vitamin A; see Figure 6.15) and **xanthophylls** (oxygenated carotenes). Phycobilins are porphyrins in which the ring system has opened up to form a chain of pyrroles. Accessory pigments assist in the transfer of light energy to reaction centers (see below). They absorb light in regions of the spectrum at which chlorophylls *a* and *b* do not absorb well (see Figure 15.6) and extend the spectrum of light that photosynthetic organisms can use. Collectively, chlorophylls and accessory pigments absorb radiant energy across the spectrum of visible light.

15.3.4. Photosystems I and II

Light-absorbing pigments and their associated proteins are assembled into a functional unit called a **photosystem.** Green plants and cyanobacteria possess two such assemblies, **photosystem I (PSI)** and **photosystem II (PSII).** Photosystem I responds to light below 700 nm; photosystem II responds to light below 680 nm. Accordingly, the *reaction centers* of these two photosystems are designated **P700** and **P680,** respectively, where P stands for pigment. Both photosystems function in the light reactions of photosynthesis. Photosystem I leads to production of reducing power in the form of NADPH. Photosystem II leads to evolution of oxygen through *photolysis* (splitting brought about by light) of water.

Absorption of light by these photosystems involves both resonance energy transfer and electron transfer. When a photon is absorbed by a photosystem, it excites a chlorophyll molecule, boosting its electron to a higher energy level. The excited chlorophyll molecule passes its ex-

citation energy by resonance energy transfer to surrounding chlorophyll molecules. The excitation energy passes in stepwise fashion and in a random manner from one chlorophyll molecule to another. We call the molecules that serve as conduits for the excitation energy **antenna chlorophylls.** Much as antennas function to receive incoming radio or TV signals, so antenna chlorophylls serve to gather the incoming energy of light.

Ultimately, the excitation energy becomes channeled by antenna chlorophylls to a reaction center (Figure 15.7), where it brings about an excitation of chlorophyll molecules called **specialized chlorophylls.** Reaction centers containing excited specialized chlorophylls are designat-

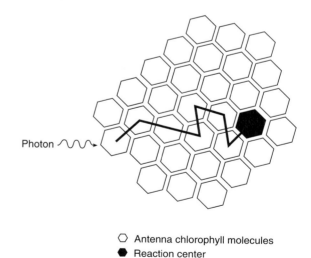

○ Antenna chlorophyll molecules
⬡ Reaction center

Figure 15.7.   Diagram showing the flow of energy through a series of antenna chlorophylls to a photosynthetic reaction center.

ed **P700*** and **P680*,** respectively. Specialized chlorophylls are chemically identical to antenna chlorophylls
but have different properties because of their location and
immediate environment. They have *lower* excited-state
energies than antenna chlorophylls and, therefore, *trap* the
excitation energy and prevent it from being passed along
by further resonance energy transfer. Instead, the excitation energy is dissipated by electron transfer. A specialized
chlorophyll molecule transfers its excited electron to an
appropriate acceptor, thereby reducing the acceptor while
the specialized chlorophyll is converted to a cation (Chl^+).
The chlorophyll ultimately returns to its ground state by
replacing its lost electron from another source.

Most chlorophyll molecules are antenna chlorophylls. There are usually several hundred antenna chlorophylls for every one specialized chlorophyll. Antenna
chlorophylls transfer excitation energy with great efficiency (over 90%) and convey it rapidly (in less than 10^{-10} s)
to the reaction center. Once the excitation energy reaches
the specialized chlorophylls in the reaction center, the actual photochemistry of the light reactions begins.

15.4. LIGHT REACTIONS

The **light reactions** include those parts of the photosynthetic process that are directly dependent on light and do not
proceed without it. All components of the light reactions—
light-absorbing pigments, electron carriers, and ATP synthase—are located in the thylakoid membrane. The light reactions produce NADPH and ATP, both of which
subsequently provide the energy to "fix" CO_2 in the dark reactions (Figure 15.8). The light reactions comprise four individual reactions, which we will discuss in the order listed:

- *Photooxidation of chlorophyll*—photochemical
  excitation of chlorophyll
- *Photoreduction of NADP+*—production of NADPH
- *Photooxidation of water*—splitting of water
  (photolysis); evolution of O_2
- *Photosynthetic phosphorylation*—synthesis of ATP

In plants and cyanobacteria, we describe the set of reactions by an energy diagram called the **Z-scheme** (because of its resemblance to the letter Z, looked at side-
ways; Figure 15.9). In this scheme, photosystems I and II
are linked in series.

15.4.1. Photooxidation of Chlorophyll

In a *photochemical excitation* step, photosystem I absorbs
a photon of light that boosts an electron to a higher ener-

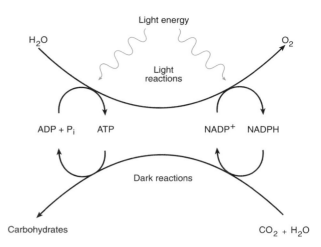

Figure 15.8. The light and dark reactions of photosynthesis. The for-
mer include water photolysis, ATP synthesis, and NADPH production.
The latter convert CO_2 to carbohydrate, using ATP and NADPH formed
by the light reactions.

gy level, thereby forming a molecule of excited chlorophyll. The excitation energy passes, by resonance energy
transfer and via antenna chlorophylls, to a reaction center
(P700), where it leads to formation of an excited specialized chlorophyll (P700*). The P700* loses its excited
electron by electron transfer to an appropriate acceptor,
believed to be a chlorophyll molecule (A_0). In the process,
the acceptor is reduced.

Recall that a half-reaction with a smaller reduction
potential represents a stronger reducing agent than one
with a larger reduction potential (Section 12.1). Thus,
P700 is a weaker reducing agent than A_0 (see Fig. 15.9).
Ordinarily, P700 could not reduce A_0; on the contrary,
A_0 would reduce P700. Yet here P700 is made to reduce
A_0. This constitutes an "uphill" reduction in which a
weaker reducing agent (P700) is made to reduce a
stronger one (A_0). The "uphill" reduction occurs only
because the energy of the absorbed photon excites P700
to P700*, and this *excited form of chlorophyll is a
stronger reducing agent* than A_0. The reduction potential of P700* is smaller than that of A_0, and hence P700*
can reduce A_0.

15.4.2. Photoreduction of NADP+

From the reduced electron carrier (A_0), the electron passes through a chain of electron carriers constituting a *photosynthetic electron transport system*. In principle, the
coupled oxidation–reduction reactions are similar to those
of the mitochondrial electron transport system.

Phylloquinone (vitamin K_1) is the electron carrier
(A_1) that follows A_0 (Figure 15.9). From there, the elec-

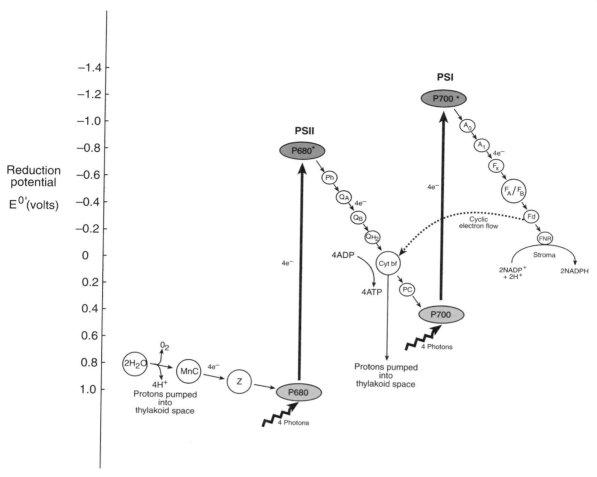

Figure 15.9. The Z-scheme of photosynthesis. Photosystems I and II (PSI and PSII) are linked in series and include two "uphill" and three "down-hill" reductions.

tron passes through a series of three *membrane-bound ferredoxins* (F_x, F_A, F_B). Ferredoxins are nonheme, iron–sulfur proteins that have low reduction potentials. From membrane-bound ferredoxins, the electron passes to a *soluble* ferredoxin (Fd) and from there to *Fd:NADP+ reductase* (FNR). This enzyme is a flavoprotein, containing FAD as a prosthetic group. The enzyme catalyzes reduction of $NADP^+$ by transfer of a hydride ion (H^-) from $FADH_2$ (see Section 11.1). $NADP^+$ serves as the terminal electron acceptor in this photosynthetic electron transport system and is reduced to NADPH:

$$NADP^+ + H^- \rightarrow NADPH$$

All of the steps from P700* to $NADP^+$ constitute ordinary "downhill" reductions. Each carrier has a smaller reduction potential than the one following it, so each carrier reduces the subsequent one in the series.

15.4.3. Photooxidation of Water

Photooxidation of chlorophyll results in formation of an "electron hole" in the reaction center P700—an electron from the center was used to reduce $NADP^+$. The lost electron must be replaced by obtaining an electron from some other source. That source is water, the splitting of which produces electrons:

$$2H_2O \rightarrow 4H^+ + 4e^- + O_2$$

However, although water constitutes the ultimate source of the electrons, the splitting of water is carried out in an indirect way involving photosystem II. Water splitting is coupled to absorption of a photon (*photolysis*) by photosystem II. The complete photosynthetic system, therefore, requires *absorption of two photons, one by each of the two photosystems.*

Absorption of a photon by photosystem II, as in photosystem I, leads to production of an excited specialized chlorophyll, in this case P680*. P680* loses its electron, via electron transfer, to a *pheophytin (Ph)*. Pheophytins are identical to chlorophylls in their structure but have the centrally bound Mg^{2+} replaced by two protons. The reduction of pheophytin by P680 constitutes another "uphill" reduction, again made possible by absorption of a photon that excites P680 to P680*.

From pheophytin, the electron passes through a set of electron carriers constituting a second photosynthetic electron transport system and ultimately fills the electron hole in photosystem I. After pheophytin, the electron passes through a series of *plastoquinones* (Q), compounds structurally similar to coenzyme Q (ubiquinone). A *cytochrome/iron–sulfur complex (cytochrome bf complex)* constitutes the next electron carrier in the sequence. The complex consists of *cytochrome b_6, cytochrome f,* and *iron–sulfur clusters.* From the cytochrome/iron–sulfur complex, the electron passes via *plastocyanin (PC)* to P700. Plastocyanin is a copper-containing peripheral membrane protein located on the surface of the thylakoid membrane. All steps from P680* to P700 are ordinary "downhill" reductions. Transfer of the electron from plastocyanin to P700 effectively fills the electron hole in that reaction center.

However, loss of an electron from P680* produces an electron hole in photosystem II. This hole is filled by an electron derived from the splitting of water. Thus, water constitutes the ultimate source of the electrons used to replenish those lost by photochemical excitations resulting from the absorption of two photons. A *manganese-containing protein complex (MnC)* mediates water photolysis. The complex contains four protein-bound manganese ions that cycle through a series of oxidation states comprising various combinations of Mn^{3+} and Mn^{4+}. The sequence of steps allows for the formation of O_2 without generating hazardous partially reduced forms of oxygen. Water photolysis, like the reverse reaction catalyzed by cytochrome oxidase, is a four-electron reaction. The MnC-catalyzed photolytic sequence constitutes an oxidation, involving the loss of four electrons. By contrast, the cytochrome oxidase catalyzed conversion of oxygen to water constitutes a reduction, involving the gain of four electrons.

From MnC, the electron transfers to an electron carrier designated Z and identified recently as a tyrosine residue located on a polypeptide chain of phototosystem II. From Z, the electron passes to P680. Transfer of the electron from water to P680 is an ordinary "downhill" reduction.

Note that both photosystems carry out oxidation–reduction reactions but with different end results. Photosystem I generates a strong reducing agent (P700*) capable of reducing $NADP^+$. Photosystem II generates a strong oxidizing agent (P680) capable of oxidizing water. (In Figure 15.9, four electrons flow through the Z-scheme. We will explain the reason for this below.)

15.4.4. Photosynthetic Phosphorylation

Operation of the electron transport system that links photosystems I and II is *coupled* to synthesis of ATP. We refer to this mode of ATP synthesis as **photosynthetic phosphorylation** or **photophosphorylation.** It represents a third mode of ATP synthesis that differs from *oxidative phosphorylation* and *substrate-level phosphorylation,* discussed earlier. In principle, the mechanism of photophosphorylation is similar to that of oxidative phosphorylation. In both cases, electron transport and ATP synthesis become coupled by means of an electrochemical proton gradient (*chemiosmotic coupling*) across a membrane containing *ATP synthase* (Figure 15.10). However, photophosphorylation differs from oxidative phosphorylation in the way in which the electrochemical gradient is generated, in the type and sequence of electron carriers, and in the dependence on absorption of light.

In photophosphorylation, the transmembrane proton gradient that drives ATP synthesis develops due to two reactions that yield protons. One is the photolysis of water:

$$2H_2O \rightarrow 4H^+ + 4e^- + O_2$$

The other involves the cytochrome *bf* complex that reacts with plastoquinone, which shuttles between an oxidized and a reduced form (Q and QH_2), much like CoQ and $CoQH_2$:

$$QH_2 + 2(Cyt\ bf)_{ox} \rightleftarrows Q + 2(Cyt\ bf)_{red} + 2H^+$$

These proton-producing reactions pump protons into the *thylakoid space (intrathylakoid space; fluid-filled lumen).* As a result, the pH decreases (high $[H^+]$) in the thylakoid space and increases (low $[H^+]$) in the stroma. The difference that develops amounts to about 2–3 pH units. Due to the gradient, protons move from the thylakoid space into the stroma through the membrane-bound ATP synthase, thereby driving ATP synthesis. Much like the mitochondrial ATP synthase (F_0-F_1 ATPase), the chloroplast enzyme consists of two factors and is also known as **CF_0-CF_1 ATPase,** where C stands for chloroplast.

We can demonstrate the chemiosmotic coupling of

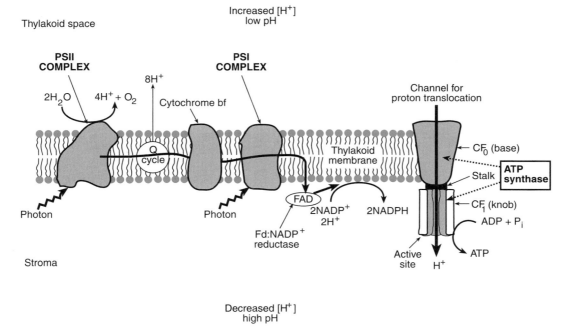

Figure 15.10.   Localization of Z-scheme components and ATP synthase in the thylakoid membrane. PSI, PSII, and the cytochrome *bf* complex constitute three large assemblies interconnected by mobile carriers, plastocyanins (PC) and plastoquinones (Q). [Adapted from D. R. Ort and N. E. Good, *Trends Biochem. Sci.* 13:467–469 (1988) with kind permission of Elsevier Science-NL, Sara Burgerhartstraat 25, 1055 KV Amsterdam, The Netherlands.]

photophosphorylation experimentally. We first soak chloroplasts for several hours in a buffer at pH 4 so that their internal pH becomes stabilized at that value. Then we change the external pH abruptly by transferring the chloroplasts to a buffer at pH 8, containing both ADP and P_i. When we do this, there occurs a sudden burst of ATP synthesis accompanied by disappearance of the pH gradient across the chloroplast membrane. ATP synthesis from ADP and P_i proceeds in the dark; no light is required. This experiment was originally performed by André Jagendorf in 1966. The results provide strong support for the general chemiosmotic coupling hypothesis.

Electron carriers of the photosynthetic electron transport system are largely organized in three complexes that resemble the clustering of mitochondrial electron carriers to form respiratory complexes. The three complexes (photosystem I, photosystem II, and the cytochrome *bf* complex) need not be very close to each other because they are effectively interconnected by highly mobile electron carriers, plastoquinones and plastocyanins. Photosystem I and ATP synthase are located primarily in the stroma lamellae (unstacked regions) while photosystem II is located almost exclusively in the grana (stacked regions). The cytochrome *bf* complex is distributed throughout both the stacked and unstacked regions.

15.4.5. Balance Sheet of the Light Reactions

Reduction of $NADP^+$ to NADPH requires transfer of a hydride ion, equivalent to a proton plus *two electrons:*

$$H^- = H^+ + 2e^-$$

Accordingly, *two* photons must be absorbed by photosystem I. Since reduction of $NADP^+$ results in *two* "electron holes" in photosystem I, *two* electrons must be derived from photosystem II to fill these holes. Thus, photosystem II must also absorb *two* photons, and a total of *four* photons is required to bring about the reduction of one molecule of $NADP^+$.

However, four photons cannot account for the electron balance imposed by water, which constitutes the ultimate electron source for filling the "electron holes" in the two photosystems. Water oxidation must yield molecular oxygen (O_2, not $\frac{1}{2}O_2$) so that two molecules of water must be oxidized by photolysis, releasing four electrons:

$$2H_2O \rightarrow 4H^+ + 4e^- + O_2$$

Therefore, *four* photons must be absorbed by photosystem I, and *four* photons must be absorbed by photosystem II, resulting in the reduction of *two* molecules of NADP$^+$. Consequently, Figure 15.9 shows a flow of *four* electrons through the Z-scheme, so that *eight photons must be absorbed for every two molecules of NADP$^+$ reduced.*

Photosystem I:

$$2NADP^+ + 2H^- + 2H^+ \xrightarrow{4h\nu} 2NADPH + 2H^+$$
$$(4H^+ + 4e^-)$$

Photosystem II:

$$2H_2O \xrightarrow{4h\nu} 4H^+ + 4e^- + O_2$$

Overall reaction for photosystems I and II:

$$2NADP^+ + 2H_2O \xrightarrow{8h\nu} 2NADPH + 2H^+ + O_2$$

The overall reaction is strongly endergonic, having a negative $\Delta E^{\circ\prime}$ and a positive $\Delta G^{\circ\prime}$:

$$2NADP^+ + 4H^+ + 4e^- \rightarrow 2NADPH + 2H^+ \quad (1)$$
$$\underline{2H_2O \rightarrow 4H^+ + 4e^- + O_2 \quad\quad (2)}$$
$$2NADP^+ + 2H_2O \rightarrow 2NADPH + 2H^+ + O_2$$

$$E_1^{\circ\prime} = -0.32 \text{ V}$$
$$\underline{E_2^{\circ\prime} = -0.82 \text{ V}}$$
$$\Delta E^{\circ\prime} = -1.14 \text{ V}$$
$$\Delta G^{\circ\prime} = +220 \text{ kJ mol}^{-1}$$

This reaction cannot go in the absence of light and is made feasible only by absorption of photons, resulting in "uphill" reductions. The light energy absorbed drives the endergonic synthesis of NADPH and the endergonic photolysis of water.

15.4.6. Efficiency of the Light Reactions

15.4.6A. NADPH Formation.
We can evaluate the efficiency of the light reactions by considering the two products formed, NADPH and ATP. As you just saw, the energy expenditure for synthesis of a mole of NADPH in the light reactions is 220 kJ. Therefore, the energy required to form two moles of NADPH equals 440 kJ. This energy is provided by absorption of eight moles of photons (eight einsteins). Each einstein has an energy of about 171 kJ so that the total absorbed energy is

$$8 \text{ einsteins} \times 171 \text{ kJ/einstein} = 1368 \text{ kJ}$$

This constitutes a more than sufficient amount of energy to drive NADPH formation. On the basis of these values, reduction of NADP$^+$ to NADPH has an efficiency of

$$\frac{440 \text{ kJ}}{1368 \text{ kJ}} \times 100 \approx 32\%$$

similar to that calculated for CO_2 fixation (see Section 15.2) and other metabolic pathways.

15.4.6B. ATP Production.
To evaluate the efficiency of ATP synthesis, we can proceed in two ways. First, let us consider the entire process comprising the light reactions. Experiments indicate that translocation of about 12 H$^+$ is associated with evolution of one molecule of oxygen (O_2) and synthesis of four molecules of ATP. Thus, photophosphorylation leads to

$$4ADP^{3-} + 4P_i^{2-} + 4H^+ \rightarrow 4ATP^{4-} + 4H_2O$$

When you combine this equation with those given above for photosystems I and II, you obtain an overall equation that sums up all four of the composite light reactions:

$$2NADP^+ + 4ADP^{3-} + 4P_i^{2-} + 2H^+ \xrightarrow{8h\nu}$$
$$2NADPH + 4ATP^{4-} + 2H_2O + O_2$$

To calculate the efficiency of ATP synthesis, recall that ATP hydrolysis has a $\Delta G^{\circ\prime}$ of -30.5 kJ mol^{-1}. According to the overall equation, four molecules of ATP are synthesized for every eight photons absorbed. In other words, energy equivalent to four *moles* of ATP ($4 \times 30.5 = 122$ kJ) becomes trapped when light energy equivalent to eight *moles* of photons ($8 \times 171 = 1368$ kJ) is expended. This indicates an efficiency of energy conservation of approximately 9%, low in comparison with that of other metabolic pathways.

However, free energy has also been trapped in the form of NADPH synthesis. The free energy *potentially derivable* from oxidation of NADPH via the mitochondrial electron transport system is like that for oxidation of NADH, and oxidation of one molecule of NADPH generates three molecules of ATP. Because the overall reaction yields 2 NADPH, an additional 6 ATP could be produced per molecule of O_2 evolved. Accordingly, 8 moles of photons potentially yield a total of 10 moles of ATP, resulting in an efficiency of

$$\frac{10 \text{ mol} \times 30.5 \text{ kJ/mol}}{1368 \text{ kJ}} \times 100 \approx 22\%$$

For our second estimate of the efficiency of ATP synthesis, we consider just the chain of electron carriers that links photosystems I and II. The potential drop of this system amounts to about 1.2 V per $2e^-$ transferred (see Figure 15.9). Because the drop is traversed twice, by $2e^-$ in each pass, the total potential used equals 2.4 V. Recall that a potential drop of 0.16 V is required for synthesis of ATP when $2e^-$ pass through a chain of electron carriers (Section 12.3). Accordingly, ATP synthesis represents a trapping of $4 \times 0.16 = 0.64$ V or an efficiency of

$$\frac{4 \times 0.16 \text{ V}}{2.4 \text{ V}} \times 100 \approx 27\%$$

On this basis, photophosphorylation appears to be somewhat less efficient than oxidative phosphorylation.

15.4.7. Cyclic Electron Flow

We refer to the flow of electrons through the Z-scheme of photosynthesis as *noncyclic electron flow*, and to ATP synthesis coupled to it as **noncyclic photophosphorylation.** The two constitute the normal mode of operation during photosynthesis. A different mode of operation can, however, occur. It involves a cyclic electron flow in which the excited electron returns to the molecule from which it originated (Figure 15.11).

In *cyclic electron flow*, the excited electron is first transferred from P700* to the initial acceptor (A_0) as before. Following that transfer, the electron passes through the standard noncyclic electron flow carriers up to ferredoxin (Fd). From ferredoxin, the electron is shunted via the cytochrome *bf* complex and plastocyanin to P700 rather than to NADP$^+$. This cyclic set of reactions in-

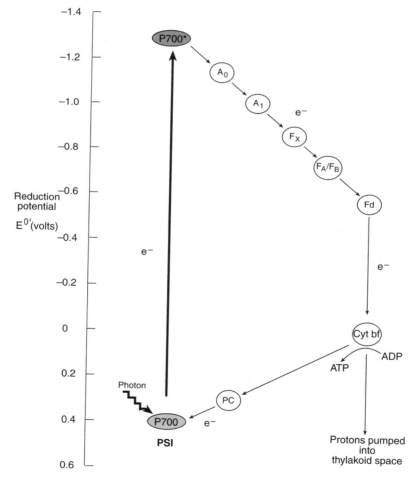

Figure 15.11.   Cyclic photophosphorylation. The process involves only photosystem I, features no net oxidation or reduction, and results in ATP synthesis.

volves *only photosystem I* but uses electron carriers from both photosynthetic electron transport systems and leads to production of ATP.

We refer to ATP synthesis coupled to cyclic electron flow as **cyclic photophosphorylation.** Investigators believe it to be somewhat more efficient than noncyclic photophosphorylation and to result in synthesis of two molecules of ATP for every four protons translocated and three photons absorbed:

$$2ADP^{3-} + 2\,P_i^{2-} + 2H^+ \xrightarrow{\ 3\ h\nu\ } 2\,ATP^{4-} + 2H_2O$$

Cyclic photophosphorylation occurs when cells have a high NADPH/NADP+ ratio but require ATP for various metabolic reactions. Since there exists insufficient NADP+ to accept electrons from reduced ferredoxin, ATP is generated by the cyclic electron flow. Cyclic photophosphorylation may also help to maintain the balance between production of NADPH and ATP. As you will see shortly, the dark reactions require large quantities of ATP. Under some conditions, noncyclic photophosphorylation may not provide enough ATP to drive both the dark reactions and other ATP-requiring reactions of the cell. In that case, cyclic photophosphorylation may supply the additional needed ATP.

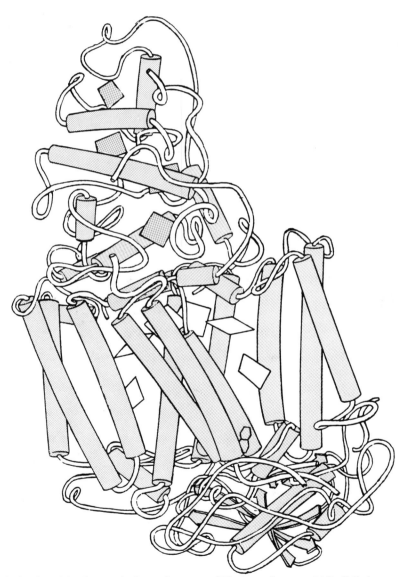

Figure 15.12.   Schematic drawing of the photosynthetic reaction center of *Rhodopseudomonas viridis.* Cylinders represent α-helices (11 are transmembrane), and flat arrows represent β-sheets. Dark rectangles designate heme groups, and gray rectangles designate four bacteriochlorophylls and two bacteriopheophytins. (Courtesy of Dr. J. S. Richardson.)

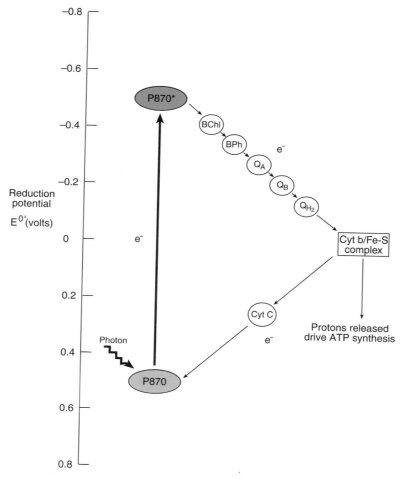

Figure 15.13. Cyclic electron flow in *Rhodopseudomonas viridis*. BChl, Bacteriochlorophyll; BPh, bacteriopheophytin.

Photosynthetic bacteria that do not use oxygen also use a cyclic electron flow. The most extensively studied system is that of the purple sulfur bacterium *Rhodopseudomonas viridis*. The molecular structure of its reaction center was elucidated by Johann Deisenhofer, Hartmut Michel, and Robert Huber in 1984 (Nobel Prize, 1988). The center contains two specialized bacteriochlorophyll molecules ("special pair") embedded in the hydrophobic region of a transmembrane protein and surrounded by other bacteriochlorophylls and accessory pigments (Figure 15.12). A cytochrome, carrying four heme groups, is bound to the reaction center on the external side of the cell membrane (in the periplasmic space).

The special pair becomes excited collectively by light of longer wavelengths (870 nm); hence it is called P870 (Figure 15.13). Loss of an electron by the excited special pair (P870*) to a pheophytin occurs very rapidly, within about four picoseconds (4×10^{-12} s). The ex-tremely rapid transfer prevents the electron from returning to its ground state. It appears that a molecule of bacteriochlorophyll functions in the transfer between P870* and pheophytin.

From pheophytin, the electron passes through a series of plastoquinones and from there to a cytochrome/iron–sulfur complex similar to that linking photosystems I and II in plants. The electron then returns to P870 via cytochrome *c*. ATP synthesis is associated with the cyclic electron flow.

As opposed to plants and cyanobacteria, which obtain their reducing equivalents from photolysis of water and convert them to usable forms by producing NADPH, photosynthetic bacteria must obtain their reducing equivalents from other sources. This they achieve by using inorganic compounds (H_2S, HSO_3^-, $Na_2S_2O_3$) or organic compounds (succinate, lactate, acetate) as reducing agents.

15.5. DARK REACTIONS

The **dark reactions** follow the light reactions. The term "dark reactions" is somewhat misleading. It does not mean that the reactions occur only in the dark; rather, the term emphasizes that these reactions *do not require direct participation of light.* In the dark reactions, or **CO_2 fixation,** CO_2 is converted to carbohydrate. The process consists of a large number of reactions that take place in the stroma and use the ATP and NADPH produced by the light reactions (Figure 15.8). Light, although not directly required, affects the dark reactions since it activates several of the enzymes involved in CO_2 fixation (see Section 15.5.3). In simplified fashion, the following (unbalanced) equation describes the dark reactions:

$$6CO_2 \xrightarrow[\text{enzymes}]{\text{NADPH, ATP}} C_6H_{12}O_6$$

The path that the carbon of CO_2 takes during the dark reactions was elucidated by Melvin Calvin (Nobel Prize, 1961), James Bassham, and Andrew Benson between 1945 and 1953 and is known as the **Calvin cycle** or **reductive pentose phosphate cycle.** These investigators determined the initial steps of the cycle by passing a suspension of unicellular green algae *(Chlorella)* through an illuminated glass tube. They injected radioactive carbon dioxide ($^{14}CO_2$) into the tube and ran the emerging suspension into hot alcohol to inactivate enzymes and stop the reaction. They varied the time elapsed between injecting $^{14}CO_2$ and stopping the reaction to allow the labeled carbon to appear in a larger or smaller number of metabolic intermediates.

After stopping the reaction, the researchers extracted the algae and analyzed the extract by two-dimensional paper chromatography and autoradiography. When they allowed the suspension to carry out photosynthesis for a minute or longer, the extract contained a complex mixture of many ^{14}C-labeled compounds including carbohydrates, amino acids, lipids, and nucleotides. When they reduced the reaction time to 30 s, the extract contained only a limited number of labeled compounds, and when they decreased the time allowed for photosynthesis to 5 s, the extract yielded an autoradiogram that contained only a single spot. Analysis of the material corresponding to this spot showed it to be *3-phosphoglycerate.* The experimenters concluded that 3-phosphoglycerate must be the first *stable* compound formed in the dark reactions of photosynthesis.

15.5.1. Ribulose 1,5-Bisphosphate Carboxylase (Rubisco)

Subsequent work showed that the initial acceptor of carbon dioxide is a five-carbon compound, *ribulose 1,5-bisphosphate,* which is cleaved to two three-carbon compounds (3-phosphoglycerate) so that the flow of carbon is given by

$$C_5 \xrightarrow{C\ (CO_2)} 2C_3$$

The reaction whereby CO_2 combines with ribulose 1,5-bisphosphate to produce two molecules of 3-phosphoglycerate is catalyzed by **ribulose 1,5-bisphosphate carboxylase (rubisco)** and proceeds through formation of an enediol intermediate (Figure 15.14).

Ribulose 1,5-bisphosphate carboxylase of plants and most photosynthetic microorganisms is a large oligomer,

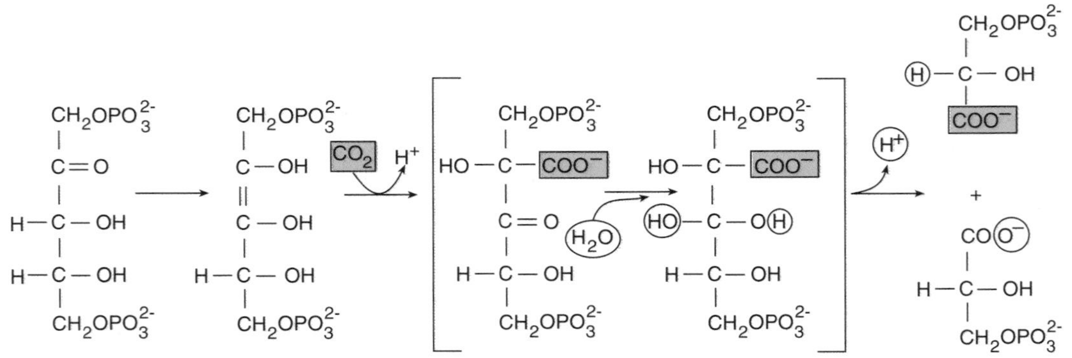

Ribulose 1,5-bisphosphate | Enediol intermediate | Intermediates | 2 Molecules of 3-phosphoglycerate

Figure 15.14. The reaction catalyzed by ribulose 1,5-bisphosphate carboxylase (rubisco) when the enzyme functions as a carboxylase in CO_2 fixation.

having a molecular weight of about 560,000. It consists of 16 subunits, eight large ones (MW = 56,000 each) and eight small ones (MW = 14,000 each). The large and small subunits are specified, respectively, by a gene of chloroplast DNA and by a gene of nuclear DNA. Each large subunit contains both a catalytic and a regulatory site. The small subunits enhance the activity of the large subunits.

Because so much photosynthesis occurs all over the globe, scientists consider rubisco to represent the most abundant protein on Earth. The enzyme typically accounts for more than 50% of soluble leaf proteins and constitutes about 15% of all chloroplast proteins. Researchers have estimated that there are about 40×10^6 tons of rubisco in the world and that the enzyme is being synthesized at a rate of about 4×10^{13} g per year! Because the enzyme is so abundant, some scientists have proposed using it as a dietary protein supplement.

15.5.2. Calvin Cycle

We commonly divide the Calvin cycle into two phases (Figure 15.15):

   A.   *Production phase*
      1. Carboxylation
      2. Phosphorylation
      3. Reduction
      4. Carbohydrate formation
   B.   *Regeneration phase*

15.5.2A. Production Phase.

The **carboxylation stage,** the first part of the production phase, consists of the rubisco-catalyzed reaction whereby ribulose 1,5-bisphosphate is carboxylated to form two molecules of 3-phosphoglycerate.

Carboxylation is followed by a **phosphorylation stage** that requires ATP and in which 3-phosphoglycerate becomes converted to *1,3-bisphosphoglycerate.* A **reduction stage** follows the phosphorylation and leads to reduction of 1,3-bisphosphoglycerate to *glyceraldehyde 3-phosphate* by means of NADPH. Both the ATP and the NADPH required in these reactions are supplied by the light reactions. We can summarize the first three stages of the production phase as follows (Figure 15.16):

$$\text{Ribulose 1, 5-bisphosphate}^{4-} + CO_2 + H_2O \rightarrow$$
$$2 \text{ (3-phosphoglycerate}^{3-}) + 2H^+$$

$$\text{3-Phosphoglycerate}^{3-} + ATP^{4-} \rightarrow$$
$$\text{1, 3-bisphosphoglycerate }^{4-} + ADP^{3-}$$

$$\text{1, 3-Bisphosphoglycerate }^{4-} + NADPH + H^+ \rightarrow$$
$$\text{glyceraldehyde 3-phosphate}^{2-} + NADP^+ + P_i^{2-}$$

In order to fix enough CO_2 for synthesis of one molecule of hexose, the first reaction must be multiplied by 6 and the remaining two by 12. Thus, 6 ribulose 1,5-bisphosphate, 6 CO_2, and 6 H_2O react to yield 12 glyceraldehyde 3-phosphate, 12 $NADP^+$, and 12 P_i.

After formation of 12 molecules of glyceraldehyde 3-phosphate, the path of carbon is split. Two molecules of

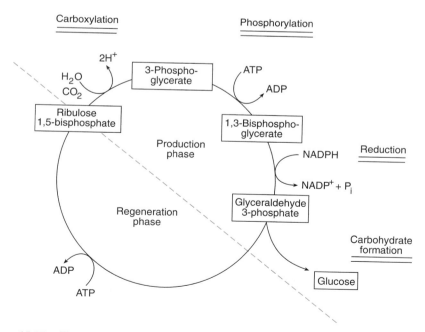

Figure 15.15.   The two major parts of the Calvin cycle—a production phase and a regeneration phase.

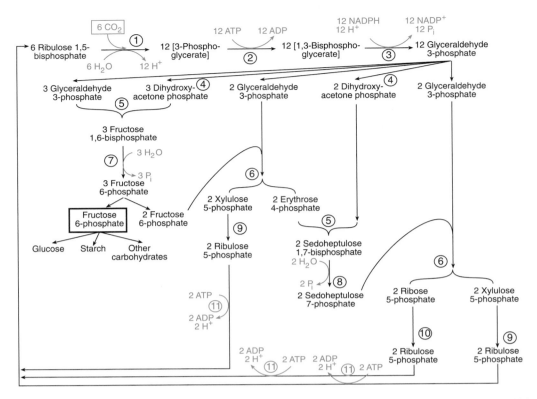

Figure 15.16. Reactions of the Calvin cycle. Six molecules of ribulose 1,5-bisphosphate yield 12 molecules of glyceraldehyde 3-phosphate. Of these, two form one molecule of fructose 6-phosphate, and 10 regenerate six molecules of ribulose 1,5-bisphosphate. Numbers designate enzymes: ① Ribulose 1,5-biphosphate carboxylase; ② phosphoglycerate kinase; ③ glyceraldehyde 3-phosphate dehydrogenase; ④ triose-phosphate isomerase; ⑤ aldolase; ⑥ transketolase; ⑦ fructose 1,6-bisphosphatase; ⑧ sedoheptulose 1,7-bisphosphatase; ⑨ ribulose 5-phosphate epimerase; ⑩ ribulose 5-phosphate isomerase; ⑪ phosphoribulose kinase.

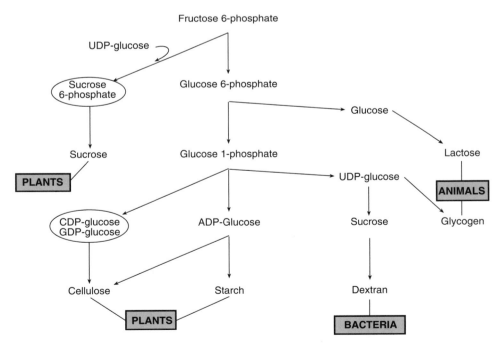

Figure 15.17. Some pathways for carbohydrate synthesis from fructose 6-phosphate.

glyceraldehyde 3-phosphate serve for synthesis of one molecule of hexose as part of the **carbohydrate formation stage** of the production phase. The remaining 10 molecules of glyceraldehyde 3-phosphate are converted to six molecules of ribulose 1,5-bisphosphate in the regeneration phase of the cycle.

Synthesis of fructose 6-phosphate is initiated with six molecules of glyceraldehyde 3-phosphate; three of these are used directly, and three are first converted to the isomeric form dihydroxyacetone phosphate. These precursors lead to formation of three molecules of fructose 1,6-bisphosphate, which undergo dephosphorylation to yield three molecules of fructose 6-phosphate. Two molecules of fructose 6-phosphate (equivalent to four molecules of glyceraldehyde 3-phosphate) become diverted to the regeneration phase. The third fructose 6-phosphate can lead to formation of glucose, starch, or other carbohydrates. These reactions include some of the steps of gluconeogenesis and constitute the carbohydrate formation stage of the production phase (Figure 15.17).

15.5.2B. Regeneration Phase.
Ten of the 12 molecules of glyceraldehyde 3-phosphate serve to regenerate six molecules of ribulose 1,5-bisphosphate in the regeneration phase of the cycle. This phase is initiated with six molecules of glyceraldehyde 3-phosphate; four are used directly, and two are first converted to the isomeric form, dihydroxyacetone phosphate. The equivalent of four molecules of glyceraldehyde 3-phosphate become siphoned off from the pathway of hexose synthesis as two molecules of fructose 6-phosphate.

Reactions of the regeneration phase resemble those of the pentose phosphate pathway in that they involve a reshuffling of carbon skeletons (Figure 15.18) and action of *transketolase* (see Section 10.4). Transaldolase does

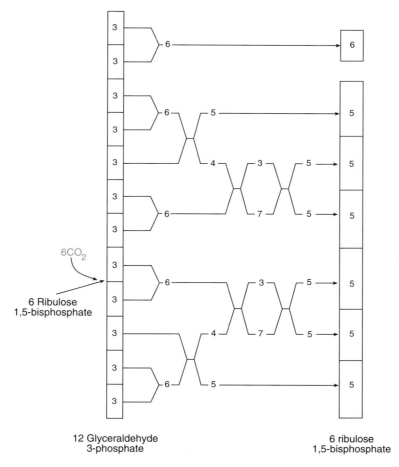

6CO$_2$

6 Ribulose
1,5-bisphosphate

12 Glyceraldehyde
3-phosphate

6 ribulose
1,5-bisphosphate

Figure 15.18. Schematic diagram of the rearrangements of carbon skeletons in the Calvin cycle. Twelve molecules of glyceraldehyde 3-phosphate regenerate six molecules of ribulose 1,5-bisphosphate and form one molecule of fructose 6-phosphate. Numbers indicate the number of carbon atoms per molecule.

not participate in the Calvin cycle; instead, two reactions are catalyzed by *aldolase*.

15.5.2C. Overall Reaction.

To arrive at the overall reaction of the Calvin cycle, consider the stoichiometric relationships involved:

Reactants

CO_2: 6 CO_2 for rubisco reaction

ATP: 12 ATP for phosphorylation of 12 molecules of 3-phosphoglycerate
6 ATP for phosphorylation of 6 molecules of ribulose 5-phosphate

NADPH: 12 NADPH for reduction of 12 molecules of 1,3-bisphosphoglycerate

H_2O: 6 H_2O for rubisco reaction
5 H_2O for hydrolysis of 3 molecules of fructose 1,6-bisphosphate and 2 molecules of sedoheptulose 1,7-bisphosphate

Products

ADP: 18 ADP from phosphorylation by 18 ATP

P_i: 12 P_i from reduction of 12 molecules of 1,3-bisphosphoglycerate
5 P_i from hydrolysis of 3 molecules of fructose 1,6-bisphosphate and 2 molecules of sedoheptulose 1,7-bisphosphate

$NADP^+$: 12 $NADP^+$ from reduction by 12 NADPH

H^+: 6H^+ from phosphorylation of 6 molecules of ribulose 5-phosphate

Other: 1 molecule of fructose 6-phosphate

Combining these values yields the overall reaction:

$$6CO_2 + 18ATP^{4-} + 12NADPH + 11H_2O$$
$$\downarrow$$
$$\text{Fructose 6-phosphate}^{2-} + 18ADP^{3-} + 17P_i^{2-} + 12NADP^+ + 6H^+$$

Thus, the Calvin cycle leads to net synthesis of one molecule of hexose as *fructose 6-phosphate*. Ribulose 1,5-bisphosphate does not appear in the overall equation. Although six molecules of ribulose 1,5-bisphosphate enter the cycle, *all six* are ultimately regenerated. Subsequent reactions can convert fructose-6-phosphate to glucose, starch, or other carbohydrates as you saw in Figure 15.17.

15.5.3. Control of the Cycle

The two phases of photosynthesis—the light and dark reactions—are interconnected. Because ATP and NADPH are products of the light reactions and reactants in the dark reactions, the light reactions affect the rate and extent of the dark reactions. As soon as ATP and NADPH start to form, the dark reactions begin. Dark reactions start while light reactions are still in progress and terminate some time after illumination has stopped.

There is a second linkage between the two photosynthetic phases in addition to that of product/reactant. Light is not only a required ingredient in the light reactions but also functions as an activator of several Calvin cycle enzymes. The activity of phosphoribulose kinase, for example, increases 100-fold on illumination. The mechanism of **light activation** involves *ferredoxin* (Section 15.4) and *thioredoxin* (Section 14.4) and is diagrammed in Figure 15.19.

When chloroplasts are subjected to strong illumination, reduced ferredoxin accumulates because of the activity of photosystem I. Under these conditions, ferredoxin donates some of its electrons to thioredoxin rather than to $NADP^+$. Reduction of thioredoxin is catalyzed by *ferredoxin-thioredoxin reductase* and converts a disulfide bond of thioredoxin to two sulfhydryl groups. Thioredoxin's sulfhydryl groups then undergo a *disulfide exchange* with a disulfide bond in the Calvin cycle enzyme being light-

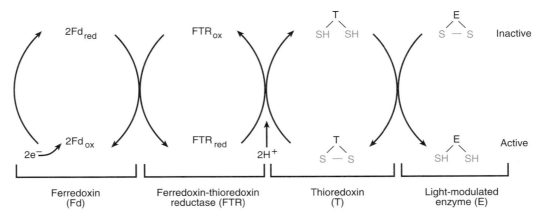

Figure 15.19. Light activation of some Calvin cycle enzymes. Electrons generated by photosystem I reduce ferredoxin. The mini electron transport system leads to reduction of a disulfide bond in a susceptible enzyme.

activated. Reduction of the enzyme's disulfide bond results in activation of the enzyme. Reduced thioredoxin also stimulates ATP synthase (CF_0-CF_1 ATPase) of chloroplasts. This ensures a high rate of ATP synthesis when chloroplast illumination is intense.

Light modulates the activity of several enzymes of carbohydrate catabolism via a similar mechanism except that there the effect is an opposite one, resulting in **light inactivation.** *Phosphofructokinase* and *glucose 6-phosphate dehydrogenase,* for example, are inhibited when their disulfide bonds are reduced by thioredoxin. Recall that these two catalysts constitute key enzymes of glycolysis and the pentose phosphate pathway, respectively.

Consequently, light stimulates carbohydrate synthesis in plants by activating Calvin cycle enzymes but prevents carbohydrate degradation by inhibiting enzymes of glycolysis and the pentose phosphate pathway. In the dark, these effects are reversed, and plants assume some metabolic characteristics of animals. Rather than carry out photosynthesis and produce and store carbohydrates, plants draw on their carbohydrate reserves for growth and respiration and degrade carbohydrates via glycolysis, the citric acid cycle, and the pentose phosphate pathway.

In addition to the regulatory effects of light, the dark reactions are controlled by means of the enzyme rubisco, located in the stroma. Rubisco is stimulated by increases in pH, [Mg^{2+}], and [NADPH], all of which are produced by illumination of chloroplasts.

Light reactions cause pumping of protons from the stroma into the thylakoid space. Hence, the [H^+] in the stroma decreases and the stroma becomes more alkaline, thereby activating rubisco. Movement of the protons is accompanied by movement of magnesium ions in the opposite direction to preserve electrical neutrality. As a result, the [Mg^{2+}] increases in the stroma, which again serves to activate the enzyme. Lastly, as the light reactions progress, NADPH accumulates. NADPH serves as the terminal electron acceptor of the light reactions and as a positive allosteric effector of rubisco. Increasing concentrations of NADPH result in enzyme activation.

15.6. PHOTORESPIRATION

We have known since the 1960s that illuminated plants carry out reactions whose essence is the reverse of photosynthesis: they consume O_2 and evolve CO_2. This process, called **photorespiration,** becomes especially pronounced at low levels of CO_2 and high levels of O_2. Photorespiration differs from oxidative phosphorylation and results from rubisco's unusual ability to function as *either* a carboxylase or an oxygenase. Oxygen competes with carbon

dioxide as a substrate for the enzyme. Because of this dual activity, we also refer to rubisco as *ribulose 1,5-bisphosphate carboxylase-oxygenase.*

When rubisco functions as an oxygenase, it binds O_2 at its active site instead of CO_2 (the two molecules are similar in size). Binding of oxygen initiates photorespiration. In this process, ribulose 1,5-bisphosphate undergoes conversion to 3-phosphoglycerate and *phosphoglycolate* (Figure 15.20). 3-Phosphoglycerate can enter the Calvin cycle, but phosphoglycolate is dephosphorylated to *glycolate,* followed by a series of reactions that convert it back to ribulose 1,5-bisphosphate. These reactions lead to evolution of CO_2, require reducing power in the form of NADH, and require energy input in the form of ATP. The entire pathway involves three types of organelles—*chloroplasts, peroxisomes,* and *mitochondria* (Figure 15.21). Photorespiration competes with and decreases the efficiency of photosynthesis owing to:

- loss of both CO_2 and ribulose 1,5-bisphosphate that could have fed into the Calvin cycle;
- evolution of CO_2 and consumption of O_2, which reverses the effects of CO_2 fixation;
- diversion of reducing power from CO_2 fixation, since NADH, expended in photorespiration, could have been converted to NADPH and used in the Calvin cycle; and
- loss of carbon, since not all of the carbon ultimately returns to the chloroplast

Photorespiration has no known metabolic function, so we consider it a wasteful process. In addition to decreasing photosynthetic efficiency, its operation requires both ATP and NADH. Consumption of NADH effectively depletes the supply of NADPH because the two reducing compounds are interconvertible by action of the enzyme *transhydrogenase:*

$$NADPH + NAD^+ \rightleftarrows NADP^+ + NADH$$

Thus, photorespiration uses up both ATP and NADPH produced by the light reactions, thereby undermining the reactions' usefulness. Additionally, as opposed to ordinary cellular respiration, photorespiration is not coupled to an oxidative phosphorylation; reduction of O_2 is *not* accompanied by synthesis of ATP. Therefore, photorespiration results in a threefold wastefulness: O_2 is needlessly reduced, and both NADPH and ATP are needlessly expended.

Photorespiration leads to a tremendous loss of photosynthetic products. In some plants, it results in as much as a 50% reduction of these products. If photorespiration could be avoided, the food supply to the growing world

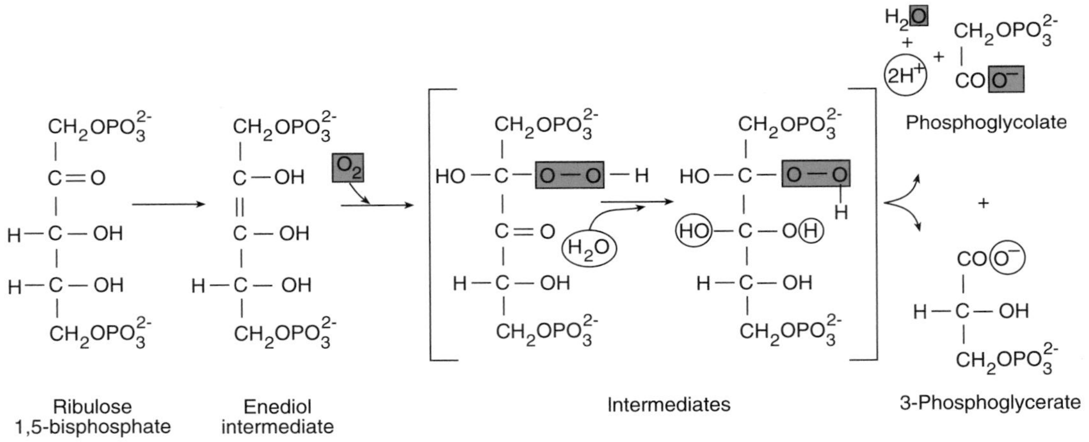

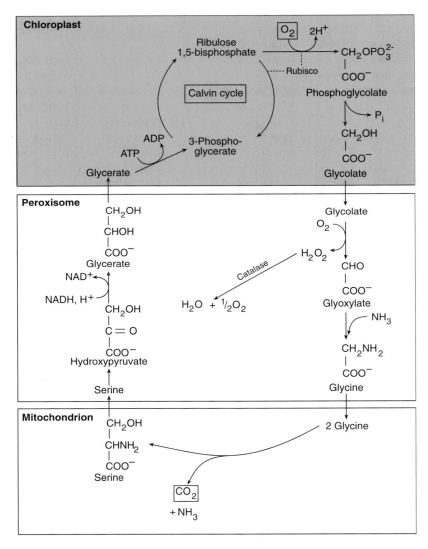

Figure 15.20.   The reaction catalyzed by ribulose 1,5-bisphosphate carboxylase when the enzyme functions as an oxygenase in photorespiration.

Figure 15.21.   The pathway of photorespiration. Peroxisomes are organelles that produce and use hydrogen peroxide.

population would greatly increase. Because of this, efforts are under way to use genetic engineering to either modify the specificity of rubisco or to introduce the C_4 cycle (see below) into plants that do not have it.

15.7. THE C_4 CYCLE

Fixation of CO_2 via the Calvin cycle occurs in all photosynthetic plants, but the mechanism of fixation varies. Plants that grow in *temperate* climatic zones convert CO_2 directly to stable 3-phosphoglycerate, a *three-carbon compound*. These plants are called C_3 **plants.**

Other plants, however, fix CO_2 by a different mechanism. In these plants, a number of reactions *precede* operation of the Calvin cycle. Because these reactions include an initial fixation of CO_2 to stable oxaloacetate, a *four-carbon compound,* these plants are called C_4 **plants.** Following its initial fixation, CO_2 is released and fixed a second time via the Calvin cycle. We term the complete mechanism of CO_2 fixation in these plants the C_4 **cycle** or the **Hatch–Slack pathway** in honor of its discoverers, Marshall Hatch and Roger Slack. These investigators worked on the elucidation of the C_4 cycle between 1966 and 1970.

C_4 plants include not only some crop species, such as sugarcane, sorghum, and corn, but also desert plants, crabgrass, and Bermuda grass. These plants all have two properties in common: they thrive in hot and sunny environments, including tropical and desert zones, and they have evolved a unique leaf anatomy coupled with specific metabolic properties.

Plant leaves contain two types of cells, **mesophyll cells** and **bundle-sheath cells** (Figure 15.22). In C_3 plants, mesophyll cells contain chloroplasts and carry out the reactions of the Calvin cycle; bundle-sheath cells are devoid of chloroplasts.

In C_4 plants, both types of cells contain chloroplasts, but all of the rubisco is concentrated in bundle-sheath cells so that the reactions of the Calvin cycle are carried out only by these cells. Mesophyll cells have become specialized and contain all of the enzymes required for operation of the C_4 cycle.

Mesophyll cells, in C_4 plants, surround bundle-sheath cells in a close concentric packing, separating them effectively from the epidermis and the stoma. This minimizes the leaf's water loss by transpiration and allows bundle-sheath cells to retain their water and function in photosynthesis even in hot, dry environments.

In C_4 plants, the initial acceptor of CO_2 is *phosphoenolpyruvate (PEP)*. CO_2 fixation, catalyzed by *PEP carboxylase,* occurs in mesophyll cells and yields oxaloacetate (Figure 15.23). Oxaloacetate subsequently undergoes conversion to malate, which diffuses into bundle-sheath cells, where it is decarboxylated and converted to pyruvate. The CO_2 produced is fixed via the Calvin cycle, and pyruvate returns to the mesophyll cells, where it is converted to phosphoenolpyruvate.

Carbon dioxide is thus *fixed twice;* once in meso-

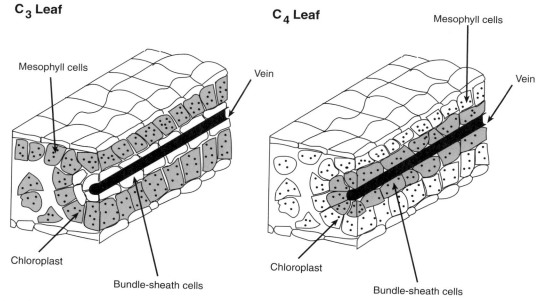

Figure 15.22.   Leaf structure of C_3 and C_4 plants. A vein comprises a bundle of tubes transporting water, inorganic ions, and organic compounds through the leaf. Stoma are minute openings in the leaf through which gas exchange takes place. Colored cells have the capacity to carry out normal CO_2 fixation via the Calvin cycle.

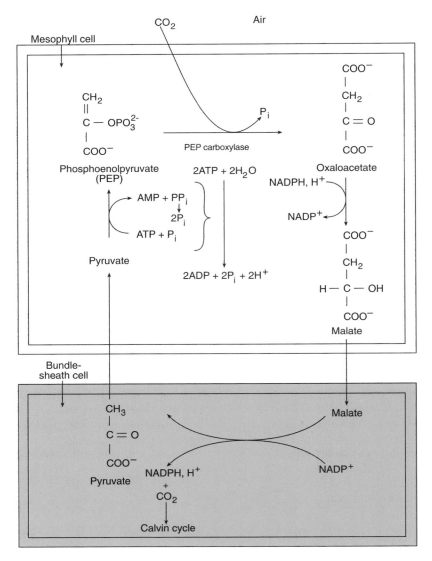

Figure 15.23.   The C_4 cycle (Hatch–Slack pathway).

phyll cells to oxaloacetate (C_4 cycle) and once in bundle sheath cells to 3-phosphoglycerate (Calvin cycle). Double fixation increases the efficiency of the Calvin cycle. The initial CO_2-fixing enzyme (PEP carboxylase) has much greater affinity for CO_2 than rubisco. PEP carboxylase serves as an effective scavenger of CO_2, so that much more CO_2 is ultimately "fixed" by rubisco than would otherwise be the case. High [CO_2] also minimizes the possibility that rubisco will use O_2, rather than CO_2, as a substrate and initiate photorespiration.

Double CO_2 fixation in C_4 plants costs more in terms of ATP expended than the single fixation that occurs in C_3 plants. Conversion of pyruvate to PEP uses *pyruvate phosphate dikinase,* an unusual enzyme that catalyzes phosphorylation of *both* pyruvate and P_i by ATP:

$$Pyruvate^- + ATP^{4-} + P_i^{2-} \rightarrow$$
$$PEP^{3-} + PP_i^{3-} + AMP^{2-} + H^+$$

As in other reactions producing PP_i, pyrophosphatase catalyzes the hydrolysis of PP_i to $2P_i$. Consequently, the energy expenditure becomes *equivalent* to hydrolysis of two energy-rich bonds (ATP $\rightarrow$ AMP + PP_i; $PP_i \rightarrow 2P_i$). The energy requirement is large because synthesis of PEP by a simple transfer of a phosphate group from ATP is thermodynamically unfavorable. The overall reaction of the combined C_4 and Calvin cycles is

$$6CO_2 + 30ATP^{4-} + 12NADPH + 23H_2O$$
$$\downarrow$$
$$Fructose\ 6\text{-}phosphate^{2-} + 30ADP^{3-} + 29P_i^{2-} + 12NADP^+ + 18H^+$$

This equation differs from that of the Calvin cycle by increased numbers of ATP, ADP, P$_i$, H$_2$O, and H$^+$. The increases result from the added energy expenditure in the pyruvate phosphate dikinase reaction of the C$_4$ cycle. That reaction requires an *effective* expenditure of 2 ATP per molecule of CO$_2$ fixed:

$$2ATP^{4-} + 2H_2O \rightarrow 2ADP^{3-} + 2P_i^{2-} + 2H^+$$

Multiplying this equation by 6 (for fixation of 6 CO$_2$) and adding it to the Calvin cycle equation yields the overall stoichiometry for the combined operation of the two cycles.

Photosynthetic efficiency of C$_4$ plants is much greater than that of C$_3$ plants. C$_4$ plants synthesize carbohydrate faster and grow faster than C$_3$ plants. Three reasons explain the increased efficiency:

1. [CO$_2$] for the rubisco reaction is increased by the prior fixing of CO$_2$ via PEP carboxylase, which has a much greater affinity for CO$_2$ than rubisco.
2. Water loss is minimized due to the leaf's unique anatomical structure. Hence the plants can function even in dry environments and can fix CO$_2$ even when the water supply is low.
3. Wasteful photorespiration is greatly minimized, or eliminated entirely, by the increased [CO$_2$], which prevents rubisco from using O$_2$ as a substrate. Prevention of photorespiration exacts a price from C$_4$ plants; they must expend extra energy to form PEP. However, the high energy price is well worth the photosynthetic efficiency gained.

SUMMARY

Photosynthesis—synthesis of carbohydrate from CO$_2$ and H$_2$O in the presence of light and pigments—constitutes essentially the reverse of respiration and is carried out by plants, algae, and bacteria. Most photosynthetic organisms use water as electron donor and produce oxygen. Some organisms use other compounds as electron donors and do not evolve oxygen. Plants and cyanobacteria carry out photosynthesis in chloroplasts by means of two photosystems. Photosystem I (PSI) and photosystem II (PSII) respond, respectively, to light below 700 nm and below 680 nm. The two photosystems function in series, an arrangement called the Z-scheme.

Absorption of a photon excites a molecule of chlorophyll in a photosystem by boosting an electron to a higher energy level. The excitation energy passes via resonance energy transfer through antenna chlorophylls to a reaction center containing specialized chlorophylls. Excited specialized chlorophylls transfer an electron to a suitable acceptor, thereby reducing the acceptor. In PSI, the electron passes through a chain of electron carriers to NADP$^+$, which is reduced to NADPH. In PSII, the electron passes through another set of electron carriers to PSI, where it fills the "electron hole" produced when PSI lost an electron. The electron hole produced in PSII is filled with an electron derived from the splitting of water (photolysis). ATP synthesis is coupled to operation of the chain of electron carriers linking PSI and PSII.

Light reactions comprise four individual reactions, require absorption of eight photons for reduction of two molecules of NADP$^+$, and lead to synthesis of four molecules of ATP by photophosphorylation. Photosynthetic bacteria use only one photosystem, and ATP synthesis is coupled to a cyclic electron flow.

Dark reactions (CO$_2$ fixation) convert carbon dioxide to carbohydrate via the Calvin cycle. The cycle uses ATP and NADPH produced in the light reactions and results in net synthesis of hexose. Ribulose 1,5-bisphosphate carboxylase (rubisco) initiates the reactions of the Calvin cycle by catalyzing the combination of CO$_2$ with ribulose 1,5-bisphosphate to yield two molecules of 3-phosphoglycerate. Rubisco has both carboxylase and oxygenase activity. When the the enzyme functions as an oxygenase, it leads to photorespiration, a wasteful process of unknown metabolic function that decreases photosynthetic efficiency.

Certain plants (C_4 plants) use a double fixation of CO_2, once into oxaloacetate and once into 3-phosphoglycerate. The C_4 pathway is more efficient than the single fixation pathway in ordinary plants (C_3 plants).

SELECTED READINGS

Bassham, J. A., and Calvin, M., *The Path of Carbon in Photosynthesis,* Prentice-Hall, Englewood Cliffs, New Jersey (1957).

Deisenhofer, J., and Michel, H., The photosynthetic reaction center from the purple bacterium *Rhodopseudomonas viridis, Science* 245:1463–1473 (1989).

Golbeck, J. H., Structure and function of photosystem I, *Annu. Rev. Plant Physiol. Plant Mol. Biol.* 43:293–324 (1992).

Govindjee, and Coleman, W. J., How plants make oxygen, *Sci. Am.* 262(2):50–58 (1990).

Greenbaum, E., Lee, J. W., Tevault, C. V., Blankinship, S. L., and Mets, L. J., CO_2 fixation and photoevolution of H_2 and O_2 in a mutant of *Chlamydomonas* lacking photosystem I, *Nature (London)* 376:438–441 (1995).

Hall, D. D., and Rao, K. K., *Photosynthesis,* Cambridge University Press, Cambridge (1994).

Hartman, F. C., and Harpel, M. R., Structure, function, regulation, and assembly of D-ribulose 1,5-bisphosphate carboxylase/oxygenase, *Annu. Rev. Biochem.* 63:197–234 (1994).

Hatch, M. D., C_4 photosynthesis: A unique blend of modified biochemistry, anatomy, and ultrastructure, *Biochim. Biophys. Acta* 895:81–106 (1987).

Krauss, N., *et al.,* Three-dimensional structure of system I of photosynthesis at 6Å resolution, *Nature (London)* 361:326–331 (1993).

Kühlbrandt, W., Wang, D. N., and Fujiyoshi, Y., Atomic model of plant light-harvesting complex by electron crystallography, *Nature (London)* 367:614–621 (1994).

Ogren, W. L., Photorespiration: Pathways, regulation, and modification, *Annu. Rev. Plant. Physiol.* 35:415–442 (1984).

Vermaas, W., Molecular biological approaches to analyze photosystem II structure and function, *Annu. Rev. Plant Physiol. Plant Mol. Biol.* 44:457–481 (1993).

Yachandra, V. K., De Rose, V. J., Latimer, M. J., Mukerji, I., Sauer, K., and Klein, M. P., Where plants make oxygen: A structural model for the photosynthetic oxygen-evolving manganese cluster, *Science* 260:675–679 (1993).

REVIEW QUESTIONS

A. Define each of the following terms:

Phytol	Photon
Internal conversion	Fluorescence
Resonance energy transfer	Quantum yield
Electron transfer	Red drop
Thylakoid disk	Hill reaction
Reaction center	Photophosphorylation
Emerson enhancement effect	Rubisco

B. Differentiate between the two terms in each of the following pairs:

Photosynthesis/ photorespiration	Photosystem I/ photosystem II
Light reactions/dark reactions	Ground state/excited state
Chlorophyll/chloroplast	Stroma/grana
Antenna chlorophylls/ specialized chlorophylls	Light activation/ light inactivation
Cyclic photophosphorylation/noncyclic photophosphorylation	Carboxylation stage/ phosphorylation stage
Calvin cycle/ Hatch-Slack Pathway	Mesophyll cells/ bundle-sheath cells

C. (1) Discuss each of the four individual light reactions that occur in plant photosynthesis and explain the operation of the Z-scheme.

(2) What is the mechanism of photorespiration? Why does photorespiration decrease the efficiency of photosynthesis and why does it represent a wasteful process?

(3) What are the two phases of the Calvin cycle and what are their subdivisions? Outline the path of carbon for each phase. What are the energetics of the Calvin cycle?

(4) How are the dark reactions affected by light?

(5) Compare the photosynthetic process in C_3 and C_4 plants.

PROBLEMS

15.1. A student calculates the free energy changes for various parts of the Z-scheme from the potentials shown in Figure 15.9. Do these calculated values correctly describe the free energy changes in the chloroplast under physiological conditions? Why or why not?

15.2. Write out the equations for the reactions in plant photosynthesis that are responsible for establishing a pH gradient across the thylakoid membrane.

15.3.* A student sets up a number of experiments along the line of Priestley's classic experiment. The student uses (a) a C_3 plant that does not carry out photorespiration; (b) a C_3 plant with extensive photorespiration; and (c) a C_4 plant that does not carry out photorespiration. Everything else being equal, predict the relative lengths of time that the mouse will survive in the apparatus under these conditions.

15.4. Based on the estimated amount of carbon fixation given in the text, calculate the annual amount of glucose that can be produced via photosynthesis.

15.5. Destruction of the ozone layer by fluorocarbons used in spray cans increases the extent of ultraviolet irradiation of the Earth. Would you expect this to affect the rate of photosynthesis? Why or why not?

15.6. What is the quantum yield of ATP synthesis in cyclic photophosphorylation if the absorption of three photons leads to the synthesis of two molecules of ATP?

15.7. Based on the data in Figure 15.6, what are the likely colors of carotenoids, phycocyanins, and phycoerythrins?

15.8.* Assume that photolysis of water in photosynthesis leads to ozone (O_3) formation:

$$3H_2O \rightarrow O_3 + 6H^+ + 6e^-$$

On this basis, write the equations for the overall reactions of photosystem I, photosystem II, and their combined operation. How many photons must strike each photosystem and how many electrons must flow through the Z-scheme?

15.9.* What is the quantum yield for ATP synthesis of the system in the previous problem if we know that a total of two molecules of ATP are synthesized for every two electrons passing from P680* to P700 (Figure 15.9)?

15.10.* What is the efficiency of energy conservation for the electron transport system from P680* to P700 based on Problems 15.8 and 15.9?

15.11. Theoretically, what is the maximum number of molecules of ATP that could be obtained from the flow of one electron in bacterial photosynthesis

(Figure 15.13) for the chain of carriers (a) from P870* to the Cyt b/Fe-S complex and (b) from the Cyt b/Fe-S complex to P870?

15.12. Would you be able to demonstrate the Emerson enhancement effect by using photosynthesizing cells of the bacterium *Rhodopseudomonas viridis* and supplementing light at 870 nm with that at 700 nm? Why or why not?

15.13.* Calculate $\Delta G^{\circ\prime}$ for the excitation of P870 to P870*. What is the minimum number of photons of 870-nm light that will bring about the excitation of one molecule of P870?

15.14.* When photosynthesis is carried out in the presence of $^{14}CO_2$, what fraction of the carbons in glyceraldehyde 3-phosphate will be labeled after 12 molecules of glyceraldehyde 3-phosphate have been formed? Depending upon which two molecules of glyceraldehyde 3-phosphate react to form fructose 6-phosphate, the latter will have different degrees of labeling. What are the various fractions of ^{14}C in fructose 6-phosphate that could be produced from this pool of glyceraldehyde 3-phosphate?

15.15. Leaves of deciduous trees are green in the summer and often have brilliant red and yellow colors in the fall. Given that these leaves are known to lose their chlorophyll in the fall, what must cause the nongreen colors?

15.16. Cyclic photophosphorylation, as illustrated in Figure 15.11, leads to synthesis of two molecules of ATP for every three photons (700 nm) absorbed. On this basis, what is the efficiency of solar energy conservation?

15.17.* Calculate $\Delta E^{\circ\prime}$ and $\Delta G^{\circ\prime}$ for the reduction of $NADP^+$ by ferredoxin based on the data in Table 12.1.

15.18. You place a potted C_3 plant and a potted C_4 plant in a sealed glass container containing an adequate amount of moisture. You then illuminate the container. As time goes on, the C_4 plant thrives, but the C_3 plant shrivels up and dies. How do you explain these results?

15.19. The herbicide dichlorophenyldimethylurea (DCMU) blocks electron transfer between Q_A and QH_2 in photosystem II (see Figure 15.9). Do you expect DCMU to inhibit cyclic photophosphorylation? Why or why not?

15.20. DCMU-inhibited chloroplasts do not normally evolve oxygen (O_2), but when ferricyanide is added to the chloroplasts, they do. How do you explain this result?

15.21.* A researcher determines the kinetics of CO_2 incorporation by rubisco once in a nitrogen atmosphere and once in an oxygen atmosphere. He plots the data as $1/v$ versus $1/[CO_2]$ and obtains two straight lines (v = velocity of the reaction). Draw the lines expected, and explain how the Michaelis constants (K_m), calculated from the lines, are related.

15.22. The steady-state concentrations of ATP, ADP, and P_i in actively photosynthesizing spinach chloroplasts are $120\mu M$, $6.00\mu M$, and $700\mu M$, respectively. Under these conditions, what is the minimum amount of energy required for the synthesis of one mole of ATP? Hint: Use $([ADP][P_i])/[ATP]$ for the [products]/[reactants] term. $\Delta G^{\circ\prime}$ for ATP hydrolysis is -30.5 kJ mol^{-1}.

15.23.* The chloroplasts of the previous problem are known to lead to the synthesis of four molecules of ATP from ADP for every eight photons absorbed. Assuming that all the photons are of the same wavelength, calculate the maximum wavelength of light that can result in photosynthesis in this system, based on an efficiency of 9% for the entire process comprising the light reactions.

15.24. What is the minimal potential drop per $2e^-$ required for synthesis of ATP in actively photosynthesizing spinach chloroplasts, based on your calculations in Problem 15.22?

15.25. What is the maximum increase in reduction potential per e^- that can be brought about by a photon of blue light (500 nm)?

15.26.* Consider the following schematic representations of photosynthetic systems in *hypothetical* organisms. Each half-reaction ties into the scheme via its oxidant or reductant as shown. For each representation, indicate whether the scheme, as outlined, is theoretically *possible* or *impossible*. If the latter applies, explain why this is so.

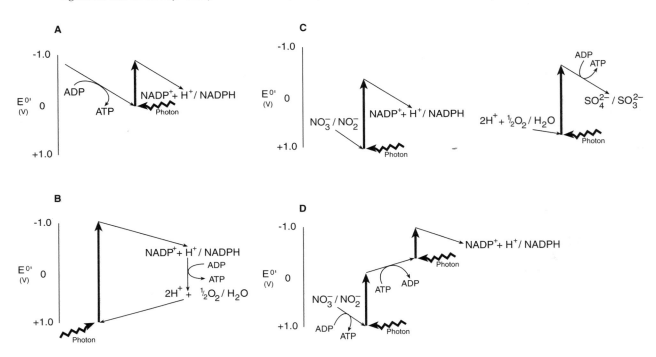

Library of Congress Cataloging-in-Publication Data

Stenesh, Jochanan.
    Biochemistry / Jochanan Stenesh.
        p.   cm.
    Includes bibliographical references and index.
    ISBN 0-306-45732-6 (hardcover). -- ISBN 0-306-45733-4 (pbk.)
    1. Biochemistry.   I. Title.
QP514.2.S635 1998
572--dc21   97-41014
                                                             CIP

ISBN 0-306-45732-6 (Hardbound)
ISBN 0-306-45733-4 (Paperback)

© 1998 Plenum Press, New York
A Division of Plenum Publishing Corporation
233 Spring Street, New York, N.Y. 10013

http://www.plenum.com

10 9 8 7 6 5 4 3 2 1

Printed in the United States in America

Biochemistry

Metabolism

This part of the book deals with the multitude of reactions whereby bio-molecules are synthesized, degraded, and interconverted in living systems. Energy transformations constitute integral aspects of these processes. Some reactions produce usable energy, whereas others require an input of energy to proceed. To provide an overall view, we consider both synthesis and degradation of a given type of biomolecule in a single chapter. We will cover major pathways of carbohydrate, lipid, amino acid, and nucleotide metabolism. Because of their unique links to DNA, we will discuss the metabolism of proteins and nucleic acids separately in Part IV.

PLENUM PRESS • NEW YORK AND LONDON

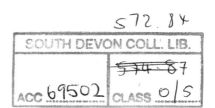